Chemistry and Physics of Coal Utilization—1980

(APS, Morgantown)

AIP Conference Proceedings
Series Editor: Hugh C. Wolfe
Number 70

Chemistry and Physics
of Coal Utilization—1980
(APS, Morgantown)

Editors
Bernard R. Cooper
West Virginia University
and
Leonidas Petrakis
Gulf Research and Development Company

American Institute of Physics
New York
1981

PROGRAM COMMITTEE

B. R. Cooper (Chairperson), West Virginia University
C. Alexander, Jr., University of Alabama
Y. Chen, Oak Ridge National Laboratory
A. Davis, Pennsylvania State University
W. A. Ellingson, Argonne National Laboratory
J. Gethner, EXXON Research & Engineering Company
C. L. Herzenberg, Argonne National Laboratory
W. D. Jackson, Energy Consulting
I. S. Jacobs, General Electric Research & Development Center
T. J. O'Brien, Morgantown Energy Tech. Ctr. - U.S.D.O.E.
R. B. Oder, Gulf Research and Development Company
L. Petrakis, Gulf Research and Development Company
M. Poutsma, Oak Ridge National Laboratory
H. L. Retcofsky, Pittsburgh Energy Tech. Ctr. - U.S.D.O.E.
P. C. Scott, U. S. Department of Energy
P. G. Shewmon, Ohio State University
M. Siskin, EXXON Research & Engineering Company
S. E. Stein, West Virginia University
P. L. Walker, Jr., Pennsylvania State University
D. D. Whitehurst, Mobil Research and Development Center
W. H. Wiser, University of Utah
D. E. Woodmansee, General Electric Research & Development Center

Proceedings Editors:

B. R. Cooper and L. Petrakis

Local Arrangements (West Virginia University)

W. E. Vehse (Chairperson)   B. R. Cooper
M. D. Aldridge   R. Koppelman
P. M. Schollaert   M. L. Jones
            W.V.U. Conference Staff

PREFACE

The Conference on the Chemistry and Physics of Coal Utilization was designed as an interdisciplinary conference centered on the physics and chemistry phenomena involved in coal utilization, including the chemistry and physics of coal itself. The speakers were asked to bear in mind the bridging to practical technology systems, and discussion of this bridging was the objective of the closing panel session. The Conference was designated as a Topical Conference of the American Physical Society, and was held on June 2-4, 1980, at the Lakeview Inn in Morgantown, West Virginia. It was primarily intended as a working conference for those already engaged in, or at least technically informed on, coal conversion and utilization research.

The Conference was attended by approximately 200 scientists and engineers. The program consisted of lectures by 23 invited speakers and two panel discussions by invited experts. In addition there were two sessions at which a total of 49 poster contributions were presented. This volume contains the invited papers (with one exception), reports of the panel discussions, and abstracts for the poster contributions.

We appreciate the continuing efforts of the American Physical Society, through the Committee on the Applications of Physics, to foster the interest of physical scientists in research relevant to coal utilization. We gratefully acknowledge the financial support of West Virginia University, the U.S. Department of Energy, and the Gulf Oil Foundation, in sponsoring the Conference.

Bernard R. Cooper

Leonidas Petrakis

TABLE OF CONTENTS

Chapter 1   COAL STRUCTURE

Coal Structure... 1
    John W. Larsen
Crosslinked Macromolecular Structures in Bituminous
Coals: Theoretical and Experimental Considerations........... 28
    Lucy M. Lucht and Nikolaos A. Peppas
Relationship of Coal Maceral Composition to Utilization....... 49
    R. R. Thompson

Chapter 2   COAL STRUCTURE - RESONANCE TECHNIQUES

C-13 NMR on Solid Samples and Its Application to
Coal Science.. 66
    G. E. Maciel, M. J. Sullivan, N. M. Szeverenyi,
    and F. P. Miknis
^{29}Si NMR: A New Tool for Coal Liquids Characterization........ 82
    K. D. Rose and C. G. Scouten
Electron Spin Resonance (ESR) Characterization of Coal and
Coal Conversions.. 101
    L. Petrakis and D. W. Grandy

Chapter 3   COAL STRUCTURE - EXPLORATORY TECHNIQUES

Understanding Coal Using Thermal Decomposition and
Fourier Transform Infrared Spectroscopy....................... 121
    P. R. Solomon and D. G. Hamblen
Acoustic Microscopy for the Characterization of Coal.......... 141
    C. F. Quate
Low Temperature Chemical Fragmentation of Coal................ 154
    N. C. Deno, K. Curry, A. D. Jones, R. Minard,
    T. Potter, W. Rakitsky, K. Wagner, and R. J. Yevak

Chapter 4   COAL STRUCTURE PANEL DISCUSSION

Panel Discussion - A Critique of Determinations of
Coal Structure... 167
    H. L. Retcofsky, P. H. Given, R. H. Schlosberg, and
    S. E. Stein

Chapter 5   INSTRUMENTATION AND MONITORING

Electrical Properties of Coal at Microwave Frequencies
for Monitoring... 175
    Constantine A. Balanis
Instrumentation for Transport and Slurries................... 198
    Nancy M. O'Fallon

Mossbauer Spectroscopy for Pyrite Analysis in Coal............ 209
    Lionel M. Levinson

Chapter 6   SURFACE PHENOMENA: CATALYSIS

Spectroscopic Studies of CO Chemisorption on Transition
Metals and Studies of the Kinetics of CH_4 Synthesis........... 235
    John T. Yates, Jr., D. W. Goodman, R. D. Kelly,
    and T. E. Madey
Mechanisms of Catalyzed Gasification of Carbon............... 236
    D. W. McKee
Chemical and Physical Aspects of Refining Coal Liquids........ 256
    Y. T. Shah, G. J. Stiegel, and S. Krishnamurthy
Mineral Matter Effects in Coal Conversion.................... 291
    B. Granoff and P. A. Montano

Chapter 7   INTERFACES AND COMBUSTION

Fundamental Research in Coal Combustion: What Use
Is It?... 309
    Robert H. Essenhigh
Heat Capacity and NMR Studies of Water in Coal Pores.......... 332
    S. C. Mraw and B. G. Silbernagel
Surface Chemical Problems in Coal Flotation.................. 344
    S. R. Taylor, K. J. Miller, and A. J. Deurbrouck
Physical and Chemical Coal Cleaning......................... 357
    T. D. Wheelock and R. Markuszewski

Chapter 8   ELECTRICAL PROCESSING OF COAL

Electrochemical Coal Gasification - Operating Conditions,
Variables, and Practical Implications....................... 388
    J. Hickey, S. Lalvani, and Robert Coughlin
Selective Magnetic Enhancement of Pyrite in Coal by
Dielectric Heating at 27 and 2450 MHz....................... 417
    Delwyn D. Bluhm, Glenn E. Fanslow,
    Stephen Beck-Montgomery, and Stuart O. Nelson

Chapter 9   CLOSING PANEL DISCUSSION

Panel Discussion - Interaction of Research with Coal
Technology Development...................................... 438
    U. Merten, I. Wender, W. H. Wiser, D. E. Woodmansee,
    and B. R. Cooper (compiled by B. R. Cooper)

10 ABSTRACTS FOR CONTRIBUTIONS PRESENTED AT POSTER SESSIONS

Introduction.. 442
Chemical Characteristics of American Coals.................. 443
    S. Raj

A New Procedure for the Separation of Coal Macerals........... 443
    G. R. Dyrkacz, C. A. A. Bloomquist, L. H. Fuchs,
    and E. P. Horwitz
ENDOR Study of Bituminous Coal............................... 444
    I. Miyagawa and C. Alexander
Matrix ENDOR Studies of the Carbonization of Western
Canadian Coking Coals.. 444
    P. R. West and S. E. Cannon
Small Angle X-Ray Scattering Study of the Porosity
in Coals... 445
    P. W. Schmidt, M. Kalliat, and C. Y. Kwak
X-Ray Diffraction Studies of Framboids....................... 445
    A. S. Stiller, A. S. Pavlovic, and J. M. Cook
Coal Characterization by X-Ray Diffraction................... 446
    A. S. Pavlovic, J. M. Cook, and J. J. Renton
Trace Element Analysis of Coal by Proton-Induced X-Ray
Emission (PIXE).. 446
    D. C. Buckle and G. C. Grant
A Rapid, Direct Method for Determining Organic Sulfur
Content in Coal.. 447
    T. D. Davies and R. Raymond, Jr.
A Comparative Study of the Properties of Marcasite and
Pyrite... 448
    M. S. Seehra and M. S. Jagadeesh
Coal Mineral Analyses at the Coal Research Bureau Using
the Multiple Analysis Technique.............................. 448
    W. Grady, D. Gierl, and T. Simonyi
Isothermal Studies of the Plastic State of Coal.............. 449
    L. P. Yates, H. E. Francis, and W. G. Lloyd
The Impermeability Characteristic of Caking Coals
Upon Heating... 449
    Z. Wang and J. K. Shou
Investigation of Structural Deformation of Coal Particle
in Pyrolysis... 450
    M. Chiou and H. Levine
The Electrical Properties of Coal Slag...................... 451
    R. Pollina and R. Larsen
Electron Beam Ionization for Coal Ash Precipitators.......... 451
    R. H. Davis, W. C. Finney, and L. C. Thanh
Observation of Particle Trajectories and Inertial Effects
in a Regular Lattice of Magnetized Fibers.................... 452
    W. Lawson, R. Treat, and F. Zeller, III
Magnetic Separation, Thermo- and Magnetochemical
Properties of Coal Liquid Residues........................... 452
    E. Maxwell, D. R. Kelland, I. S. Jacobs, and
    L. M. Levinson
Microspectroscopy of Coal.................................... 453
    J. S. Gethner
V^{4+} Paramagnetic Resonance as a Probe of the Structure and
Dynamics of Coal Related Fuels............................... 454
    N. S. Dalal

On the Rapid Estimation of % Ash in Coal from Silicon
Content Obtained via FNAA, XRF, or Slurry-Injection AA...... 454
    D. G. Hicks, J. E. O'Reilly, and D. W. Kopenaal
Composition and Properties of Jet and Diesel Fuels
Derived from Coal and Shale................................. 455
    J. Solash and R. N. Hazlett
Electrochemical Behavior of Coals, H-Coal Liquids
and Fe^{++} Ion.. 455
    R. P. Baldwin, K. F. Jones, J. T. Joseph,
    and J. L. Wong
Mossbauer Effect Study of Victorian Brown Coal.............. 456
    J. D. Cashion, B. Maguire, and L. T. Kiss
Desulfurization of Hot Coal Gases by Regenerative
Sorption... 457
    V. M. Jalan and D. Wu
Air/Water Oxidative Desulfurization of Coal and Sulfur-
Containing Compounds....................................... 457
    R. P. Warzinski, S. Friedman, and R. B. LaCount
Liquid Sulfur Dioxide Treatment of Coal: Comminution,
Extraction, and Desulfurization............................ 458
    D. F. Burow and R. K. Sharma
Reaction of Bituminous Coal with Inorganic Solvents........ 459
    T. S. Miller and A. P. Hagen
Bromination of Anthracite.................................. 459
    C. G. Woychik and D. D. L. Chung
Characterization of the Oxidation of Bituminous
Coal by ^{13}C n.m.r. 460
    J. A. MacPhee and B. N. Nandi
Refractory Oxides for High-Temperature Coal-Fired MHD
Air Heaters.. 460
    R. J. Pollina and R. R. Smyth
Application of a Multiple-Beam Laser Doppler Velocimeter
to Measure Velocity Distributions of a Fluidized Bed....... 461
    E. J. Johnson and M. E. McDonnell
Frontier Orbital Interactions in Model Coal Reactions...... 462
    P. S. Virk and D. J. Ekpenyong
The Hydrogasification of Lignite and Sub-Bituminous Coals... 462
    B. Bhatt, P. T. Fallon, and M. Steinberg
A Discussion of Physical and Chemical Problems Pertinent
to UCG 463
    T. L. Eddy and C. I. Anekwe
An X-Ray Diffraction Study of the Reactions of Zinc
Compounds with Coal.. 463
    H. Beall and R. J. Wadja
Model Compound Studies of the Mineral Matter Catalysis
in Coal Liquefaction....................................... 464
    B. C. Bockrath and K. T. Schroeder

Shock Activation of Catalysts................................. 464
    R. A. Graham, B. Morosin, P. M. Richards, F. V. Stohl,
    and B. Grancff
Improved Methanation Catalysts............................... 465
    C. S. Brooks, G. S. Golden, and F. D. Lemkey
Modeling of Fischer-Tropsch Synthesis........................ 466
    S. Novak
A Fundamental Chemical Kinetics Approach to Coal Conversion... 466
    R. E. Miller and S. E. Stein
A General, Quantitative, Kinetic Model for Coal
Liquefaction... 467
    T. Gangwer
Liquefaction of Kentucky and Illinois Coals in a
Batch Microreactor... 468
    D. C. Cronauer, R. G. Ruberto, and D. C. Young
Ball Valve Design for Solid Withdrawal Service in
Coal Liquefaction.. 468
    A. J. Patton
Aging of SRC Liquids... 469
    T. Hara, L. Jones, K. C. Tewari, and N. C. Li
Liquefaction Reactivity of Extruded Coal..................... 469
    S. Mori, B. H. Davis, A. W. Fort, and W. G. Lloyd
Summary Report of the Rawlins 1 for Gasification of Steeply
Dipping Coal Beds.. 470
    B. E. Davis and J. E. Miranda
Hydration Process for Enhanced Calcium Utilization in
Fluidized-Bed Combustion.-................................... 470
    D. S. Moulton, E. B. Smyk, and J. A. Shearer
Author Index 471

COAL STRUCTURE

John W. Larsen
Department of Chemistry, University of Tennessee
Knoxville, Tennessee 37916
and
Chemistry Division, Oak Ridge National Lab.
P.O. Box X, Oak Ridge, Tennessee 37830

ABSTRACT

Some aspects of the structure of coal are discussed. Emphasis
is placed on the changes occurring during coalification and on the
macromolecular structure of coal. The various forms of oxygen in
coals are described, and the limitations to our knowledge of heteroatom functional groups are discussed. Some aspects of coal structure
especially relevant to the chemical reactivity of coals are described. Much more must be learned about coal structure, both physical
and chemical before realistic attempts to develop structure-reactiv-
ity relationships are possible.

I. INTRODUCTION

A topic as vast as the chemical structure of coal cannot be
covered in a single lecture. A number of topics have been selected
on the following basis: an organic chemist thinks they may prove
interesting to physicists. Specifically, the way a few properties
change with something we call coal rank will be discussed. Coal rank
can be loosely associated with the relative age of coals within a
geological province. It is the degree to which the coalification
process has progressed. Some chemical structural aspects will be
discussed, evidence for a macromolecular gel structure of coal will
be provided, and something known in my research group as the 86 %
carbon anomaly will be described. Many apparently unrelated properties of coal go through some sharp changes at about 86 % carbon. The
relation between them and the reason for the changes will be discussed.

The biggest problem with coal is that the stuff is a rock. It
is not a substance whose purity can be controlled. It is something
that started out as a swamp and went through a terrific number of
complicated chemical stages and aged for a few hundred million years.
It is an enormously complex material. This is one of the reasons to
be interested in it. It is a worthwhile research area, not just
because the stuff is of enormous practical importance, but because
working with a material this complicated is a fascinating and very
difficult intellectual challenge. Coal contains inorganic materials
covering an enormous size range. We have found mineral matter particles which pass through a o.5 μ filter. There are a variety of
types of organic material in this terribly complicated mixture.

Different parts of the plant go through the coalification process in different ways, yielding materials whose chemistries are different and which in coal are all mixed up together. Coal is a very high surface area material with a complex pore structure.[1] Coals are insoluble. This means that chemistry occurs at surfaces and these surfaces are very poorly characterized. To react, reagents must penetrate a pore network about which we have a lot of information. But in many cases the information is inadequate to decide whether the chemistry is limited by the chemical reactivity or by the physical constraints on mass transport.[2] One of the themes that will run through this paper is that, very often, when we try to do chemistry on coal, it may be strongly limited by the physical structure of the material, and our understanding of the physical structure is insufficient to know when these limitations are coming into play.

II. PHYSICAL STRUCTURE

Figures 1 and 2 demonstrate clearly the heterogeneous complex mixtures to be found in many coals. All of the different regions are different chemically and the results of a chemical reaction done on that piece of coal are uninterpretable in terms of "coal structure." One of the things that one has to be very careful about is getting coal samples which are largely one maceral, or getting samples of separated macerals. Stay away from coals that look like this.

Coals are very high surface materials. The plot in Figure 3 shows surface areas of coal measured by BET adsorption isotherms. Using N_2 at 77 $^{\circ}$K, the surface areas measured are very small. Using CO_2 at 298 $^{\circ}$K, much larger surface areas are obtained. This latter technique more closely approximates the total of internal plus external surface. In this high surface area material with a very complex pore system, many of the pores are small enough so that organic reagents are going to have a very hard time diffusing through them, if indeed they can. Another factor, which is often not considered at all, is the fact that coals swell when placed in many organic solvents, as much as doubling their volume.[3] What this swelling does to the pore size distribution or surface area is unknown.

A brief discussion of the origin of coal is in order.[4] The organic starting materials are celluloses, lignins and other plant debris. (See Figure 4.) Occasionally people try to get from these structures to reasonable guesses for the structure of coal. This is an extraordinarily difficult process and does not seem to be a good way of deducing coal structure. In lignins, one can very easily pick out the kinds of structures most of us believe occur in coal, largely aromatic molecules containing oxygen. Cellulose is also oxygen rich. The coalification process, which begins with these materials and ends up eventually as coal, is a deoxygenation followed by a dehydrogenation. Coalification starts out with lignins, etc., and ends up with antracite. The process is more or less continuous and it is stopped when the coal is dug out of the ground.

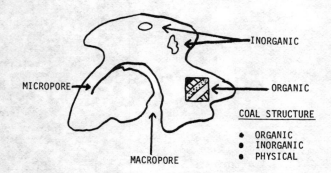

Fig. 1 Coal is a Rock
(thanks to Rick Schlosberg of Exxon for this drawing).

4

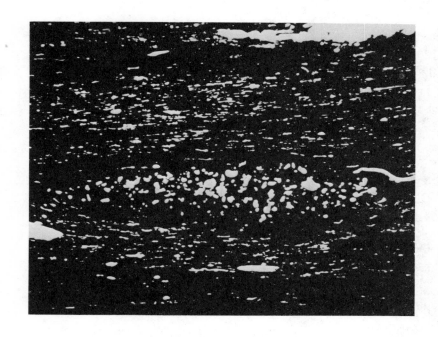

Fig. 2. Coal-A Thin Section
(thanks to Alan Davis, Penn State Univ. for this photograph).

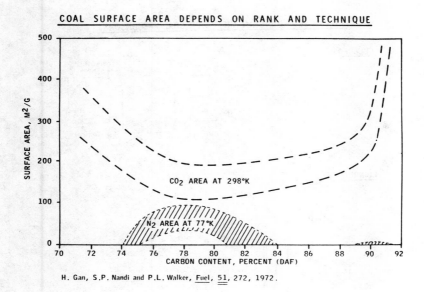

COAL SURFACE AREA DEPENDS ON RANK AND TECHNIQUE

H. Gan, S.P. Nandi and P.L. Walker, Fuel, 51, 272, 1972.

Fig. 3.   Coal Surface Area as a Function of Carbon Content.

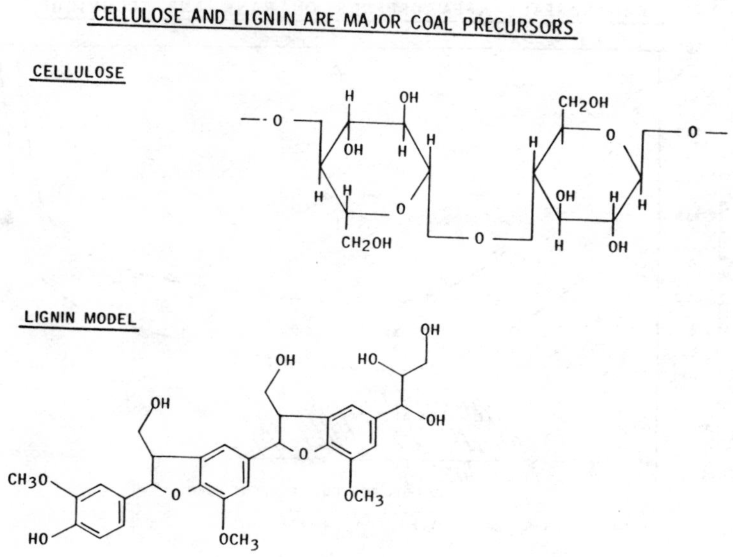

CELLULOSE AND LIGNIN ARE MAJOR COAL PRECURSORS

CELLULOSE

LIGNIN MODEL

Fig. 4. Coal Precursors.

It may be a young or an older coal, depending upon the degree to which this chemistry has progressed. From the data in Figure 5, it is clear what is happening: the amount of oxygen is decreasing fairly continuously, hydrogen decreases much less rapidly, and a big drop in hydrogen content does not come until the bituminous stage is nearly past. It is only with anthracties that what might be called chicken wire structures for coal begin to be important. In most lignites and coals, the size of the polynuclear aromatic units is not that large, and does not change much with rank. Oxygen is lost, and the way the oxygen functional groups change is shown in Figure 6.

COALIFICATION IS A DEOXYGENATION/AROMATIZATION PROCESS

		WT.% (DAF BASIS)		
		C	H	O
INCREASING SEVERITY OF GEOLOGICAL CONDITIONS	WOOD	49.3	6.7	44.4
	PEAT	60.5	5.6	33.8
PRESSURE TEMPERATURE TIME	LIGNITE	69.8	4.7	25.5
	SUB-BITUMINOUS	73.3	5.1	18.4
	BITUMINOUS	82.9	5.7	9.9
	ANTHRACITE	93.7	2.0	2.2

INCREASING AROMATIZATION
- DEHYDRATION
- LOSS OF "O" FUNCTIONALITIES
- ALKYLATION
- OLIGOMERIZATION

Fig. 5 Change in Elemental Analysis with Coalification
(thanks to Mike Siskin of Exxon for this figure).

8

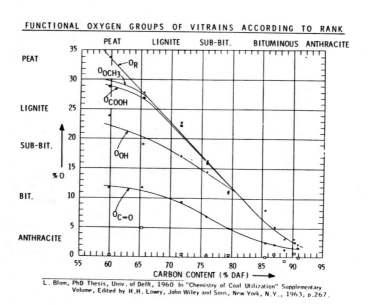

FUNCTIONAL OXYGEN GROUPS OF VITRAINS ACCORDING TO RANK

L. Blom, PhD Thesis, Univ. of Delft, 1960 In "Chemistry of Coal Utilization" Supplementary
Volume, Edited by H.H. Lowry, John Wiley and Sons, New York, N.Y., 1963, p.267.

Fig. 6.   Oxygen Functionality as a Function of Rank.

Almost all of the X-ray work on coal is fairly old. The X-ray
folks I know think it would be a good idea if much of this was redone
using modern X-ray technology and computer programs, but none of them
seem to be willing to do it themselves. There are some real problems
with the X-ray structures for coal. One is that the material is so
complicated that there is no unique solution for the scattering pattern. The problem is worked backwards; does a coal structure give
the observed scattering patterns? Not surprisingly, the picture
which emerges fits in very nicely with chemical intuition. Hirsch's
conclusions are summarized in Figure 7.[5] A mid rank coal consists of
layer structures. These are thought to be about 20 A° across and
partially oriented. In a mid rank coal, there is some stacking and
the stacks are twisted and at a variety of angles to each other, but
a fairly parallel kind of arrangement is emerging. In a higher rank
coal, the parallel kind of arrangement is much more apparent. As expected for this model, anthracites have very small pores. In the low
rank coal the layers have not begun to stack in a parallel fashion,

yielding a much more amorphous, much opener structure. The number of atoms in a planar aromatic layer, based primarily on Hirsch's analysis of the X-ray data, is not very large, around 18 per unit (Figure 8). It is interesting to note that there is no break in the curve around 86 %carbon. The point will come up again. Figure 9 shows a 20 A$^\circ$ square and will give you an idea of the size of the molecules occuring in a lamella structure detected by X-ray diffraction.

SCHEMATIC MODEL OF COAL STRUCTURE ACCORDING TO HIRSCH

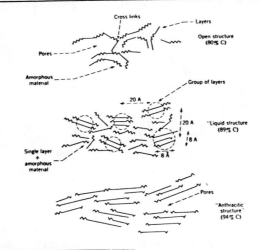

P.B. Hirsch, Proc. Royal Soc. (London), A226, 143 (1954).

Fig. 7. Coal Structure from X-Ray Evidence.

10

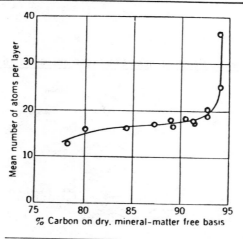

Variation of the average number of
planar atoms in an aromatic
layer with carbon content of coal.

P.B. Hirsch, Proc. Conf. Science in the Use of Coal,
Inst. Fuel, London, 1958, pp. A29–33.

Fig. 8. Average Number of Atoms per Planar
Structural Unit Based on X-Ray Data.

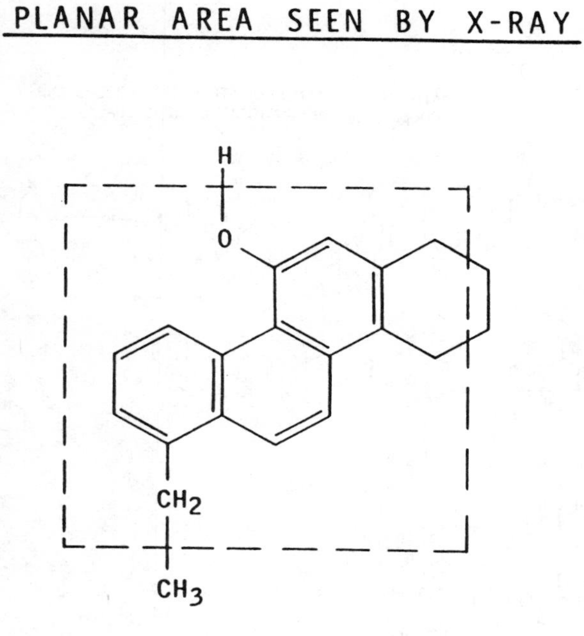

PLANAR AREA SEEN BY X-RAY

Fig. 9. Molecule Illustrating a 20 A° Plane
(thanks to Mike Siskin of Exxon for this drawing).

III. CHEMICAL STRUCTURE

Figure 10 is the "coal structure" currently in vogue. It is
derived from a large number of chemical and physical structural investigations of coals and represents the groups thought to be present in a bituminous rank vitrinite. It is worth emphasizing that
this is not a coal structure. It is a collection of the types of
structures thought to be present and written down together in a way
which satisfies the bulk analytical data. It is worth considering
what must be done before real coal structures can be written. I
believe that the organic chemist's approach to structure will not
work with coal. Organic structural techniques and representations
were developed for describing pure compounds. Coal is a complex
mixture. It must be described in terms of averages, average positions or average structures. I suggest that the best way of des-

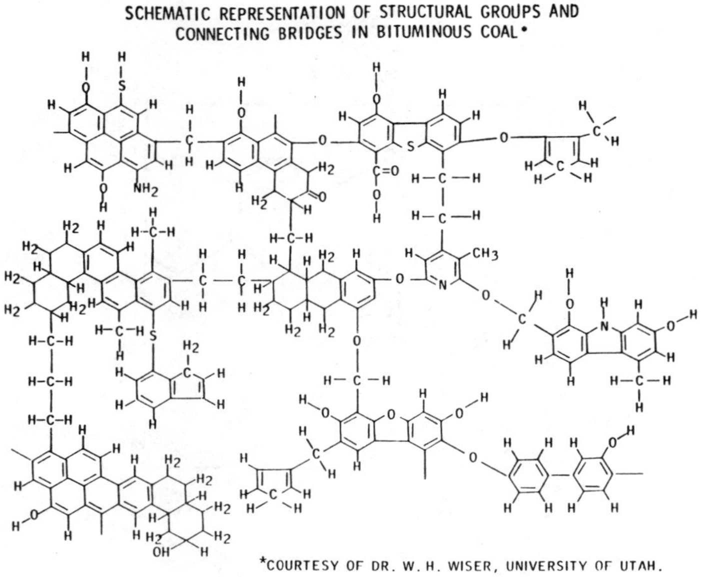

12

*COURTESY OF DR. W. H. WISER, UNIVERSITY OF UTAH.

Fig. 10. A Representation of the Structural Groups
and Connecting Linkages in Bituminous Coal

cribing coal would be to define the radial distribution functions
from a single point of all the important groups in a coal. This
would not only define quantity (i.e., how many OH groups), but also
would define average relative positions (i.e., how often a hydroxyl
and a benzylic methylene were adjacent). Wiser's structure gives a
nice overall picture. There are large aromatic clusters functional-
ized in reasonable ways linked together by short chains of methylene
groups and occasional ethers. It is a model which seems to hold up
fairly well. But it has flaws as pointed out by its originator.
For example, it is not known whether hydroxyl groups are apart or
close together. The spatial distribution of any functional group in
coal is not known. As suggested above, this complex material might
be best described using the techniques that chemical physicists use
to describe glasses, liquids and disordered solids. There is much
more information and accuracy in that kind of approach to coal
structure than the organic chemical structures now being written.

Chemically the chief difficulty in working with coal is not its unreactivity, but the fact that it is so very reactive. Try to do one reaction and there are a dozen occuring and they have to be sorted out. Another problem is reagent access. Reagents must be delivered into a polymer. Both of those problems are a long way from being understood.

My research group is interested in structure reactivity relations. Frank Mayo (SRI International) used to tell me constantly that coal must be a macromolecular solid. After all, it will swell up to double its volume in a good solvent without dissolving. It behaves like a rubber, like a three dimensionally cross-linked solid. Coal is also a viscoelastic material; with the right coals the elastic response is quite impressive. Again this is consistent with a three dimensional crosslinked structure. Nick Peppas'

TABLE 1. SWELLING OF FOUR PYRIDINE EXTRACTED COALS WITH A SERIES OF ORGANIC SOLVENTS

COAL	TOLUENE	ETHANOL	ACETONE	1,4 DIOXANE	PYRIDINE
N.D. LIGNITE	1.20	1.35	1.44	1.49	1.82
WYODAK	1.33	1.43	1.54	1.71	2.16
ILL. NO. 6	1.44	1.41	1.54	1.83	2.18
BRUCETON	1.43	1.39	1.46	1.78	2.03

Fig. 11.   Swelling of Pyridine Extracted Coals (wt %)
by Various Solvents.

14

excellent paper describes one way of investigating the macromolecular nature of coals. Figure 11 contains some swelling data for a
few coals in five solvents. The coals all have been exhaustively
extracted with pyridine to remove all of the small molecules. In
pyridine, the swelling is around 100 %; the coal has doubled in
weight. It has not quite doubled in volume because of some of that
solvent has gone to fill up the pores. One can use data based on
solvent swelling and the interaction of the swelling molecules with
coals together with some statistical mechanics, to calculate a very
significant number: the number average molecular weight per cross-
link ($\bar{M}_c$).[6] The data used in Figure 12 were measured by Sanada and
Honda, and by Kirov, O'Shea and Sergeant.[7] They used two very
different techniques but in both cases measured solvent swelling.
Two different techniques were used to obtain the Flory χ perameter,
the other quantity needed to calculate $\bar{M}_c$.

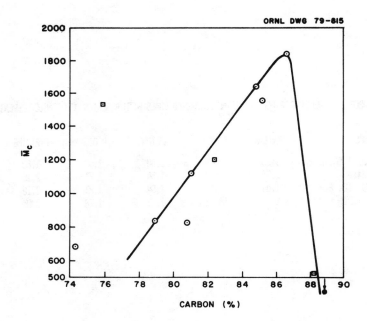

Fig. 12. Number Average Molecular Weight per Cross-Link
as a Function of Carbon Content.

The number average molecular weight per cross-link is a measure of the cross-link density. Consider relative coal reactivity. A loosely linked coal should be more reactive than a coal which is more highly cross-linked. Figure 12 shows a very sharp minimum in the cross-link denisty near 86 % carbon. If true, this is fascinating, and significant. Other work supports the existence of this minimum. In particular, the results of Van Krevelen's treatment of the coalification process as a condensation polymerization also leads to a minimum in cross-link density at 86 % carbon.[8] Van Krevelen argued as follows. The organic materials which are coal and lignite precursors are soluble in aqueous base and therefore do not have a cross-linked macromolecular structure. If they had a cross-linked gel structure they would not be soluble. Van Krevelen then treated the coalification process as a condensation polymerization of these materials with gel formation. This polymerization was subjected to the statistical mechanical treatment of condensation polymerization that was derived by Flory.[9] He used as a model the polymerization of two trifunctional monomers of equal reactivity. It seems absurdly simple, but the results are impressive and significant. The progress of the model polymerization is shown in Figure 13. At the start, there is 100 % monomer. The concentrations of the oligomers rise and then fall, and gel formation begins at 50 % reactions. If a particular reaction is obeying these statistics, the extent of reaction can be discovered by extracting from the polymer all of the soluble material, the sol. All of the molecules which have not been incorporated into the gel are soluble. For example, an extractability of 20% demonstrates a bit more than 50 % reactions. Knowning the extent of reaction, the molecular weight distribution of soluble material can be calculated. Van Krevelen exhaustively extracted four coals with pyridine and used the amount of extracted material to calculate the extent of the reaction. He then selectively precipitated molecules from those pyridine solutions by adding ligroin. The molecular weight of each of the precipitates was measured generating the molecular weight distribution of the sol. This was compared to that predicted by his model. This is shown in Figure 14. The agreement is quite remarkable. Now if this treatment is valid, the amount of material extracted can be associated with the degree of coalification, i.e., with the extent of reaction. The more highly cross-linked is the coal, the less material will be extractable from it. The less highly cross-linked is the coal, the more material will be extractable. Much information is available on the extractibility of coals into pyridine. It has a tendency to produce suspensions rather than true solutions and cannot be totally removed from the coal. This causes experimental problems and uncertainties. The data in Figure 15 are from Van Krevelen, as is the line on the Figure 15. The scattering is significant, but the shape of the curve-gradual up, sharp down with a maximum at 86%, is identical to that obtained from solvent swelling studies. This treatment of coalification is quite consistent with the existence of minimum in the cross-link density at about 86 % carbon.

16

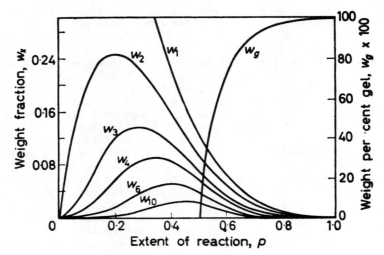

Weight fractions of various molecular species in a trifunctional polycondensation[5]

Fig. 13. Product Distribution as a Function of Extent of Reaction of Trio Trifunctional Monomers (from Ref. 8).

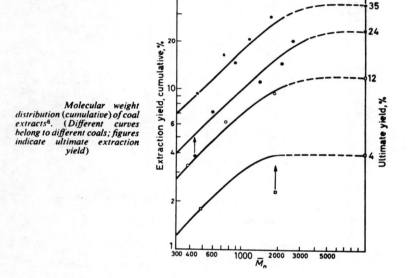

Molecular weight distribution (cumulative) of coal extracts[6]. (Different curves belong to different coals; figures indicate ultimate extraction yield)

Fig. 14. Molecular Weight Distribution of the Pyridine Extract from Four Coals (from Ref. 8).

18

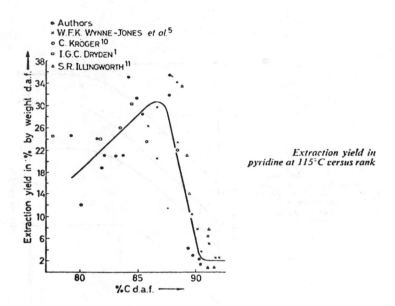

*Extraction yield in
pyridine at 115°C versus rank*

Fig. 15.   Yield of Pyridine Extract as a Function
of Carbon Content (from Ref. 9).

IV. MISCELLANEOUS PROPERTIES AND REACTIVITY

Next, some coal properties will be examined. Coal reactivity
under mild conditions passes through a maximum at 86 % carbon.
Figure 16 shows the conversion of a series of coals to pyridine
solubles on heating at 800 °F for three minutes in the Mobil synthetic SRC solvent.[10] The principle process is the transfer of
hydrogen from tetralin to the coal producing pyridine solubles. A
thermal depolymerization of coal in the presence of a hydrogen
donor is occuring (see Figure 17). Carried out for a short time
this is a mild depolymerization. Our hypothesis is that the
reaction time is so short that the cross-link density may be the
factor controlling reactivity. In a highly cross-linked coal, many
bonds must be broken to get soluble material out. Less soluble
product is obtained from a highly cross-linked coal than is obtained from a coal of low cross-link density. Figure 16 shows the

Mobil data and $\bar{M}_c$ plotted together. Mobil has also converted coal by heating it in pyrene. The pyrene serves to shuttle hydrogen from one place to another in the coal during the depolymerization. Again, there is a reactivity maximum at roughly 86-87 % carbon (Figure 18). The fluidity of coals measured with a Geisler Plastimiter also shows a maximum at 86 % carbon, although the shape of the curve is different (Figure 19). It is not unreasonable to expect a maximum in fluidity to occur at a minimum in cross-link density. Figure 20 shows a plot of spin-lattice relaxation time measured by nmr vs % carbon. Again, a maximum at 86 % carbon is visible. Originally interpreted in terms of network flexibility,[11] an explanation based on the diffusion rate of paramagnetic centers is more reasonable.[12]

More examples of coal properties which undergo changes at 86 % carbon can be given. These will suffice. Figures 21 and 22 show that nothing untoward is happening to the bulk of analytical data at 86 % carbon. We must look at the details of coal structure to find an explanation. Obviously, we would like to associate the changes in properties with a change in the degree of cross-linking. Indeed the agreement seems to be much more than coincidental.

In this paper we have tried to make two main points. One is that coal is extraordinarily complex. The other is that, in spite of its complexity, it can be treated in semi-quantitative ways and relationships between structure and behavior can be discerned and tested experimentally.

Acknowledgement

Support of our work in this area by the Department of Energy, Dow Chemical USA, and the Exxon Research Foundation is gratefully acknowledged.

20

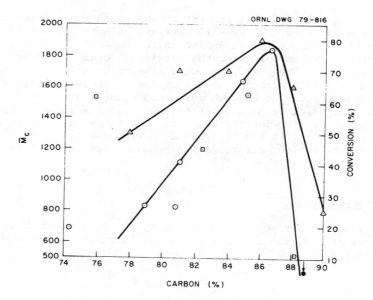

Fig. 16.   Conversion of Coals at Short Contact Time.

21

Pott-Broche Process

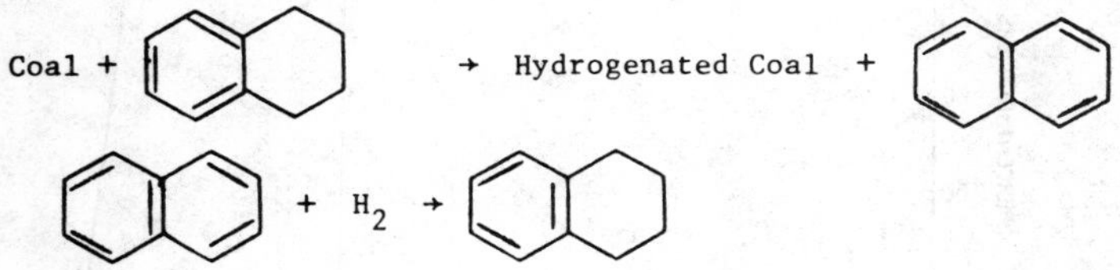

Coal + → Hydrogenated Coal +

+ H$_2$ →

Solvent Refined Coal Process Is Quite Similar

Fig. 17.   Schematic of a Hydrogen-Donor Liquefaction.

22

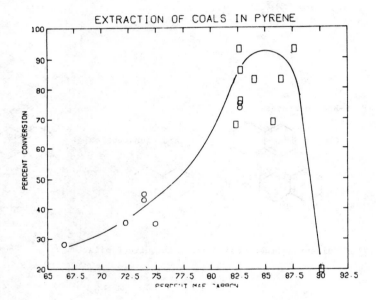

Fig. 18.   Conversion of Coal on Heating in Pyrene.

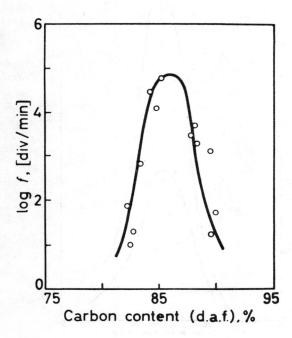

Fig. 19. Fluidity as a Function of
Carbon Content (from Ref. 3).

24

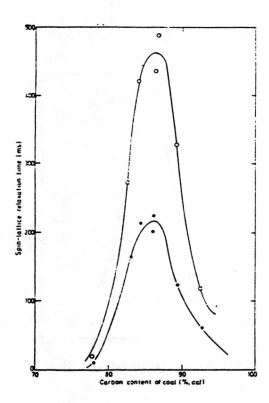

Fig. 20. Spin Lattice Relaxation Time as a
Function of Carbon Content (from Ref. 11).

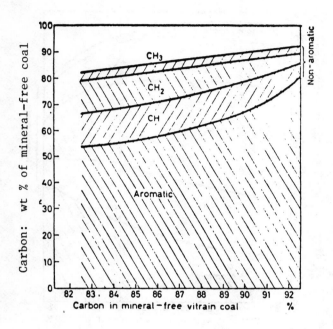

Fig. 21 Distribution of Carbon in Coals
(thanks to IGC Dryden for permission to use this figure).

26

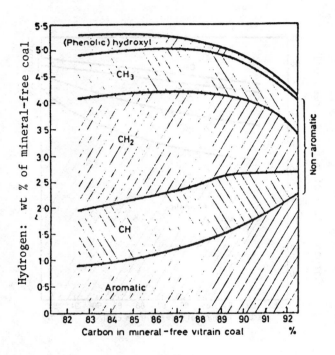

Fig. 22. Distribution of Hydrogen in Coals
(thanks to IGC Dryden for permission to use this figure).

REFERENCES

1. H. Gan, S. P. Nandi, and P. L. Walker, Jr., Fuel, 51, 272 (1972).
2. J. W. Larsen, P. Choudhury, T. K. Green and E. W. Kuemmerle, ACS Adv. in Chem., Vol. 000, M. L. Gorbaty Ed., 1980.
3. Y. Sanada and H. Honda, Fuel, 45, 295 (1966).
4. D. W. van Krevelen, "Coal", Elsevier, New York, 1961.
5. P. B. Hirsch, Proc. Roy. Soc. (London), A226, 143 (1954).
6. J. W. Larsen and J. Kovac, ACS Symposium Series, 71, 36 (1978).
7. N. Y. Kirov, J. M. O'Shea, and G. D. Sergeant, Fuel, 47, 415 (1968).
8. D. W. van Krevelen, Fuel, 45, 229 (1966).
9. H. N. M. Dormans and D. W. van Krevelen, Fuel, 39, 273 (1960).
10. D. D. Whitehurst, M. Farcasiu and T. O. Mitchell, "The Nature and Origin of Asphaltenes in Processed Coals," EPRI Report #AF-252, February 1976.
11. T. Yokono and Y. Sanada, Fuel, 57, 334 (1978).
12. T. Yokono, K. Miyazawa, Y. Sanada and H. Marsh, Fuel, 58, 896 (1979).

CROSSLINKED MACROMOLECULAR STRUCTURES IN BITUMINOUS COALS: THEORETICAL AND EXPERIMENTAL CONSIDERATIONS

Lucy M. Lucht and Nikolaos A. Peppas*
School of Chemical Engineering, Purdue University
West Lafayette, Indiana 47907

ABSTRACT

Ample evidence from physicochemical experiments suggests that bituminous coals can be described as highly crosslinked and entangled networks of macromolecular chains of irregular structure. Theoretically these structures can be analyzed by statistical mechanical models considering non-Gaussian distribution of the macromolecular chains along with departure from the Flory theories of crosslinked macromolecules. The models of Kovac (1978) and Peppas and Lucht (1979) have been developed in order to describe non-extractable coal matrices and their behavior during swelling in appropriate swelling agents. The molecular weight between crosslinks $\overline{M}_c$ and the crosslinking density ρ_x can be determined for various solvents and equilibrium swelling ratios. Few experimental data are available to which these models can be applied. Thus, in view of these new theoretical models, experimental research must be directed towards the reexamination of extraction and swelling behavior of bituminous coals. Some of the important parameters to be determined for characterization of the physical structure of coals include the thermodynamic interaction parameter χ, the crosslinking parameters $\overline{M}_c$ and ρ_x and the molecular weight distribution of the extractable coal portion.

INTRODUCTION

Investigation of the molecular structure of coals and study of the relationships between their structure and properties remain important problems in coal characterization despite considerable developments in recent years [1]. Since coal is a high molecular weight natural product exhibiting polydispersity and morphological inhomogeneity, the study of its structure is associated with a number of unique theoretical and experimental problems, requiring the careful accumulation of scientific information obtained by use of a wide range of experimental methods.

Coal structure is conceived at various levels and the scale of coal characterization is important in the development of structure-properties relationships. For example, one may look at the macroscopic or microscopic level, or investigations may be carried out for the establishment of the petrographic or macro-

* Author to whom correspondence should be addressed.

molecular structure. Recently, Larsen and Kovac [2] presented an interesting break-down of the coal structure in three levels related to various aspects of the macromolecular structure.

In this review paper we will discuss the macromolecular structure of bituminous coals. From the point of view of experimental investigations most of the work in recent years has concentrated on two aspects of this structure, that is the chemical nature of coal and the molecular weight and molecular weight distribution of coal matrices. However, very little information is available concerning the physical structure and the way the various macromolecular chains are distributed in the coal solid state [3].

From the coal processing point of view this last type of structural analysis may be the most important one for the prediction of coal properties directly related to specific types of coal liquefaction, and other related processes.

Considerable evidence in the literature, based mainly on extraction and swelling experiments suggests that bituminous coals consist of a highly crosslinked macromolecular network and other uncrosslinked macromolecular chains of size varying from 400 to 4000 depending on the experimental technique and determination method used [2], [4-7]. An analysis of the scientific evidence supportive of the crosslinked structure of bituminous coals is presented here.

EVIDENCE OF CROSSLINKED STRUCTURE OF COAL

Interpretation of the physical structure of macromolecular chains in coal with theories and models based on crosslinked macromolecular networks has gained popularity in recent years, although reference to these models has existed for more than twenty years in the European, Japanese and especially Russian literature [4]. Unfortunately some of this analysis is obscured by incomplete molecular theories, rather simplistic experiments and incorrect reference to unacceptable terminology such as the "polymeric nature" of bituminous coals.

The term "polymeric nature" of bituminous coals is both incorrect and deceiving. If one excludes mineral matter, ash and other impurities naturally found in coal, its structure can be described by two distinct but closely related phases. The first one includes small and large molecules of (only partially) determined molecular weight which are uncrosslinked and occupy a considerable fraction of the whole coal sample. These macromolecular chains may be either intact uncrosslinked molecules formed from the original matter during diagenesis [1], or partially degraded chains that were formed under appropriate temperature and pressure conditions through depolymerization reactions during the metamorphic stage of coal development.

The main part of the coal structure consists of a highly crosslinked macromolecular phase forming a crosslinking three-dimensional network (coal matrix). Present evidence suggests that most of this crosslinked phase is observed in vitrinite and inertinite with

less available in exinite [4]. Figure 1 presents a simplified
scheme of the physical structure of coal, showing various forms of
macromolecular chains and defects thereof.

The crosslinked structure cannot be extracted or dissolved at
low temperatures (unless the solvent is acting by combined diffusion
and reaction, i.e., depolymerization) and it is characterized by both
physical and chemical crosslinks. Physical crosslinks refer to
highly entangled macromolecular chains which, due to their inherent rigidity, exhibit restrictive mobility. Consequently, dis-
entanglement is highly unlikely to happen. Chemical crosslinks
have been formed by chemical reaction of two or more coal chains
leading to tetrafunctional crosslinks or crosslinks of higher or
lower functionality ($\phi \neq 4$).

Coal researchers have looked upon models of various "recon-
structed" chemical structures of coal as a means of supporting
their theories [1]. Thus, model networks such as the Wiser [8] or the
Given [9] models can be used only as indicative of the type of
chemical crosslinks one would expect in coal, but not as proof of
the physical structure of coal and the size of macromolecular
chains between crosslinks. Chain ends, unreacted functionalities,
chain loops, multifunctional crosslinks, etc. (see Fig. 1) are other
types of defects rendering an accurate analysis of the coal network
improbable.

In addition to all these general problems, macromolecular
chains in coal cannot be described by a repeating unit as in
conventional polymers. Such a repeating unit does not exist,
making the use of the term "polymer structure" unacceptable. Highly
irregular aromatic structures described in recent reviews [1,10] and
known to all coal researchers require that a hypothetical repeating
unit be established if one is to proceed with a statistical mechanical analysis of the coal matrix structure.

Invasive Techniques

Invasive experimental techniques supporting the existence of
crosslinked structures in coal are mainly based on extraction,
sorption, diffusion and swelling of fine coal particles, especially
with good "solvents" such as pyridine and tetrahydrofuran. The
term "solvent" is used here rather loosely, since these techniques
use organic solvents as swelling agents.

Early coal literature includes many references to extraction
of coal using pyridine, where one refers to extractable and non-
extracted matter [11]. The latter is believed to be a three-dimensional crosslinked coal network, which swells but does not dissolve. The extractable fraction is readily soluble and therefore
uncrosslinked. Specific research on the swelling behavior of coal
in pyridine and other solvents by Sanada and Honda [12,13], in methanol by Malherbe and Carman [14], in chloroform by Brown and Waters [15]
and in various solvents by Kirov et al [16] has shown beyond doubt
that a crosslinked structure is present at 25°C and that this
structure does not "dissolve" under any extraction conditions, as

long as the temperature is kept low and other degradation conditions are absent. None of these authors has attempted to describe the chemical nature of the network chains, although some
speculations were presented by Gadyatskii et al [17]. Even when
solvents with high electron donor thermodynamic capbility are used
[18,19] and appreciable solvation of coal at moderate temperatures
is observed there is still a considerable fraction of crosslinked
matter.

The two major references to the crosslinked structure of coal
are those of Van Krevelen [20,21] and Larsen and Kovac [2], and are
based on the findings of others [11-13,15] as well as partial unpublished data of the same authors [22]. Actually Van Krevelen [23]
seems to be the first one to recognize that regardless of the final
structure of depolymerized coal and its initial chemical structure,
its physical and chemical properties depend on the way the various
macromolecular chains are arranged and linked in its structure, i.e.
on its "polymeric character" (sic).

Other results from invasive experimental techniques supportive
of crosslinked structures include sorption and diffusion experiments on various types of coal [24,25].

In fact, East European researchers have supported the idea of
a crosslinked macromolecular coal structure for some thirty
years [4]. Three dimensional networks have been observed directly
or indirectly [26] as early as 1950. However, major questions arise
as to the validity of other of their beliefs, such as the alleged
chemical similarity of coal matrix and coal extracts and the
obvious but unsuccessful effort to "define" a repeating unit for
coal [4].

Non-Invasive Techniques

Non-invasive experimental techniques for the evaluation of the
crosslinked structure of coals include mostly mechanical testing
such as stress-strain behavior at low deformations [27,28] and
ultrasonic determination of shear and compressive moduli in coal
samples [29,30]. Some of these techniques and their findings have
been reviewed recently by Larsen and Kovac [2]. Through studies of
the elasticity of coal and the recoverable strain in stress-strain
experiments, general comments can be made about the existence of
crosslinked structures in coal. However, it is to be noted that
these data can be used only for qualitative purposes. Efforts to
determine molecular parameters from these data using elasticity
theories usually fail due to the high inhomogeneity and the
porous structure of the materials.

Chemical Nature of the Crosslinks

The question of the nature of the crosslinks in coal is a
point of considerable scientific dispute. Most of the work
related to the identification of coal crosslinks is based on
analysis of the products of depolymerization/degradation reactions.

In fact one of the traditional methods of structural investigation of high molecular weight organic substances in coal is based on their degradation and the study of the decomposition products. Because of the complexity of the system and the strong intermolecular interactions, the analysis of these results is rather difficult and often leads only to speculations about the chemical structure [1,31].

The most promising technique of analysis of the degradation products (coal liquids) appeared last year and it is based on mass-analyzed ion kinetic energy spectrometry (MIKES), a technique modified for use in coal liquids by Cooks and his collaborators [32,33]. Numerous other techniques have been used including elemental analysis and identification of decomposition and extraction products from coal [34-38], IR spectroscopy [39], high resolution NMR [41] and GC/mass spectrometry [38] just to name a few of the important investigations. Based on these results we have now a clearer picture of the types of chemical bonds involved in crosslinking. A good review of this subject is presented by Lazarov and Angelova [4]. Etheric, methylene or sulfidic bonds are the main types of crosslinks observed [4-6,37,42]. Recently Gibson [43] has attributed with some confidence these crosslinks to either oxygen or non-aromatic carbon bridges between neighboring aromatic groups, while Gorbaty et al [3] have discussed further research needs in this area.

Before any discussion of experimental characterizations of crosslinked parameters, some further analysis of the crosslinked networks will have to be presented.

THEORETICAL MODELS FOR COAL NETWORKS

As discussed earlier, the macromolecular networks observed in bituminous coals are highly crosslinked and entangled systems of chains of highly irregular structure. The actual spatial configuration of the macromolecular chains is not known. However, based on recent observations of the swelling behavior [2], equilibrium swelling ratio, chemical structure [1], and several references to the molecular weight distribution of extracted and degraded chains [44], we can postulate highly rigid chains exhibiting only minor flexibility, mainly where the aromatic structures are linked through methylene or etheric groups. Rotation is rather restricted and macromolecular models developed by Flory [45], such as the freely rotating chain and the freely jointed chain cannot describe coal chains.

Once the extractable macromolecular chains have been removed by an appropriate technique, the remaining three-dimensional macromolecular network can be described by a highly simplified network structure as shown in Figure 1. This representation facilitates the use of statistical mechanical theories and physicochemical analysis usually employed in the interpretation of the structure of crosslinked polymers. Since a repeating unit is needed to describe the macromolecular chains, we can define a hypothetical unit of molecular weight m for which the only necessary assumption is that it consists of a highly aromatic structure contributing

33

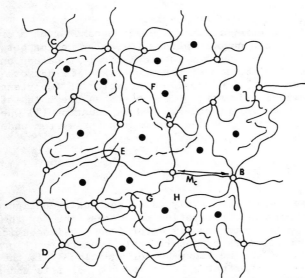

Fig. 1: Simplified representation of the crosslinked structure of coal including possible defects. ——— : Chains participating in network structure; — — — : extractable (unreacted or degraded) chains; $\bigcirc$: crosslinks (junctions); ● molecules of swelling agent; $\overline{M}_c$: molecular weight between crosslinks; A: tetrafunctional crosslink; B: multifunctional crosslink; C: unreacted functionalities; D: chain end; E: entanglement; F: chain loop; G: effective network chain; H: mesh size.

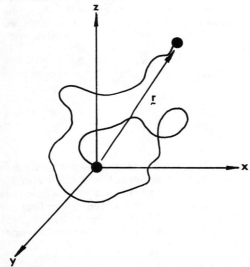

Fig. 2: Spatial configuration of macromolecular chain with end-to end vector $\underset{\sim}{r}$.

to the overall rigidity of the chain.

The single most important molecular parameter of a cross-linked network is the number average molecular weight between crosslinks $\overline{M}_C$. This parameter is statistical in nature and depends on the distribution of the mesh sizes of a network as defined by deGennes [46]. A derived parameter of importance is also the crosslinking density ρ_x expressed as the moles of crosslinks per unit volume and defined by equation (1) where υ is the specific volume of the macromolecular system under consideration.

$$\rho_x = \frac{1}{\upsilon \overline{M}_c} \tag{1}$$

Statistical mechanical analysis of the chain configuration in crosslinked networks and derivation of expressions for the calculation of $\overline{M}_c$ in terms of readily determined physicochemical parameters have been subjects of considerable research by Flory [45,47] and his coworkers [48]. The Flory-Rehner equation [48] is the mode of analysis of most polymer networks in terms of elastic and swelling behavior. Unfortunately this theory inherently assumes that the chains between crosslinks are flexible and long enough so that a Gaussian distribution function can be assumed. In most polymer systems the spatial configuration of a macromolecular chain taking the origin of coordinates at one end of the chain (see Figure 2) is expressed by the probability $W(x,y,z)$ dxdydz that the components of the end-to-end vector $\underset{\sim}{r}$ are x to x + dx, y to y + dy and z to z + dz or that the other end of the macromolecular chain lies in the volume element of size dxdydz which is located at x, y and z. The Gaussian distribution is given by equation (2) where β is a parameter which for freely jointed chains is expressed by equation (3) where ℓ is the bond length and n is the number of links of the macromolecular chain.

$$W(x,y,z)dxdydz = \frac{\beta^3}{\pi^{3/2}} e^{-\beta^2 r^2} dxdydz \tag{2}$$

with

$$\beta = (\frac{3}{2n\ell^2})^{1/2} \tag{3}$$

The Gaussian distribution is unsatisfactory at high extensions, and when the end-to-end vector approaches full extension of the chain. The latter is a case similar to the one observed with coal macromolecules. Then one can use modified distributions derived by James and Guth [49], Kuhn and Grün [50] and other investigators [51]. One such approach is through the inverse Langevin function L^*, when the distribution becomes

$$W(r)dr = C \exp[- \int_o^r L^* (\frac{r}{n\ell}) \, dr/\ell] 4\pi r^2 dr \tag{4}$$

and where the expression in brackets can be analyzed in terms of convenient series expansions.

Obviously analysis of highly crosslinked networks such as those expected in coal structures cannot follow Flory's theories. It must be based either on a different treatment of the statistics of chains or on incorporation of specific structural characteristics of these networks. For example non-Gaussian statistics [51,52] have been used by some investigators. Others have incorporated theories of nonaffine deformation of the crosslinks, restricted junction fluctuations [53], multifunctional junctions [54], etc. There are even recent analyses using the reptation theory [55]. This model is of particular interest when one considers the physical crosslinks (entanglements) of the networks and their mobility with respect to the whole network. In coal matrices, there is no information as to the existence of entanglements, especially because of the lack of rheological data on this system. Some of these data will be obtained in our laboratory with sensitive rheological equipment such as the rotary rheometer or the mechanical spectormeter. A reasonable hypothesis proposed by Larsen [2] and supported also by these authors is that very few entanglements could exist in coal mainly due to the short chains observed. It is also assumed that most of the entanglements would be permanent due to the high crosslinking of the whole network.

A typical procedure for the description of the behavior of highly crosslinked networks has been recently reported by Kovac [51]. The macromolecular chain is defined as a set of bond vectors and the distribution and partition functions can be written in terms of the end-to-end vector and the equilibrium force f to which the macromolecular chain is subjected. The distribution function is then expanded according to a specific series function, giving the modified distribution which is used to calculate exact expressions of the end-to-end distance as a function of applied force. In addition, the entropic change can be calculated from this expression as $T\Delta S_{e\ell}$.

This statistical mechanical analysis is meaningless unless applied to a specific physicochemical experiment, whereby one can calculate M_c and ρ_x through the swelling behavior. In a fashion similar to the one followed by Flory and Rehner [47,48], recent analyses [51,56] consider the change of free energy during swelling ΔF as the summation of the change of free energy due to dilution of the macromolecular chains ΔF_{mix} and the change of free energy of the elastic chains $\Delta F_{e\ell}$.

$$\Delta F = \Delta F_{mix} + \Delta F_{e\ell} \tag{5}$$

The free energy due to thermodynamic interactions is conveniently expressed by the Flory-Huggins equation (6),

$$\Delta F_{mix} = RT[\ell n(1-\upsilon_2) + \upsilon_2 + \chi\upsilon_2^2] \tag{6}$$

while the change of free energy due to the elastic forces in the network is expressed by the $- T\Delta S_{e\ell}$ term derived earlier, where

$\Delta H_{e\ell}$ is assumed as negligible.

Based on this type of analysis three general expressions (equations 7,8 and 9) can be used to describe the swelling behavior of coal networks and to determine $\overline{M}_c$. The first is a direct result of the Flory-Rehner equation[48] as modified by Peppas and Merrill[57]. This equation applies to loosely crosslinked networks and is thus not recommended for coal structures although it is presented here for the sake of comparison.

$$\frac{1}{\overline{M}_c} = \frac{\frac{\overline{\upsilon}}{V_1}[\ln(1-\upsilon_{2,s}) + \upsilon_{2,s} + \chi_1 \upsilon_{2,s}^2]}{[\frac{1}{2} \upsilon_{2,s} - \upsilon_{2,s}^{1/3}]} \tag{7}$$

The Kovac model[51] describes swelling of a modified isotropic Gaussian network with $\Delta H_{e\ell} = 0$

$$\frac{1}{\overline{M}_c} = \frac{-\frac{\overline{\upsilon}}{V_1}[\ln(1-\upsilon_{2,s})+\upsilon_{2,s}+\chi_1 \upsilon_{2,s}^2][1-\frac{3}{N}\upsilon_{2,s}^{-2/3}+ \frac{3}{N^2} \upsilon_{2,s}^{-4/3} - \frac{1}{N^3}\upsilon_{2,s}^{-2}]}{\upsilon_{2,s}^{1/3} [1 + \frac{2}{N} \upsilon_{2,s}^{-2/3} + \frac{1}{N^2} \upsilon_{2,s}^{2/3}]} \tag{8}$$

A similar expression for a modified isotropic Gaussian network is derived from the Peppas-Lucht model[56].

$$\frac{1}{\overline{M}_c} = \frac{\frac{\overline{\upsilon}}{V_1}[\ln (1-\upsilon_{2,s}) + \upsilon_{2,s} + \chi_1 \upsilon_{2,s}^2][1 - \frac{1}{N} \upsilon_{2,s}^{2/3}]^3}{[\frac{1}{2} \upsilon_{2,s} - \upsilon_{2,s}^{1/3}][1 + \frac{1}{N} \upsilon_{2,s}^{1/3}]^2} \tag{9}$$

In all these equations $\overline{M}_c$ is the number average molecular weight between crosslinks, $\overline{\upsilon}$ and V_1 are the specific volume of the macromolecule and the molar volume of the swelling agent, $\upsilon_{2,s}$ is the final equilibrium swelling volume fraction of polymer in the swollen macromolecular network and χ is the macromolecule/swelling agent thermodynamic interaction parameter according to Flory[47], which depends on temperature T and $\upsilon_{2,s}$. It is to be noted that whereas the Flory equation is independent of the number of repeating units between two crosslinks, since it was derived for long flexible chains, the other two models include the parameter N which expresses the number of links in short-chain models. Both models are expressed in terms of N since this parameter defines the deviation from Gaussian (Flory) behavior.

Comparison of the three models shows that values predicted for $\overline{M}_c$ using the non-Gaussian models[51,56] vary considerably from Flory's Gaussian model[47]. For example Table I shows computer calculations of $\overline{M}_c$ for two coal samples of density $\rho = 1.3$ gr/cm^3 in two hypothetical solvents having interaction parameters of

χ = 0.79 and 0.51 and equilibrium swelling volume fractions of $\upsilon_{2,s}$ = 0.62.

TABLE I
Computer Simulation of $\overline{M}_c$

Model	Molecular Weight Between Crosslinks	
	Sample I χ = 0.79, $\upsilon_{2,s}$= 0.62	Sample II χ = 0.51 $\upsilon_{2,s}$= 0.62
Flory [47] [eqn.7]	1300	400
Kovac [51] [eqn.8]	2450	750
Peppas–Lucht [56] [eqn.9]	2200	700

Use of these models for the determination of $\overline{M}_c$ in coal networks is not without problems. For one thing, assumptions made in the derivation of these equations include short but linear chains between crosslinks, affine network behavior, lack of defects such as loops and multifunctional crosslinks, negligible entanglements that cannot disentangle during swelling, and swelling behavior that can be described by the Flory-Huggins thermodynamic theory. In addition this analysis must be applied to network structures free of impurities (such as mineral matter), pore structures and unreacted chains.

In applying these equations for the determination of $\overline{M}_c$, the molar volume of the swelling agent V_1 is readily available and the specific volume of the coal sample under experimental investigation $\overline{\upsilon}$ can be calculated as the reciprocal of the density of the coal matrix. The average molecular weight $\overline{M}_c$ depends on the equilibrium volume swelling ratio Q and/or the coal volume fraction at swelling equilibrium $\upsilon_{2,s}$ defined as

$$\upsilon_{2,s} = \frac{1}{Q} = \frac{V_c}{V_{s,c}} \qquad (10)$$

where V_c and $V_{s,c}$ are the volumes of "dry" and swollen coal. During application of this equation care must be taken to determine and exclude the quantity of swelling agent absorbed in the porous structure and therefore not effectively contributing to the swelling of the coal network.

The most sensitive dependence of $\overline{M}_c$ is on the value of the thermodynamic interaction parameter χ. Computer simulation and calculation of $\overline{M}_c$ for various values of χ factors (Figure 3) shows the effect of a thermodynamically "good" solvent on the accurate determination of crosslinking parameters. Obviously, the

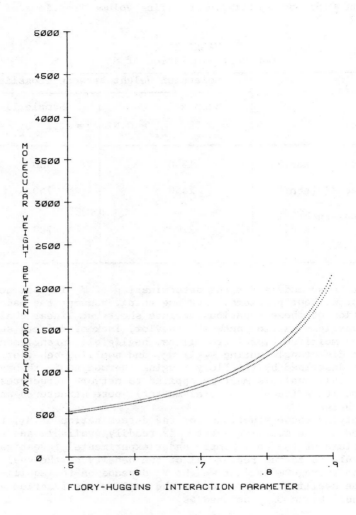

Fig. 3: Effect of χ-factor on the determination of $\overline{M}_c$, the number
average molecular weight between crosslinks, using equation (9).
Data obtained by computer simulation using a hypothetical coal of
$\rho=1.3$ gr/cm^3, $\upsilon_{2,s}=0.62$ and two solvents with $V_1=88.7$ cm^3/mole
(upper curve) and $V_1=85.3$ cm^7/mole (lower curve).

accurate determination of $\overline{M}_c$ depends on the selection of the proper swelling agent.

ATTEMPTS TO CHARACTERIZE THE NETWORK STRUCTURE

Some previous work has been done to characterize the network structure of bituminous coals. Two groups in particular, Sanada and Honda [12,13], and Kirov et al [16] have calculated $\overline{M}_c$ for coal networks by applying the Flory-Rehner equation to swelling data. These results are presented in Table II. The data of Sanada and Honda have been corrected and recalculated for the apparent omission of the term V_1 in eqn (7). Also, the data of Kirov et al were recalculated as molecular weights $\overline{M}_c$ instead of the reported volume of crosslinks $\overline{V}_c$, a term not usually used by polymer scientists. Qualitatively, the results do not agree with each other. Kirov et al contend that the molecular weight between crosslinks M_c should drop continuously as the rank of the coal goes up. They support this belief by likening the coalification process to a trifunctional polycondensation, as proposed by Van Krevelen [21]. However, these data are not sufficient to justify this conclusion.

Sanada and Honda's data show that the molecular weight between crosslinks reaches a maximum at approximately 85 - 88%C (daf basis), dropping rapidly thereafter. It is our belief that the negative values of M_c result from the technique used in determining the thermodynamic interaction parameter χ. Since the creep compliance at 350°C and the maximum fluidity of coals as measured by the Gieseler plastometer also have maxima at approximately 85%C, there is a further possible connection between the physical response and the macromolecular structure of coal. More support for the trends of Sanada and Honda's data comes from the average molecular weights of depolymerized or alkylated coal products which show the same qualitative behavior [58]. The molecular weights of the degraded products are equal to or smaller than $\overline{M}_c$ and they reflect the location of functionalities where crosslinks can occur.

Aside from the qualitative disagreement of these data, there are flaws common to both analyses. First, the Flory-Rehner equation was developed for an affine network composed of Gaussian chains. Certainly the chains in the coal matrix are too short and stiff to follow a Gaussian distribution. This problem is tractable, however, using the modified Gaussian model developed by Kovac [51] or these authors [56]. The second problem lies in the determination of the thermodynamic coal-solvent interaction parameter χ. Sanada and Honda calculated χ from the osmotic pressure data of Wynne-Jones et al [11] for coal extracts in pyridine solution. This technique assumes that: i) coal extracts interact with a solvent in the same way that the coal matrix does; ii) χ factors are the same for a given coal rank, regardless of other parameter values; and iii) χ is independent of the concentration of extract in the solvent. Under the circumstances, these were not bad assumptions to make; the first one is necessary for most

TABLE II

Recalculated and corrected values of number average molecular weight between crosslinks $\bar{M}_c$ using data of Sanada and Honda [12,13] and Kirov et al [16].

Carbon content %C, daf	Coal density ρ_2 (gr/cm³)	Swelling Volume fraction $\upsilon_{2,s}$	Chi (χ) factor	Recalculated $\bar{M}_c$ from Kirov et al [16]	Recalculated $\bar{M}_c$ from Sanada and Honda [12,13]
65.1	----	0.524	----	----	1185
74.3	1.39	0.602	0.5	----	685
75.9	1.30*	----	----	2595	----
79.0	1.35	0.516	0.5	----	1140
80.9	1.33	0.549	0.57	----	1160
81.1	1.32*	0.512	0.63	----	2110
82.4	1.30*	----	----	2100	----
84.9	1.24	0.529	0.78	----	15130+
85.2	1.28	0.556	0.78	----	5746
86.6	1.26*	0.507	0.78	----	negative+
88.2	1.30*	----	----	940	----
88.7	1.31	0.979	0.80**	----	50 x
93.0	1.41	0.986	0.80**	----	45 x

*   Assumed values of ρ_2

**   Assumed values of χ^2

\+   Unreasonable values of $\underline{M_c}$, attributable to incorrect determination of χ factor.

x   Unreasonable values of $\underline{M_c}$, attributable to lack of values for χ factor.

methods of determining χ (except GLC). The second assumption
requires proof considering the crucial dependence of $\bar{M}_c$ on χ.
The third one is a reasonable approximation for most polymer
systems, but not for coal networks.

Kirov et al [16] used a semi-empirical method to determine δ_2,
the solubility parameter of coal. They started out with the
assumption that χ is a sum of the entropic and enthalpic mixing
interactions in swelling:

$$\chi = \chi_s + \chi_H \simeq \beta + \frac{V_1}{RT} A_{12} \tag{11}$$

where A_{12} is the interchange energy density and β is an entropic
contribution defined in terms of the coordination number of the
liquid lattice model [47]. Then, the χ factor can be expressed as

$$\chi \simeq \beta + \frac{V_1}{RT} (\delta_1 - \delta_2)^2 \tag{12}$$

A rearrangement of this last equation was used to graphically
calculate δ_2 in a trial and error procedure. The key problem with
this technique is that the equation ignores solvent-specific effects.
Nonetheless, we used this equation to calculate values of χ versus
carbon content using the values of δ_2 presented in Kirov's work [16].
These values are shown in Table III and can be used for preliminary
determinations of $\bar{M}_c$ using the previously developed models. It
remains to be seen whether these values are in agreement with our
experimental data presently under generation in our laboratory.

Besides swelling experiments it has been claimed that it is
also possible to determine the crosslinking density through mechanical tests. Larsen[2] has calculated $\bar{M}_c$ using data for Young's
moduli. These calculations are singularly unsuccessful, often
yielding unrealistic values of $\bar{M}_c$. These may reflect many phenomena; possibly, as they suggest[2c], there is an increase in inter-
molecular interactions due to the increased order in unidirectional
stress, or hindrance of the chain movement because of molecules in
the interstices of the unextracted coal. However, we believe that
use of mechanical testing data of solid porous coals to determine
molecular parameters of coal networks is not an acceptable method,
and it is bound to show significant deviations from data obtained
through swelling experiments. A possible solution of this problem
may be found in the determination of the dynamic mechanical properties of swollen coal networks in an appropriate rheological
equipment. Efforts to characterize coal matrices in view of these
clarifications are being made in our laboratory, as well as by
J. Larsen [22].

EXPERIMENTAL CONSIDERATIONS

The ultimate purpose of the procedures outlined here is to
elucidate the statistical average network structure and to determine the molecular weight between crosslinks $\bar{M}_c$ that can represent
an average three-dimensional network of a specific coal sample.

TABLE III

Determination of χ-factors from solubility parameter data using Kirov et al [16] values for three coal samples.

Solvent	δ_1	V_1 cm³/mole	$\chi \equiv$ Thermodynamic Interaction Parameter		
			$\delta_2=9.8$ Hebe/75.9%C	$\delta_2=9.5$ Greta/82.4% C	$\delta_2=9.3$ Bulli/88.2%C
Acetone	10.0	73.53	.300	.331	.360
Aniline	10.4	91.12	.355	.424	.585
Benzene	9.1	88.9	.373	.324	.306
Carbon Tetrachloride	9.5	96.5	.315	.300	.306
Chloroform	9.3	80.5	.334	.305	.300
Cyclohexane	8.2	108.1	.764	.606	.519
DMF*	11.5	77.4	.675	.819	.929
Dioxan	9.9	85.3	.301	.323	.351
Ethanol	12.7	58.3	1.123	1.302	1.431
Methanol	14.5	40.5	1.801	1.999	2.137
*Methylethyl Ketone	9.8	89.5	.300	.313	.337
Nitrobenzene	10.0	102.2	.307	.343	.384
N-Propanol	11.9	74.8	.853	1.023	1.148
*Pyridine	10.6	80.56	.386	.463	.528
*Quindine	10.7	118.2	.461	.586	.689
Toluene	8.9	106.4	.445	.364	.329
Xylene	8.8	123.5	.507	.402	.352

* Solvents considered to have specific effects

There are three main experimental steps for the determination of $\overline{M}_c$:
- i. "idealization" of the raw coal;
- ii. determination of χ, the thermodynamic interaction parameter;
- iii. swelling of the "idealized" coal matrix.

Idealization is essentially a series of separations performed to make the coal more like the system described by the swelling equations presented previously.

Flotation is a process designed to separate the minerals, heavy metals and other non-ideal coal constituents from the coal matrix. A solution whose specific gravity is less than 1.3 is made, frequently from benzene and carbon tetrachloride. The coal sample is placed in solution and the portion which floats is separated and taken as the vitrinite portion. The sinks contain the minerals, heavy metals, and other heavier macerals. The vitrinite sample received contains (typically) less than six percent ash [1] and it is the maceral thought to be most like the ideal three-dimensional network.

Partical size separation follows, after drying of the coal matrix. Since coal particles of different sizes have been shown to imbibe different quantities of solvent, the particle size and particle size distribution may be important in liquefaction. As the particle size decreases imbibition increases initially, and then decreases for very small particles. Also the process of coal grinding creates localized high temperature and pressure regions which may cause further crosslinking and/or degradation. Therefore a particle size difference could be indicative of a chemical difference.

Extraction separates the coal matrix from material which is loosely held or free in the interstices. This loose or extractable portion also acts as a filler, which would interfere with the analysis of swelling data. One of the most common extraction techniques is Soxhlet extraction, which is basically a continuous reflux of pure solvent through the material to be extracted. This procedure is usually done at the atmospheric boiling point of the solvent, but possible degradation of the coal matrix can be avoided by lowering the boiling point temperature with vacuum application. Other extraction techniques include trickling solvent through a column packed with coal or simply placing coal in the solvent at the desired temperature and pressure conditions under stirring. The last method is the least desirable as the sample is not contacted with pure solvent except initially. For good extraction solvents, the extract yield ranges from up to 5 wt % for anthracites, to 10-20 wt % for bituminous coals and 40-50 wt % for lignites [1]. The coal matrix and coal extracts obtained after extraction are separated in preparation for the next step of the experimental procedure.

Techniques for the determination of χ include membrane and vapor osmometry, solvent freezing point depression, light scattering, intrinsic viscosity and gas-liquid chromatography (GLC) [59]. Since the volume fraction of coal in the swollen state $\upsilon_{2,s}$

ranges usually from 0.5 to 1.0, membrane osmometry, light scattering
and intrinsic viscosity are not widely used, because of their
inherent trend to determine properties at the limit of $\upsilon_2 \to 0$.
However, freezing point depression and vapor osmometry lead to
readily measurable quantities even at low non-zero polymer volume
fractions while GLC covers the range of $\upsilon_2 = 0.5 - 1.0$.

Thermodynamically the χ factor is an expression of the residual
chemical potential [47] normalized by RT υ_2^2, as described by equation
(13).

$$\chi = \frac{(\mu_1 - \mu_1^0)^r}{RT \; \upsilon_2^2} = \frac{(\mu_1 - \mu_1^0)}{RT \; \upsilon_2^2} - \frac{[\ln (1 - \upsilon_2) + \upsilon_2 (1 - 1/x)]}{\upsilon_2^2}$$

(13)

where μ_1 and μ_1^0 refer to the change of free energy resulting from
the addition of one mole of solvent to an infinite amount of
solution or to the pure component respectively, both at constant
temperature and pressure; the superscript r refers to the residual
thermodynamic quantities and x is a parameter characteristic of the
size of the macromolecule with respect to the solvent in the liquid
lattice model as defined in Flory [47]. The residual chemical
potential shows the deviation of the actual thermodynamic quantity
from the reference state quantity. Usually the reference state is
described by the heat of mixing being equal to zero and the
entropy of mixing being determined by the Flory-Huggins lattice
model. Thus, it is obvious that the χ factor is dimensionless,
temperature and composition dependent, and that it increases as the
$\mu_1 - \mu_1^0$ term increases at a fixed temperature and polymer volume
fraction.

Having established a rigorous thermodynamic definition of
this interaction parameter we can determine its behavior for coal/
solvent system using one or more of the experimental methods
mentioned above. Convenient expressions for freezing point depression and membrane osmometry are available elsewhere [59]. Membrane
osmometry evaluations will require (i) appropriate selection of a
membrane with cutoff molecular weight of 2000 and (ii) verification of the assumption that the chemical structure of the
extractable coal portion is "similar" to the structure of the
crosslinked phase. Only then can it be assumed that the thermodynamic interaction parameter determined from extractable coal
samples can be applied to the crosslinked network as well. The
same assumption will have to be used when analyzing data from vapor
osmometry or freezing-point depression [59]. For these reasons,
these techniques are recommended only as alternatives to the
technique of gas-liquid chromatography (GLC).

In GLC, beads are usually coated with the macromolecular
sample and packed as the stationary phase in a chromatographic
column. The low molecular weight solvent is introduced into the

column with an inert carrier gas and its tendency to be absorbed
by the macromolecular system is measured in terms of the retention volume V_g, which is a function of the χ factor. At the limit
of $\upsilon_2 = 1$, the χ factor is determined according to equation (14)

$$\chi = \ln \frac{RTV_2}{\rho_1 V_1 V_g} - (1 - \frac{1}{x}) - P_1^0 \frac{B_{11} - V_1}{RT} \qquad (14)$$

Here V_1 and V_2 are the specific volumes of the solvent and the
macromolecular system, P_1^0 is the vapor pressure of the pure solvent
and B_{11} is the gaseous second virial coefficient for the solvent.
The other parameters have been defined earlier. This technique
has been extended to determination of χ at volume fractions of
$\upsilon_2 = 0.5 - 1.0$ by Brockmeier et al [60].

Although use of this technique for coal samples initially
requires coating of the beads with coal extracts, it may be possible to pack the chromatographic column directly with extracted coal
particles, so that the χ factor of the network can be determined
without any unnecessary assumptions. The porous nature of coal
samples complicates the determination of the retention volume
as some of the solvent will undoubtedly be entrapped in the pores.
Therefore, knowledge of the internal volume of the pores and
surface area of the particles is necessary.

The final step of the experimental analysis refers to the
swelling behavior of coal matrices. Experimentally an extracted
coal sample is surface cleaned by moderate vacuum (usually 10^{-4}
to 10^{-6} torr) in an electromicrobalance or a related apparatus
and it is exposed to the vapors of the swelling agent at constant
temperature and pressure. Thermodynamically, equilibrium is
reached when, in the presence of excess solvent, the chemical
potential of the solvent external to the coal matrix equals the
chemical potential of the solvent inside the matrix. Each set of
temperature and pressure conditions will yield a different equilibrium condition, namely a different equilibrium chemical potential
and swelling ratio.

Problems related to experimental swelling determinations
include heterogeneities within the coal sample due to maceral,
mineral and ash content as well as molecular heterogeneities
induced by multifunctional crosslinks, loops, entanglements, lack
of a distinct repeating unit etc. In addition, due to the complex
macroscopic structure of coal, swelling is not a simple phenomenon.
Simultaneous diffusion, adsorption and absorption occur, as well
as sorption and filling of the porous structure by the solvent.
Again, the void fraction and surface area are important. This
problem has been recognized by several authors and various techniques have been discussed in recent years [14, 61-63].

In addition to the three main steps of experimental characterization of the crosslinked structure of coal, other techniques can
provide information supportive of this structure. For example
recent unpublished data of our laboratory show a correlation

46

between the softening point of coal T_s, as determined by differential scanning calorimetry, and its physical structure. Determination of the molecular weight distributon of extracted and preferentially degraded chains can be used as an indication of the original crosslinked structure. Also, chemical structure analysis is needed for a better understanding of the nature and function of various functionalities.

In conclusion, the physicochemical analysis of macromolecular (polymeric) systems provides a plethora of experimental techniques, most of which can be used for the analysis of the crosslinked macromolecular structure of coal samples. Use of these techniques is not without problems, mainly due to the nonhomogeneity, anisotropy and complexity of the coal structure both in the molecular and macroscopic level. However, recognition and understanding of the differences between coal and polymers and experimental ingenuity may lead to novel techniques for coal characterization.

CONCLUDING REMARKS

Analysis of the physical behavior of macromolecular networks in bituminous coals and determination of the number average molecular weight between crosslinks and the crosslinking density of the network are important in the development of structure/properties relationships of coal samples and in the elucidation of technical problems during coal extraction, liquefaction and modification through chemical reactions. Although coal consists of a highly crosslinked and entangled structure of macromolecular chains, several tested simplifications may lead to the application of statistical mechanics and swelling theories for the determination of the physical characteristics of these networks. However, the use of these theories and experimental procedures can lead, at the best, only to an approximate quantitative analysis of the structure. For a thorough understanding of macromolecular structures in coal, additional data of molecular weights and distributions of extractable and preferentially degraded coal liquids are needed. These data will have to be compared to the M_c's determined for the coal matrix. Possible general relationships between chemical structure, elemental analysis and $\overline{M}_c$ are presently under investigation.

We believe that the results of this work and novel analytical approach will have potential for applications in coal processing, expecially in elucidation of coal liquefaction processes and the kinetics of various reactions involving coal depolymerization at higher temperatures.

Acknowledgements

This work was supported by a research grant from the Department of Energy, Contract No. ET-78-G-01-3382. Helpful discussions with Professors J.W. Larsen of the University of Tennessee, and R.G. Cooks of Purdue University and with Dr. M.L. Gorbaty of Exxon Research and Engineering Company are kindly acknowledged.

REFERENCES

1. N. Berkowitz, An Introduction to Coal Technology, Academic Press, New York, 1979.
2. J.W. Larsen and J. Kovac, in J.W. Larsen, ed., Organic Chemistry of Coal, American Chemical Society Symposium Series, Vol. 71, p. 36, 1978.
3. M.L. Gorbaty, F.J. Wright, R.K. Lyon, R.B. Long, R.H. Schlosberg, Z. Baset, R. Liotta, B.G. Silbernagel and D.R. Neskora, Science, 206, 1029, (1979).
4. L. Lazarov and G. Angelova, Khim. Tverd. Topl., 10(3), 15, (1976).
5. M.N. Zharova, N.K. Larina and A.F. Lukovnikov, Khim. Tverd. Topl., 11(3), 11, (1977).
6. M.D. Shapiro and L.S. Al'terman, Khim. Tverd. Topl., 11(3), 17, (1977).
7. M.I. Bychev, Khim. Tverd. Topl., 11(4), 55, (1977).
8. W. Wiser, ACS Fuel Prepr., 20(2), 122, (1975).
9. P.H. Given, Fuel, 39, 147, (1960).
10. D.D. Whitehurst, in J.W. Larsen, ed., Organic Chemistry of Coal, American Chemical Society Symposium Series, Vol. 71, p. 1, 1978.
11. W.F.K. Wynne-Jones, H.E. Blayden and F. Shaw, Brenn. Chem., 33, 201, (1952).
12. Y. Sanada and H. Honda, Fuel, 45, 295, (1966).
13. Y. Sanada and H. Honda, Fuel, 46, 451, (1967).
14. P.R. Malherbe and P.C. Carman, Fuel, 31, 210, (1952).
15. H.R. Brown and R.L. Waters, Fuel, 45, 17, (1966).
16. N.Y. Kirov, J.M. O'Shea and G.D. Sergeant, Fuel, 46, 415, (1967).
17. V.G. Gadyatskii, M.D. Shapiro, and A.M. Belozerov, Khim. Tverd. Topl., 5(3), 54, (1971).
18. A. Halleux and H. Tschamler, Fuel, 38, 291, (1959).
19. A. Marzec, M. Juzwa, K. Betlej and M. Sobkowiak, Fuel Process. Techn., 2, 35, (1979).
20. H.N.M. Dormans and D.W. Van Krevelen, Fuel, 39, 273, (1960).
21. D.W. Van Krevelen, Fuel, 44, 229, (1965).
22. J.W. Larsen, personal communication, October 1979.
23. D.W. Van Krevelen, Fuel, 44, 220, (1965).
24. I.L. Ettinger, R.I. Baranov, A.V. Bunin, N.V. Shulman and M.A. Oganesyan, Khim. Tverd. Topl., 8(6), 3, (1974).
25. K.A. Debelak and J.T. Schrodt, Fuel, 58, 732, (1979).
26. V.I. Kasatochkin, Dokl. Akad. SSSR, 74, 77, (1950).
27. W.T.A. Morgans and N.B. Terry, Fuel, 37, 201, (1958).
28. J.C. Macrae and A.R. Mitchell, Fuel, 36, 423, (1957).
29. J. Schuyer, H. Dijkstra and D.W. Van Krevelen, Fuel, 33, 409, (1954).
30. D.W. Van Krevelen, H.A.G. Chermin and J. Schuyer, Fuel, 38, 438, (1959).
31. J.W. Larsen and E.W. Kuemmerle, Fuel, 55, 162, (1976).
32. D. Zakett, V.M. Shaddock and R.G. Cooks, Anal. Chem., 51, 1849, (1979).

48

33. J.H. Beynon, R.G. Cooks, J.W. Amy, W.E. Baitinger and T.Y. Ridley, Anal. Chem., 45, 1023, (1973).

34. H.H. Oelert, Brennst. Chemie, 48, 362, (1967).

35. H.H. Oelert, Brennst. Chemie, 50, 178, (1969).

36. M.M. Roy, Erdöl-Kohle, Brennst. Chem., 25, 343, (1972).

37. L. A. Heredy, A.E. Kostyo and M.B. Neuworth, Fuel, 43, 414, (1964).

38. J.A. Franz, J.R. Morrey, G.L. Tingey, W.E. Skiens, R.J. Pugmire and D.M. Grant, Fuel, 56, 366, 1977.

39. G. Bergmann, G. Huck, J. Karweil and H. Luther, Brennst. Chem., 38, 193, (1957).

40. P.B. Hirsh, Proceed. Roy. Soc., A226, 143, (1954).

41. J.K. Brown, W.R. Ladner and N. Sheppard, Fuel, 39, 87, (1960).

42. K. Iwata, H. Itoh and K. Ouchi, Fuel Process Techn., 3, 25, (1980).

43. J. Gibson, J. Inst. Fuel, 67, (June 1978).

44. B.S. Ignasiak, S.K. Chakrabarty and N. Berkowitz, Fuel, 57, 507, (1978).

45. P.J. Flory, Statistical Mechanics of Chain Molecules, Interscience, New York, 1969.

46. P.G. deGennes, Scaling Concepts in Polymer Physics, Cornell University Press, Ithaca, 1979.

47. P.J. Flory, Principles of Polymer Chemistry, Cornell University Press, Ithaca, 1953.

48. P.J. Flory and J. Rehner Jr., J. Chem. Phys., 11, 521, (1943).

49. H.M. James and E. Guth, J. Chem. Phys., 11, 470, (1943).

50. W. Kuhn and F. Grün, Kolloid Z., 101, 248, (1942).

51. J. Kovac, Macromolecules, 11, 362, (1978).

52. M. Fixman and J. Kovac, J. Chem. Phys., 58, 1564, (1973); and related work.

53. P.J. Flory, Macromolecules, in press.

54. D.S. Pearson and W.W. Graessley, Macromolecules, 11, 528, (1978).

55. M. Doi and S.F. Edwards, J. Chem. Soc. Farad. Trans., 74, 1789, (1978); ibid, 1802; ibid 1818.

56. N.A. Peppas and L.M. Lucht, unpublished theory, presented in part at 87th National AIChE Meeting, Boston, Mass., August 1979.

57. N.A. Peppas and E.W. Merrill, J. Appl. Polym. Sci., 21, 1763, (1977).

58. K. Ouchi, K. Imuta and Y. Yamashita, Fuel, 44, 205, (1965).

59. R.A. Orwoll, Rubb. Chem. Techn., 50, 451, (1976).

60. N.F. Brockmeier, R.W. McCoy and J.A. Meyer, Macromolecules, 5, 130, (1972).

61. M. Shibaoka, J.F. Stephens and N.J. Russell, Fuel, 58, 515, (1979).

62. J. Guin, A. Tarrer, L. Taylor, Jr., J. Prather and S. Green, Jr., Ind. Eng. Chem., Proc. Des. Dev., 15, 490, (1976).

63. K.A. Debelak and J.T. Schrodt, J. Coll. Interf. Sci., 70, 67, (1979).

RELATIONSHIP OF COAL MACERAL COMPOSITION TO UTILIZATION

R. R. Thompson
Bethlehem Steel Corporation, Bethlehem, PA 18016

ABSTRACT

Coal as viewed microscopically is seen to be composed largely
of discrete organic components referred to as macerals. Each
maceral responds differently in carbonization, and each develops
into a particular form of carbon in the coke product and has its
own characteristic effects on coke microstructure. A classification
of macerals has been developed and a number of correlations of
this classification with coking behavior and coke properties have
been established. Examples of how this classification serves in
coal-blend formulation and for predictive purposes are given. Basic
knowledge about the chemical and physical constitution of coal and
of the mechanisms involved in the response of coal to process conditions is stressed as a requirement in the development of new cokemaking technology.

INTRODUCTION

Coal petrography, or the microscopic analysis of the maceral
composition of coal, is a technology that has gained wide acceptance
by the United States steel industry over the last two decades.
Since the first applied studies were initiated by U. S. Steel and
The Pennsylvania State University in the late 1950's, the use of
coal petrography as the primary tool in the formulation of coal
blends to produce cokes of desired properties has been adopted by
just about all steel companies in the country. Coal petrography is
also being employed by petroleum companies, coal companies, and
others who use coal as a metallurgical product or as a feedstock for
coal conversion processes.

Given the present stage of development of coal petrography and
the voluminous literature, this paper is not intended as a compre-
hensive primer on the subject. Rather, significant findings that
have led to the growth of this technology will be reviewed. The
trends and needs in future research will be pointed out. The practical focus of the paper will be on cokemaking, and illustrative
material will be drawn from the author's experience and research
done at Bethlehem Steel Corporation.

THE MACERAL CONCEPT

Any researcher who seriously investigates the field of coal
petrography can soon become discouraged by the multitude of classifications and terms used to define the organic entities, or
"macerals", of coal. Much of this confusion has resulted from a
loose definition of the term maceral, and from the application of
the term by scientific investigators from different disciplines.

Even those of us concerned with the relation of coal composition to coal utilization have been inconsistent in the application of definitions of maceral.

Most U. S. coal petrographers use Spackman's maceral definition[1]: "macerals are organic substances, or optically homogeneous aggregates of organic substances, possessing distinctive physical and chemical properties." This definition builds on Stopes' concept[2] that a maceral is the organic analog of the mineral. More precisely, macerals are analogous to a mineral series, such as the plagioclase feldspar series, where composition changes continuously from one mineral to another. Thus, the coal petrographer commonly deals with maceral groups that have different chemical and physical properties, depending on their rank or position in the coalification series. Specific macerals have not yet been clearly defined.

The maceral group terminology worked on by Spackman is employed throughout this paper. In his system, the -oid suffix designates maceral groups rather than macerals, and terms such as "vitrinoid" and "exinoid" are employed.

Maceral Classifications As Applied To Cokemaking

For the field of cokemaking, maceral groups are first classified according to their behavior in the process. Then the task is to correlate this compositional analysis with coking behavior and coke properties. Once such correlations have been established, maceral analysis can serve to predict the properties that can be expected from a coal or coal mix.[3,4]

For establishing a suitable classification of coal macerals relatable to cokemaking the most important characteristic of coking coals when heated in a reducing or coking atmosphere is their similarity to thermosetting materials. When heated to increasing temperatures under these conditions, coking coals pass through a plastic stage before resolidifying into a semicoke and finally a coke. The complex reactions and changes that take place in the plastic stage are largely responsible for the properties of the final coke products.

For producing blast furnace coke, blends of coals are formulated to achieve desired coke properties, such as high strength and low reactivity to carbon dioxide.[5,6,7] Formulation of these coal mixes on the basis of petrographic analysis requires accurate knowledge of the behavior of macerals under coking conditions and the contribution these make to the coke carbon forms and coke microstructure.

In the application of coal petrography to cokemaking the maceral groups are classified according to their degree of reaction in the coal-to-coke transformation. A typical classification, as used at the Bethlehem Steel Corporation, categorizes maceral groups in terms of whether or not they react, i. e., become plastic, in the coking process (Figure 1). Reactive macerals are those that achieve some degree of plasticity during coking and thereby contribute the bonding medium to the coke. Inert macerals are those that

REACTIVE
> REACTIVE VITRINOID
> EXINOID
> RESINOID

SEMI-INERT
> SEMIFUSINOID
> SEMIMACRINOID

SEMI-INERT TO INERT
> PSEUDOVITRINOID

INERT
> FUSINOID
> MICRINOID
> MACRINOID
> MINERAL MATTER

FIGURE 1. MACERAL GROUP CLASSIFICATION FOR COKING COALS

remain unchanged in morphology during the coking process and therefore must be bonded in the coke by the reactive macerals. Semi-
inert macerals are those that react partially during the process and
are therefore the most difficult to account for in a petrographic
analysis.

The vitrinoid group is the most important of the reactive
materials in most good North American coking coals. Generally
present in large percentages, this group provides a high degree of
plasticity to the coking mass and therefore acts as the bonding
medium. To a large extent it also determines the carbon forms and
microstructures of the coke product. In contrast, exinoid and
resinoid are reactive maceral groups that generally volatilize away
within a narrow temperature range and leave little residue in the
coke product.

Because vitrinoid provides the bonding material and controls
the kind of coke produced, one should know the degree of reaction
of different vitrinoids during coking and the resulting carbon forms
and microstructures. The major factor related to the coking
behavior of vitrinoid is identified by its rank in the metamorphic
series. The microscopic measurement of reflectance was developed
to precisely define the rank of vitrinoid in the coking coal range.[8]
The reflectance measurement relates well to other parameters of coal
rank[9,10] and has the additional advantage of being determined on
the basis of vitrinoid alone, which is the most abundant constituent
in coal and the one that changes progressively with increasing
coalification. The commonly used coking coals of North America
range in vitrinoid reflectance from about 0.60% to about 1.80%.
The high-volatile coals fall below about 1.05%, the medium-volatile

coals between about 1.05% and 1.40%, and the low-volatile coals above 1.40%. The relation of the reflectance measurement to coking properties is detailed later in the paper.

Bethlehem Steel's classification of macerals differs from other classifications in that it recognizes pseudovitrinoid as a semi-inert to inert constituent[11]. In other classifications, no category such as pseudovitrinoid was recognized, and this constituent was blanketed by the long-established conventional category of vitrinoid or vitrinite characterized as totally reactive. Bethlehem Steel's classification separates the conventional vitrinoid into two populations: (1) reactive vitrinoid, i.e., vitrinoid which is reactive in coking and controls coking behavior to a significant degree, and (2) pseudovitrinoid, which is semi-inert to inert in coking. Because of this basic distinction, the reflectance measurement should be made only on reactive vitrinoid, since it is this constituent that largely characterizes the coking behavior of the coal. Conversely, this measurement is not made on pseudovitrinoid because including the pseudovitrinoid reflectance would distort the reactivity picture based on the reactive vitrinoid. Pseudovitrinoid occurs in all coals we have examined and can occur in abundance in certain coals, making up as much as one-half of the total so-called vitrinite in the coal.

Behavior of Macerals in Carbonization

The kinds of studies we conducted to characterize the role of maceral groups in coking and to define their impact on the coke product can best be shown with a series of illustrations. Figure 2, a photomicrograph of a polished section of coal particles, shows the heterogeneity of coal. The pseudovitrinoid particle seen here

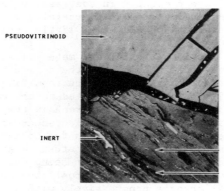

PSEUDOVITRINOID

INERT

REACTIVE VITRINOID

EXINOID

FIGURE 2. THE HETEROGENEOUS COMPOSITION
OF A HIGH-VOLATILE COAL
VIEWED MICROSCOPICALLY (315X)

is characteristic, being distinctly higher in reflectance, or brighter, than the reactive vitrinoid. Also seen are faint remnant cellular structure and some of the slits or pock marks that at times occur in the pseudovitrinoid constituent.

Figures 3 through 5 are from studies to determine the relative degree of reaction among coal constituents. These photomicrographs show samples that were heated to given temperatures in a coal plastometer, then quenched and prepared for petrographic study. Figure 3 presents views of exinoid at 400 and 420 C. At the lower temperature, exinoid has not reacted significantly. At the higher

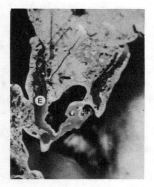 

EXINOID AT 400 C   EXINOID AT 420 C

FIGURE 3. THE BEHAVIOR OF EXINOID (E) DURING COKING OF A HIGH-VOLATILE COAL (315X)

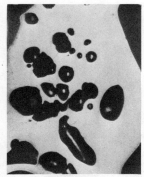 

PSEUDOVITRINOID   REACTIVE VITRINOID

FIGURE 4. REACTIONS OF PSEUDOVITRINOID AND REACTIVE VITRINOID IN A HIGH-VOLATILE COAL HEATED TO 450 C (315X)

temperature, the exinoid has melted, flowed and begun to devolatil-
ize. In this high-volatile coal, exinoid reacts and volatilizes
away within a fairly narrow temperature range, centering on
perhaps 30 C. Figure 4 provides a comparison of reactive vitrinoid
and pseudovitrinoid at 450 C, where the reactive vitrinoid has
developed in a fine anisotropic mosaic characteristic of semicoke.
The pseudovitrinoid particle exhibits a much lower degree of reaction. Studies such as those illustrated here documented the
classification of pseudovitrinoid as a semi-inert to inert material
and brought out the importance of this characterization as a guide
to the formulation of coal blends for coking. Figure 5 shows a
typical fusinoid, sometimes referred to as coalified charcoal, in
both a coal and coke. The point illustrated is that inert constituents undergo almost no change.

The impact of coal composition on the resulting coke carbon
forms and microstructures is reflected in high-temperature cokes
produced in our pilot-scale test oven facility. Figure 6 shows the
effect of reactive vitrinoid of increasing reflectance on coke-wall
carbon forms. The reactive vitrinoid of lowest reflectance (0.60%),
typical of an Illinois coal, produces isotropic carbon forms, which
tend to be highly reactive to carbon dioxide. The better coking
coals of 0.80% and 1.10% reflectance produce carbon forms having
anisotropic mosaics, which increase in size with an increase in
reflectance. These carbon forms are much less reactive to carbon
dioxide. The carbon form produced from the low-volatile coal of
1.40% reflectance is totally different, exhibiting a more graphite-
like ribbon structure. This graphitelike carbon form has a much
greater strength than other carbon forms and has a strong bonding
effect, characteristics that go far to explaining why small additions
of this type of coal add significantly to the strength of coke
produced.

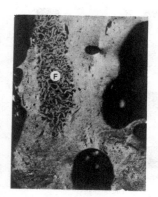

FUSINOID IN COAL   FUSINOID IN COKE

FIGURE 5. INERT NATURE OF FUSINOID (F) IN COKING (315X)

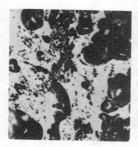

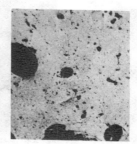

A. COKE FROM REACTIVE VITRINOID
OF 0.6% REFLECTANCE

B. COKE FROM REACTIVE VITRINOID
OF 0.8% REFLECTANCE

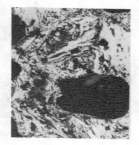

C. COKE FROM REACTIVE VITRINOID
OF 1.1% REFLECTANCE

D. COKE FROM REACTIVE VITRINOID
OF 1.4% REFLECTANCE

FIGURE 6. INFLUENCE OF VITRINOID RANK ON
COKE-WALL CARBON FORMS (315X)

Figure 7 shows the impact of pseudovitrinoid and oxidized
vitrinoid on the carbon forms in coke. Figure 7A, 7B and 7C depict
the various degrees of reaction and bonding that can result from
pseudovitrinoid constituents of different inertness. The resulting
carbon forms reflect a range in behavior from semi-inert to inert.
Figure 7D shows the isotropic and weakly bonded carbon form that
results from an oxidized reactive vitrinoid particle that otherwise
would have produced an anisotropic carbon form. The impact of
oxidized coal on coke quality is further emphasized in Figure 8,
which shows, based on pilot-scale coking tests[12], that the systematic
addition of oxidized coal to a blend greatly decreases coke strength.

Beyond these fundamental studies of maceral behavior, knowledge
of the interrelationship of coal composition, coke microstructure,
and coke quality measurements must be available before a predictive
correlation system can be established. One of the studies conducted
at Bethlehem[13] to establish some of these interrelationships is
illustrated in Figures 9 and 10. Figure 9 shows the relationship of
coke strength to weight loss by reaction with carbon dioxide when
cokes of different initial strength are used. In this reaction, the

56

A. INERT PSEUDOVITRINOID
   PARTICLE IN COKE

B. SEMI-INERT PSEUDOVITRINOID
   PARTICLE IN COKE

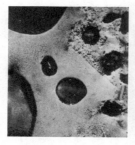

C. SEMI-REACTIVE PSEUDO-
   VITRINOID PARTICLE IN COKE

D. WEATHERED VITRINOID
   PARTICLE IN COKE

FIGURE 7. COKE-WALL CARBON FORMS FROM PSEUDOVITRINOID AND
WEATHERED VITRINOID (315X)

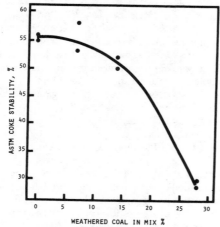

FIGURE 8. DETRIMENTAL EFFECT OF WEATHERED
COAL ON COKE STRENGTH

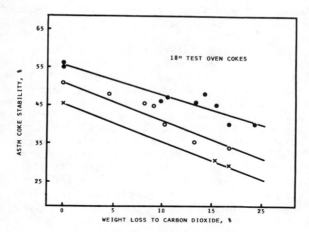

FIGURE 9. DECREASE IN COKE STRENGTH RESULTING FROM
REACTION WITH CARBON DIOXIDE

cokes are seen to lose strength uniformly, independent of the
initial strength. To prevent degradation in a blast furnace, one
should use cokes of initially high strength and low reactivity to
carbon dioxide.

Figure 10 shows the interrelationship of coke weight loss, coke
porosity as measured by an automated microscope, and initial coke
reactivity of some of the cokes in these studies. In this case,
since the low-volatile coal was a constant, the measurements reflect
the influence of different high-volatile coals. All of the cokes

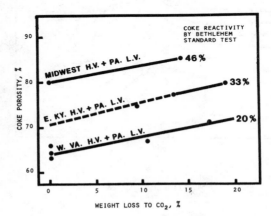

FIGURE 10. RELATIONSHIP BETWEEN COKE POROSITY AND COKE WEIGHT
LOSS TO CARBON DIOXIDE, 18" TEST OVEN COKES MADE
FROM 70% HIGH- AND 30% LOW-VOLATILE MIXES

increased in porosity with increasing weight loss. More importantly, initial porosity, initial coke reactivity and the type of high-volatile coal employed are highly related. For example, the Midwestern coal is a low-rank coal that produces an isotropic carbon form that is highly reactive and tends to produce highly porous coke. The Eastern Kentucky coal contains an abundance of pseudo-vitrinoid, which also produces isotropic carbon, whereas the West Virginia coal produces an anisotropic carbon form that is much less reactive and results in a less porous coke.

The establishing of interrelationships such as these confirmed the value of petrographic analysis and formed the basis for predictive correlation systems (Figure 11 through 13[14,15]). Figure 11 presents the correlation used at Bethlehem Steel for predicting the stability of coke from the petrographic analysis of a coal. The figure, which deals with high- through medium-volatile coals, relates the reactive vitrinoid reflectance and the inert content of coals to coke stability as measured in our pilot-scale test ovens. Thus, through a petrographic analysis of a coal or the mathematical combination of the analyses of several coals, coke stability can be predicted quite accurately from this chart.

Figure 12 shows the relationship between coal reflectance and the coking pressure generated when individual coals are coked in our test oven. Excessive coking pressure can permanently damage coke ovens and therefore coal blends must be formulated to avoid this problem. Figure 12 reveals that the major contributors to coking pressure are the low-volatile coals of high vitrinoid reflectance. Figure 13 expands on this finding and shows that coking pressure is related specifically to the reflectance and inert content of the low-volatile coals. The coals that are avoided to assure safe coking pressures for Bethlehem's operating conditions are identified.

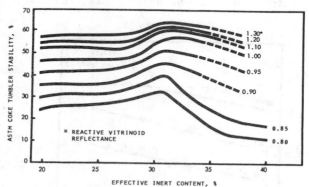

FIGURE 11. CORRELATION CURVES BETWEEN PETROGRAPHIC COMPOSITION AND ASTM TUMBLER STABILITY FOR PREDICTING THE STABILITY OF COKE PRODUCED FROM HIGH- AND MEDIUM-VOLATILE COALS

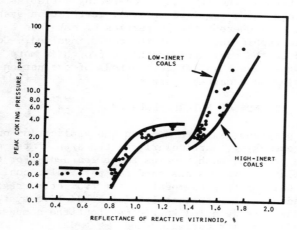

FIGURE 12. RELATIONSHIP BETWEEN REACTIVE VITRINOID
REFLECTANCE AND COKING PRESSURE

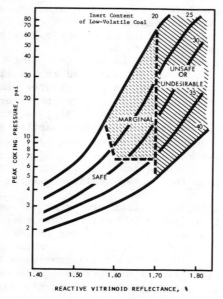

FIGURE 13. CURVES USED FOR SCREENING OUT
HIGH PRESSURE COKING COALS

To sum up to this point, a great deal of research has been
done by the steel industry to define the behavior of macerals in the
cokemaking process and to thereby establish a reliable petrographic
analysis. This work has been supplemented by carbonization research
to establish what kinds of cokes are produced from various coals and

coal mixes to determine the effects of operating variables on the product.[16] Through these combined efforts, correlation systems have been established whereby the properties of coke can be predicted from petrographic analysis of the coals involved. Coal evaluation based on the coal petrographic approach is more scientific and less expensive than earlier methods of evaluation by pilot-scale testing on a case-by-case basis.

TRENDS OF FUTURE RESEARCH

While much progress has been made on the application on coal petrography to cokemaking, particularly in the area of the conventional slot-oven process, much remains to be learned. For example, predictive correlation systems designed for particular coal supplies may need to be modified to accommodate coals from other regions. In general, broadening the applicability of predictive systems will require developing more detailed knowledge about the composition of coals and their behavior during coking.

The low-rank high-volatile coals of a reflectance of 0.6 to 0.7% can be cited as an example where more study is needed. The Midwestern coals of this rank tend to exhibit poor fluidity when coked, but certain coals of the same rank from other parts of the world exhibit very high fluidity. In such cases, more work will be needed to develop the kind of characterization that can explain why two otherwise similar coals exhibit different coking behavior.

While more work is required to broaden the application of coal petrography within the area of the conventional slot-oven process, the changing needs of the steel industry and the development of new cokemaking technology will present an even greater challenge. Some of the incentives for new technology are strictly economic, including the high capital cost of new cokemaking facilities, the need to replace old facilities, the lack of adequate capital generation, the cost of environmental control, and escalating coal costs. Other incentives are the progressive depletion of good coking coals, the inherent shortage of good coking coals in many parts of the world, and the increasing demand for higher-quality coke for the large modern blast furnaces.

New cokemaking technologies that have reached commercialization include: the charging of preheated coal to coke ovens, the densification of coal charges by partial briquetting or physical stamping of the charge, and the use of pitch binders to make coke briquettes from lower-quality coals. Many formcoking processes involving the manufacture of coke briquettes or pellets have also been investigated seriously, but none have yet progressed beyond the demonstration scale.

The emergence of these new technologies has emphasized the need for a deeper basic understanding of the coal-to-coke transformation. This kind of knowledge in combination with other types of data will be needed to develop improved or new technology. To this end, intensified study should be devoted to (1) the mechanisms involved in the coal-to-coke transformation, (2) the mechanisms of bonding in cokes and formcokes, and (3) the characterization of carbon forms and carbon products.

Mechanisms of the Coal-to-Coke Transformation

Comparatively little is known about the mechanisms by which coal transforms into coke other than that it passes through a plastic phase. The chemical and physical changes related to the plastic phase, which are so important to the development of desired products, are poorly understood, particularly at the maceral level.

A major step toward a better understanding of these mechanisms was Taylor's recognition of a liquid-crystal intermediate phase, in the transformation of coal to coke.[19] This phase, known as the "mesophase", has since been studied by others, including Marsh[20], and additional work is under way by a number of industrial laboratories and universities.[21] However, all of these reported studies have concentrated on model compounds or on petroleum and coal derivatives, rather than on coal itself.

In the area of coal, a recent breakthrough in understanding the nature of mesophase was made at Bethlehem Steel's laboratories, where the mesophase was observed during the in situ coking of vitrinite in a transmission electron microscope equipped with a hot stage. With this technique, vitrinoids from a number of coals known to have different coking behaviors, were observed as they passed through the early stages of coking. In each case, the formation of the mesophase was seen, as pictured in Figure 14 for the Pittsburgh seam coal, and the type of mesophase and the degree of its development were found to be related to the coking behavior of the coal. The results of this work, which have been submitted elsewhere for publication[22,23], confirm the important role of the mesophase in the carbonization of coal.

In connection with the coal-to-coke transformation, researchers will also need to continue improving the characterization of coals, coal macerals, and other raw materials used in cokemaking, concen-

FIGURE 14. EARLY STAGE OF MESOPHASE
DEVELOPMENT (BLACK SPHERES) IN
THE COKING OF A HIGH-VOLATILE
COAL IN A HOT-STAGE TEM

62

trating on the fundamental chemical constitution of these materials,
particularly vitrinoid. A long-range objective of this kind of
research might be learning how to manipulate the coking behavior of
coals and related materials so as to extend the range of coals that
might be usable in cokemaking.

Mechanisms of Bonding in Cokes and Formcokes

An understanding of the mechanisms of bonding is one of the keys
to improving the quality of coke. Studies in this area will need
to address: (1) the mechanisms of carbon-carbon bonding resulting
from maceral interactions, including the measurement of the strength
of those bonds, and (2) the role of binders consisting of materials
such as coaltar and pitch, petroleum pitches and solvent-refined
coals. An understanding of bonding is particularly important for
formcokes, the production of which is characterized by a large
variety of raw materials and processing conditions.

A unique problem of carbon-carbon bonding can be illustrated by
exploring the contribution of low-volatile coal to coke strength.
In most coal blends controlled amounts of low-volatile coal are used
to increase coke strength. As shown in Figure 15, the strength of
coke increases dramatically when as little as 20% low-volatile coal
is added to the blend. Yet, as an examination of the coke microstructure in Figure 16 reveals, the carbon form produced from the
low-volatile coal does not form a continuous network in the coke.
A better understanding of why low-volatile coal provides an unusually
strong bond in spite of such discontinuous networks could lead to
identifying materials that would, in one proportion or another,
substitute for the low-volatile coals.

Since studies of bonding in the coal-to-coke transformation have

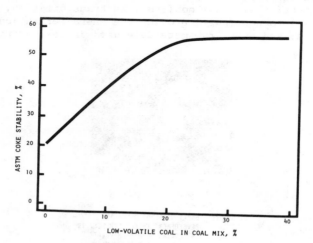

FIGURE 15. INCREASE OF COKE STRENGTH BY ADDITION OF LOW-VOLATILE
COAL TO A BLEND

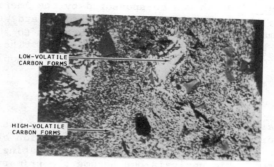

FIGURE 16. ISOLATED NATURE OF LOW-VOLATILE
CARBON FORMS IN COKE (315X)

been fostered mostly by the need to use poorer coking coals, the research has concentrated on the marriage of binders with poorly coking coals. As reported by Marsh and colleagues[24], certain binders exhibiting characteristic mesophase development are capable of transforming the carbon forms produced from poorly coking coals to carbon types characterized by strong bonding. Using this principle, Sumikin Coke Company and Sumitomo Industries have established a commercial operation to produce coke from poorly coking coals using an appropriate pitch binder. Work involving binders is most active, or course, in parts of the world lacking supplies of coal with good coking properties.

Characterization of Carbon Forms and Carbon Products

Finally, the research efforts on the mechanisms involved with the coal-to-coke transformation will require support from improved characterization of carbon forms and carbon products so that the new and emerging processes can be adequately assessed and compared. Foremost in this area will be the need to compare formcokes of different shapes, microstructure and carbon-form composition with conventional or modified slot-oven cokes.

In contrast to some of the other subjects, the characterization of carbon forms and microstructures developed in conventional cokes has received wide attention in the steel industry because of the importance of these materials to blast furnace performance. Much more of this kind of knowledge will now be needed in the area of the new nonconventional coke processes.

The research needs just identified are now widely recognized by the steel industry throughout the world. Recognition of these and

other needs by the steel industry in the United States was high-
lighted recently at the Workshop on Coal-to-Coke Transformation for
Blast Furnace Fuels which was co-sponsored by the American Iron and
Steel Institute, the National Science Foundation, and West Virginia
University. A major purpose of the conference was to stimulate
research in the basic areas related to cokemaking. The Proceedings
of the workshop[25] contain valuable information for people in our
fields.

SUMMARY AND CONCLUSIONS

Over the past three decades the steel industry has provided
the incentive for studying of the maceral composition of coal in
order to develop a more scientific basis for evaluating coking coals.
As a result petrographic analysis has become the primary tool for
this purpose and has contributed to the improvement of coke products.
In turn, this accomplishment has contributed to more efficient blast
furnace operation.

To meet the challenge of changing economic and technical conditions, the steel industry is now supporting work aimed at a deeper
understanding of the basic mechanisms underlying the coal-to-coke
transformation. Because of the complexity of coal and the carbonization process itself, these basic studies will require a broad
interdisciplinary approach and will involve a closer relationship
with the academic world as well. Such factors should also bring
the steel industry into closer contact with other industries
concerned with coal conversion.

REFERENCES

1. W. Spackman, Trans. N. Y. Acad. Sci., Ser. II, 20(5), p. 411 (1958).
2. M. C. Stopes, Fuel, 14, p. 4 (1935).
3. R. R. Thompson, J. J. Shigo III, L. G. Benedict, and R. P. Aikman, Blast Furnace and Steel Plant, 54(9), p. 817 (1966).
4. R. R. Thompson, and L. G. Benedict, Proc. Ironmaking Conference, AIME, 26, p. 91 (1967).
5. N. Schapiro, R. J. Gray, and G. R. Eusner, Blast Furnace, Coke Oven and Raw Materials Proc., AIME, 20, p. 89 (1961).
6. L. G. Benedict, R. R. Thompson, and R. O. Wenger, Blast Furnace and Steel Plant, 56(3), p. 217 (1968).
7. R. R. Thompson, A. F. Mantione, and R. P. Aikman, Blast Furnace and Steel Plant, 59(3), p. 161 (1971).
8. R. R. Thompson, and L. G. Benedict, Geol. Soc. Am. Spec. Paper 153, p. 95 (1974).
9. N. Schapiro, and R. J. Gray, Proc. Ill. Min. Inst., 68th Year, p. 83 (1960).
10. J. T. McCartney, and M. Teichmüller, Fuel, 51, p. 64 (1972).
11. L. G. Benedict, R. R. Thompson, J. J. Shigo III, and R. P. Aikman, Fuel, 47, p. 125 (1968).
12. J. C. Crelling, R. H. Schrader, and L. G. Benedict, Fuel, 58,

p. 542 (1979).

13. L. G. Benedict, and R. R. Thompson, submitted to Coal Geology (1980).

14. L. G. Benedict, and R. R. Thompson, Proc. Ironmaking Conference, AIME, 35, p. 276 (1976).

15. R. R. Thompson, and L. G. Benedict, Iron and Steelmaker, 3(2), p. 21 (1976).

16. R. O. Wenger, and V. A. Neubaum, Proc. Ironmaking Conference, AIME, 28, p. 170 (1968).

17. T. T. Gin, C. L. Dahl, and D. G. Wilson, San Fransisco AISI, Reg. Tech. Meeting preprint (1963).

18. J. A. Harrison, H. W. Jackman, and J. A. Simon, Illinois Geol. Survey Circ. 366 (1964).

19. G. H. Taylor, Fuel, 40, p. 465 (1961).

20. H. Marsh, Fuel, 52, p. 205 (1973).

21. D. S. Hoover, A. Davis, A. J. Perrotta, and W. Spackman, 14th Conference on Carbon, Extended Abstracts, University Park, Pa., p. 393 (1979).

22. J. J. Friel, S. Mehta, G. D. Mitchell, and J. M. Karpinski, accepted for publication in Fuel.

23. J. J. Friel, S. Mehta, G. D. Mitchell, and J. M. Karpinski, to be published in Proc. Ironmaking Conference, AIME, 39 (1980).

24. I. Mochida, H. Marsh, and A. Grint, Fuel, 58 (11), p. 803 (1979).

25. M. D. Aldridge, Proc. Workshop on Coal-to-Coke Transformation for Blast Furnace Fuels, Tech. Report of West Virginia University Energy Research Center (1980).

C-13 NMR ON SOLID SAMPLES AND
ITS APPLICATION TO COAL SCIENCE

G.E. Maciel*, M.J. Sullivan and N.M. Szeverenyi
Department of Chemistry, Colorado State
University, Fort Collins, Colorado 80523

F.P. Miknis
Laramie Energy Technology Center, Department
of Energy, Laramie, Wyoming 82071

ABSTRACT

^{13}C nmr spectra that approach high-resolution quality can now
be obtained on solids by using high-power ^1H decoupling (for eliminating the major line broadening due to ^1H-^{13}C dipole-dipole interactions), ^1H-^{13}C cross polarization (for circumventing the ^{13}C T_1
bottleneck) and magic-angle spinning (for eliminating line broadening due to chemical shift anisotropy). ^{13}C spectra with rather
sharp lines are obtained for crystalline samples. For coals and
other very complex mixtures, the occurrence of many different, but
very similar, chemical architectures renders for most samples two
rather broad resonances bands (the aromatic/olefinic carbon band
and the aliphatic carbon band). Additional structural information
can be obtained from the ^{13}C nmr spectrum if the coal has an unusually simple distribution of carbon types or if more sophisticated
techniques are employed. Progress, limitations and prospects for
solving important problems in coal science by the ^{13}C nmr technique
are discussed.

INTRODUCTION

A knowledge of the structures of the organic compounds in coal,
or of the relative abundances of the various structural types, is
important for at least two reasons. First, this knowledge is
essential to any realistic effort to understand the details of the
geochemical evolution of coal. Second, the utilization of coal,
e.g., conversion processes, can be optimized only if this knowledge
is available. The attainment of this kind of knowledge on samples
as complex and intractable as coal has been very difficult to
achieve. Many of the most powerful analytical techniques for organic chemical structure determination, including conventional nuclear magnetic resonance (nmr), have required liquid samples; and
typically only small fractions of a coal can be dissolved under conditions for which one can be confident of retaining the structural
integrity of the organic compounds in the sample.
Until recently the use of standard nmr approaches (e.g., the
"usual" pulse and Fourier transform techniques) on powdered or
amorphous samples of any material has yielded broad featureless

*To whom correspondence should be addressed.

resonances, with little or no useful information regarding the details of molecular structure. In obtaining a ^{13}C nmr spectrum of a coal by standard "liquid-state" Fourier transform techniques with a modern spectrometer, the broad featureless character and the low signal intensity in the spectrum obtained (if one could detect the resonance at all) are due to a combination of problems. A major line broadening influence is the magnetic dipole-dipole interaction between each ^{13}C nucleus and the magnetic moments of nearby protons. The other important ^{13}C line-broadening influence is the anisotropy of the ^{13}C chemical shift. This phenomenon arises because the resonance frequency of a given carbon nucleus depends not only on the chemical environment of the carbon (e.g., which of the five distinct carbon environments, or positions, in benzoic acid), but also on the orientation of that environment relative to the direction of the static magnetic field of the spectrometer. As all possible orientations occur randomly in an amorphous or a powdered sample, a wide range of resonance conditions can result for a given carbon type (e.g., for the carboxyl carbon of benzoic acid). This range for a given carbon type, called the chemical shift anisotropy, can for certain carbon types (e.g., carbonyl and aromatic carbons) amount to a substantial fraction (say, half) of the total ^{13}C chemical shift range ordinarily encountered with organic compounds. The result is very broad ^{13}C nmr resonance lines.

The third debilitating characteristic of common ^{13}C nmr techniques, when applied to solid samples, is the typically long ^{13}C spin-lattice relaxation times (T_1) expected for solids. (T_1 is a measure of how long it takes for the ^{13}C spin system to reattain thermal equilibrium after having been displaced from it, e.g., by a pulse of rf radiation at the Larmor frequency.) As T_1 determines how rapidly one can repeat a standard pulse and Fourier transform experiment, a regime of large T_1 values severely limits the attractiveness of the usual approach of enhancing the ^{13}C nmr signal-to-noise ratio by synchronous repetitions and computer time averaging. While ^{13}C T_1 values for coals are typically not as large as for pure crystalline materials, this still constitutes an important constraint, as well as a major impetus for the development of new techniques that are now being used for obtaining ^{13}C nmr spectra of coals.

In the liquid state the problems stated above are overcome by the rapid and random tumbling motions that are characteristic of that state. These motions average the anisotropic chemical shift to the isotropic chemical shift (and corresponding sharp resonance) of a typical liquid-state nmr experiment, and average dipole-dipole interactions to zero. These motions also provide a modulation of the energy of the spin system (e.g., a time variation of dipole-dipole interactions as molecules tumble), which gives rise to efficient spin-lattice relaxation mechanisms. For solid samples special techniques are required for elimination of the problems stated above.

CROSS POLARIZATION AND MAGIC-ANGLE SPINNING

Pines, Gibby and Waugh showed that the ^{1}H-^{13}C dipole broadening problem can be overcome in the ^{13}C nmr spectra of solids by employing high-power ^{1}H decoupling.[1,2] This approach is entirely analogous in concept to ^{1}H decoupling in solution-state ^{13}C nmr, except technically the approach is far more difficult in the solid-state work, because the ^{1}H-^{13}C couplings being overcome (dipole-dipole) are orders of magnitude larger than the ^{1}H-^{13}C couplings encountered in liquids (electron mediated J couplings, as the dipole-dipole couplings are averaged to zero by molecular motion). Thus, <u>much</u> higher power levels are required for the solid-state experiments. In carrying out their high-power ^{1}H decoupling work, Pines, Gibby and Waugh also showed that the ^{1}H, ^{13}C double-resonance experiment could be designed in a manner in which the ^{13}C spin set obtains its polarization and resultant magnetization directly from the proton spin set (at the expense of the proton polarization), rather than <u>via</u> the usual spin-lattice relaxation process. In this way, the long-$\overline{T_1}$ bottleneck is circumvented, so the ^{13}C signal-to-noise ratio can be enhanced by rapid repetitions. This phenomenon is referred to as <u>cross polarization</u> (CP).

In visualizing the CP approach, one takes advantage of the fact that the motions of the spins in the presence of more than one magnetic field can mentally be decomposed into separate motions resulting from each field, keeping in mind the fact that these "separate" motions are in reality superimposed on each other. In the CP experiment, one has the large static field $\underline{H}_0$, the (effectively) rotating rf field ($\underline{H}_{1H}$) at the ^{1}H resonance frequency (ω_{0H}) and the (effectively) rotating rf field ($\underline{H}_{1c}$) at the ^{13}C resonance frequency (ω_{0C}). It is convenient, initially to neglect the effect of $\underline{H}_0$, which is equivalent to viewing the ^{1}H spins from a frame of reference that rotates about $\underline{Ho}$ with frequency ω_{0H} and viewing the ^{13}C spins from a frame that rotates about $\underline{Ho}$ at a frequency ω_{0C}. These two frames of reference are referred to as <u>rotating frames</u>.

For the purposes of understanding the CP experiment, one focusses on the response of the ^{1}H spins to the field H_{1H} and the effect of $\underline{H}_{1c}$ on the ^{13}C spin set. The essence of the $\overline{CP}$ phenomenon is that one adjusts the amplitudes of $\underline{H}_{1H}$ and $\underline{H}_{1c}$ so the Larmor precession frequency of the ^{1}H spins about $\underline{H}_{1H}$ (ω_{1H}) is the same as the Larmor precession frequency of the ^{13}C spins about $\underline{H}_{1c}$ (ω_{1c}). Then, in their rotating frames, the ^{1}H and ^{13}C Larmor frequencies are matched, a condition referred to as the Hartmann-Hahn condition,[2] $\omega_{1c} = \omega_{1H}$. In this condition one has oscillating z-components of dipolar fields of both ^{13}C and ^{1}H spins with the same oscillation frequency, and the ^{1}H-^{13}C dipole-dipole interaction provides an efficient mechanism for the transfer of spin polarization between the ^{1}H and ^{13}C spin sets by mutual ^{1}H and ^{13}C spin flips. This is depicted in Figure 1, where the ^{1}H and ^{13}C Larmor frequency match in the rotating frames, (and consequent polarization transfer)

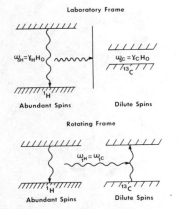

Figure 1. Larmor frequency match (Hartmann-Hahn condition in the rotating frame, contrasted with the mismatch in the laboratory frame.

is contrasted with the Larmor frequency mismatch in the laboratory frame (e.g., 60 MHz for ^{1}H and 15.1 MHz for ^{13}C).

Experimentally, the sequence of events in a typical CP experiment is shown in Figure 2, which views the ^{1}H and ^{13}C spin sets in their rotating frame(s). The first step (A) is the application of a 90° pulse to the ^{1}H spin set by the ^{1}H rf field $\underline{H}_{1H}$. Following this pulse, a 90° phase shift of $\underline{H}_{1H}$ brings it (B) into a direction that is colinear with the ^{1}H magnetization, $\underline{\mu}_H$. This is referred to as a "spin-lock" condition, and corresponds to a state of very low spin temperature for the ^{1}H spin set.[2] In the next step (C) the ^{13}C rf field, $\underline{H}_{1C}$ is applied with the Hartmann-Hahn matching condition; and ^{13}C magnetization builds up along $\underline{H}_{1C}$ by cross polarization, while there is a corresponding (but proportionately smaller) depletion of ^{1}H mag-

netization along $\underline{H}_{1H}$. When the cross polarization of ^{13}C has achieved the desired (or maximum achievable) ^{13}C magnetization, $\underline{H}_{1C}$ is turned off, and the ^{13}C free induction decay (FID) is collected (D) as the transverse ^{13}C magnetization decays, while $\underline{H}_{1H}$ remains on for the decoupling of ^{13}C-^{1}H dipole-dipole interactions. The sequence of steps depicted in Figure 2 consitutes the single-contact, Hartmann-Hahn version of the important class of experiments introduced by Pines, Gibby and Waugh.[1,2]

Schaefer and Stejskal[3,4] have shown that the line-broadening due to ^{13}C chemical shift anisotropy (CSA) which is eliminated in liquids by the randon tumbling motions of the molecules, can be eliminated in solid samples by rapid mechanical spinning of the entire sample about an axis that makes an angle of 54° 44' degrees with the direction of the static magnetic field, $\underline{H}_0$. It has been shown that rapid sample spinning, depicted schematically for an electromagnet spectrometer in Figure 3, gives residual line broadening due to CSA that vanishes when the residual terms in the orientation-dependent chemical shift expression are brought to zero via the trigonometric factor, $[1-3\cos^2\theta]$, when θ is the "magic-angle." Two basic types of magic-angle spinning devices have been used in most of the ^{13}C CP/MAS work that has been reported.[5-8] These are shown schematically in Figure 4, along with the bullet design developed in our laboratory.[7]

In 1976 Schaefer and Stejskal[3] first demonstrated that cross polarization and magic-angle spinning could be combined into a composite technique, CP/MAS, that provides ^{13}C nmr spectra on solids with resolution characteristics that approach what is typically obtained on liquid samples. Figure 5 shows the type of ^{13}C CP/MAS

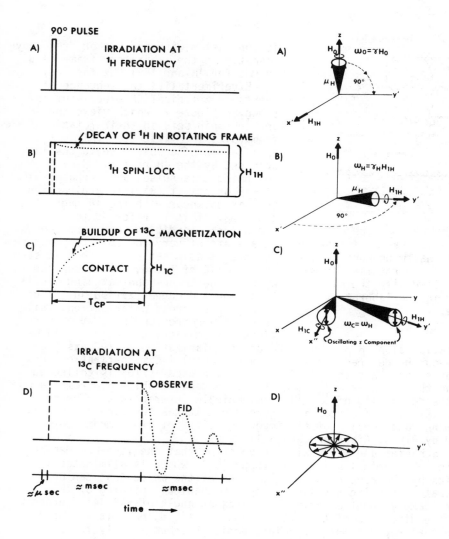

Fig. 2.   Timing of the sequence of events in a cross polarization experiment.

MAGIC ANGLE SPINNING

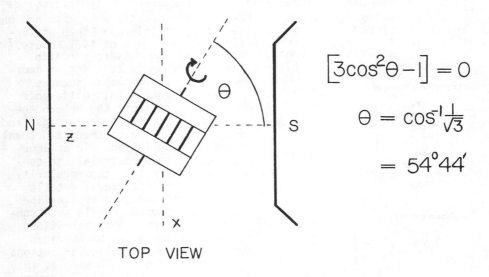

$$[3\cos^2\theta - 1] = 0$$
$$\theta = \cos^{-1}\frac{1}{\sqrt{3}}$$
$$= 54°44'$$

Fig. 3. Idealized diagram of a magic-angle spinner.

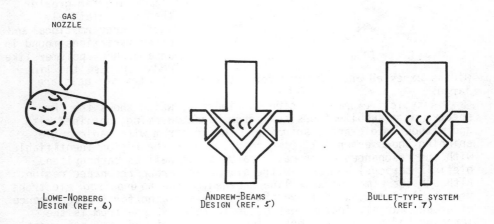

Fig. 4. Magic angle spinner designs.

spectra one can expect for pure, powdered crystalline substances (e.g., dehydroquercitin) and relatively homogeneous polymeric solids (e.g., polymethylmethacrylate, PMMA). These spectra were obtained on home-built spectrometers in our laboratory at Colorado State University, operating at a static magnetic field strength of 14.1 kgauss.[7]
The resolution apparent in the crystalline dehydroquercitin spectrum is characteristically better than what is seen for an amorphous sample such as PMMA. This difference reflects the heterogeneity of conformations (and possible other structural variations) in an amorphous material, in contrast to the regularity of structural details found in a crystalline sample. While such conformational variations can often be averaged out by molecular motions in a liquid sample, they are typically frozen out to some extent in an amorphous solid and therefore contribute to the line shape and line width. Of course, the structural variations and heterogeneity in a coal sample are far greater than the relatively subtle conformational and other variations found in an amorphous polymer like PMMA. Hence, the line widths expected and observed for ^{13}C CP/MAS spectra of coals are large.

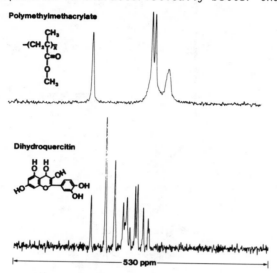

Fig. 5. [13]C CP/MAS spectra of polymeric and crystalline materials.

A typical example is seen in Figure 6, which shows the CP spectra of a subbituminous coal from Hanna, Wyoming, obtained with and without magic-angle spinning. The spectrum with magic-angle spinning shows essentially two bands, one in the region identifiable with the resonances of aromatic carbons (as well as carbonyl and olefinic carbons) and one in the aliphatic carbon resonance region. Although there may be some line broadening due to unpaired electrons present in coals, we believe that the main reason for the occurrence of broad, largely featureless bands in this ^{13}C spectrum is the large number of similar, but slightly different chemical strucures contributing to the ^{13}C spectrum in the aromatic or aliphatic region, giving rise to a multitude of closely-spaced resonances that together yield a broad band.

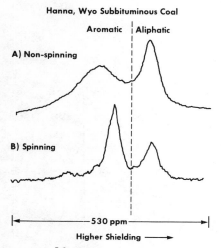

Hanna, Wyo Subbituminous Coal

Aromatic | Aliphatic

A) Non-spinning

B) Spinning

|← ——————— 530 ppm ——————— →|

Higher Shielding ——→

Fig. 6. ^{13}C CP/MAS spectrum of a subbituminous coal from Hanna, WY.

The occurrence of essentially two bands is characteristic of most coal samples, although some coals (especially lignites) with less complex variations of organic chemical structures give ^{13}C CP/MAS spectra with some additional peak definition. Thus, at the present state of standard ^{13}C CP/MAS techniques, for most coal samples one expects to distinguish mainly the aromatic from the aliphatic carbons, and in a few cases to get additional structural information directly. One can note from the spectrum of Figure 6 obtained without MAS: in order to distinguish the aromatic and aliphatic carbon resonances clearly and unequivocally, magic-angle spinning is required. The anisotropy pattern of the aromatic region of the nonspinning spectrum extends substantially into the aliphatic region.

APPLICATIONS TO COAL SCIENCE

The ^{13}C CP/MAS technique has been applied with great success to a variety of solid samples, including coals,[9-13] oil shales,[14-17] humic materials,[18-20] proteins,[21,22] cellulose[23,24] and synthetic polymers[4,25-28]. In the present paper we will confine our coverage largely to results obtained at Colorado State University.

The spectra shown in this paper were obtained on two spectrometers operating at a field of 14.1 kgauss and ^{13}C resonance frequencies of 15.1 MHz.[7] One spectrometer is completely home-built, using a 12-in magnet from a Varian HA-60 spectrometer,[29] spinning 1.5 cm^3 samples at about 2.5 KHz. The other is a modified JEOL FX-60Q spectrometer,[29] spinning 0.7 cm^3 samples at about 2.5 KHz. One should note that for coal samples the ratio of spinning speed to $\underline{H}_0$ should not be lower than what is represented here, in order to avoid interferences of the right-side spinning sideband of the aromatic carbon resonances with the aliphatic carbon resonances.

Typical ^{13}C CP/MAS spectra of coals are shown in Figure 7, which is taken from recent CP/MAS spectra,[11] and in Figure 8, which presents much earlier results.[10]

Figure 9 shows ^{13}C CP/MAS spectra[10] of a Wyoming subbituminous coal, before and after reverse combustion, and a cored sample from a field experiment. These spectra show that the reverse combustion process depletes the aliphatic carbon content, relative to the aromatics, and that the cored sample had undergone little, if any, conversion.

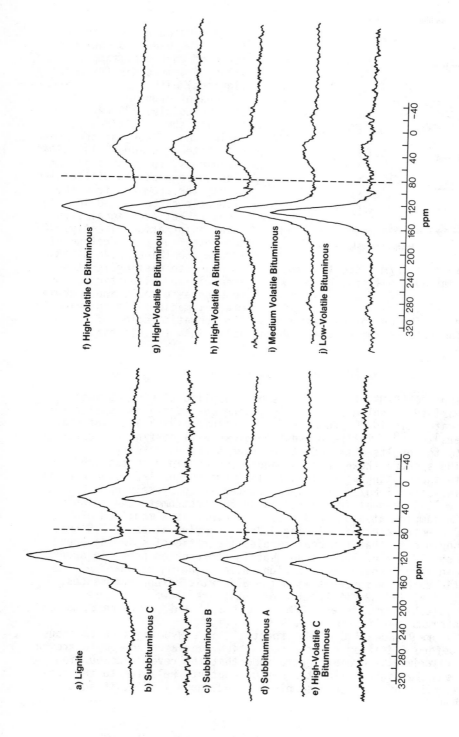

Fig. 7. ^{13}C CP/MAS spectra of coals of various rank.

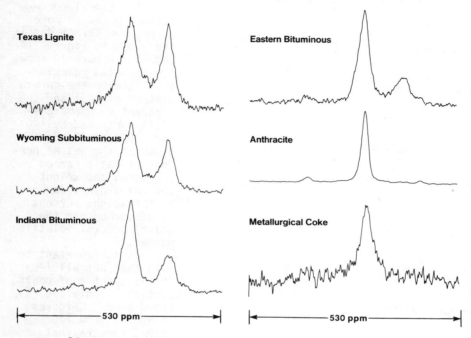

Fig. 8. ^{13}C CP/MAS spectra of various coals.

Figure 10, taken from reference 10, shows the ^{13}C CP/MAS spectra of Wyodak coal and a solvent-refined Wyodak coal product. These spectra show that the solvent-refining process also depletes the aliphatic carbons, relative to the aromatics.

Figure 11 shows the effect of high temperature on a Wyoming subbituminous coal, as seen by ^{13}C CP/MAS nmr. As with the other processes, the high-temperature heating depletes the aliphatic carbon content.

An interesting correlation based on ^{13}C CP/MAS data[10] is between the heating value of a coal and the apparent aromaticity determined from the ^{13}C nmr spectrum. This correlation is seen in Figure 12. The trend on the left side of the maximum is related to the lower molar heat of combustion of an organic compound that is partially substituted by oxygen, nitrogen or sulfur, relative to that of the corresponding hydrocarbon (e.g., C_6H_5OH relative to benzene);[6] and the trend on the right side of the maximum reflects the smaller molar heat of combustion of an aromatic compound relative to that of a related aliphatic compound (e.g., benzene and cyclohexane). Figure 13 shows the expected relationship between the apparent carbon aromaticity derived from ^{13}C CP/MAS experiments and the atomic H/C ratio (moisture and ash free).[11]

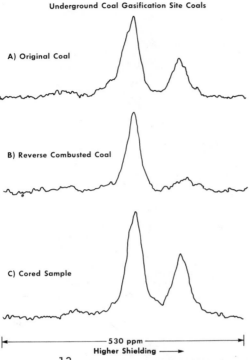

Fig. 9. ^{13}C CP/MAS spectra of Hanna, WY. subbituminous coal before and after reverse combustion, and a cored (test) sample.

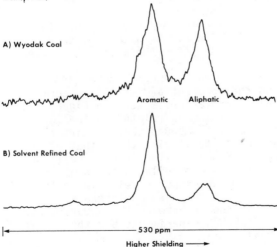

Fig. 10. ^{13}C CP/MAS spectra of a Wyodak coal before and after solvent refining.

Additional correlations based on more recent ^{13}C CP/MAS data[11] are shown in Figures 14 and 15. Figure 14 shows the expected relationship between the carbon content of coal and the apparent carbon aromaticity obtained by ^{13}C nmr. Figure 15 shows plots of the weight percent aromatic carbon (f_a^1 times the weight percent organic carbon) vs the weight percent fixed carbon and the weight percent volatile matter.

It is important to note that not all ^{13}C CP/MAS spectra of coals consist of only two broad bands. Figures 16 and 17 show ^{13}C CP/MAS spectra that have definite peaks at chemical shift values that correspond to specific carbon types (e.g., carbon-substituted aromatic carbon at about 145 ppm, unsubstituted aromatic carbon at about 128 ppm, at about 56 ppm for OCH_3 and internal methylene carbon peaks at about 31 ppm).

Numerous important questions come to mind with regard to ^{13}C CP/MAS studies of coals. These include the following:

1. How good are ^{13}C CP/MAS results on coals? (Are the different carbon types sampled completely or representatively? What is the effect of

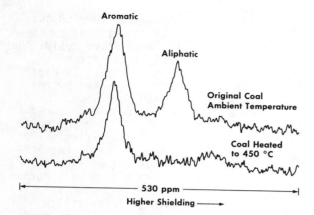

Fig. 11. ^{13}C CP/MAS spectra of a Wyoming subbituminous coal before and after heating to 450°C.

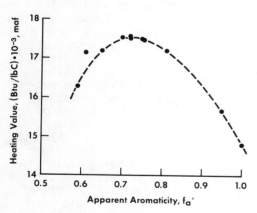

Fig. 12. Plot of heating value vs. apparent carbon aromaticity.

paramagnetic centers on the spectra?)

2. What are the best conditions for ^{13}C CP/MAS experiments on coals? (e.g., what are the "best" contact times and delay times to use?).

3. How does one best analyze ^{13}C CP/MAS experiments on coals? (What is the best way to take spinning sidebands into account? How should one account for the "low-shielding aromatic tail"?).

4. How can the efficiency of ^{13}C CP/MAS experiments on coals be improved? (How can the signal-to-noise ratio be improved per unit time?).

5. How can the structural resolution of ^{13}C CP/MAS experiments on coals be enhanced?

These and related questions have been and continue to be under study at Colorado State University and other laboratories. Such studies require systematic variations of the several pertinent experimental parameters;[11] and the answers to these questions are certainly not yet definitive. Such answers require detailed studies of the spin dynamics involved in CP/MAS experiments on coals. Studies of this type are underway. Preliminary data indicate that, for reasons not yet well understood, the answers to question 1) is favorable. Studies related to questions 2) and 4) indicate that "standard" ^{13}C CP/MAS experiments on coals can

78

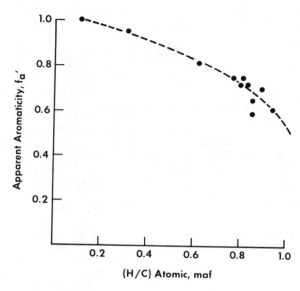

Fig. 13. Plot of apparent carbon aromaticity, f_a', vs atomic H/C ratio.

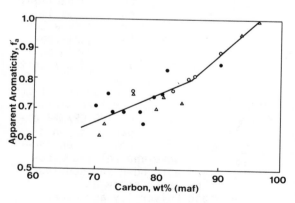

Fig. 14. Plot of apparent carbon aromaticity, f_a', vs. carbon content of coals.

usually be carried out much faster than has been customary previously. The repetition rate of the single-contact CP experiment is limited by the proton T_1 values. These values can be measured in a ^{13}C CP/MAS experiment by first applying a 180° pulse to the ^{1}H magnetization and waiting a variable time (τ), during which the proton spins undergo spin-lattice relaxation, before applying the initial 90° [1]H pulse shown in Figure 2. Measurements on ten coals of varying rank[30] yield proton T_1 values from about 10 ms to about 110 ms. The lowest values correspond to the lowest-rank coals and the highest values occur with the highest rank coals. These results indicate that substantial time savings should be possible in obtaining the ^{13}C CP/MAS spectra of most coals. Other approaches for making the experiments faster can and are being explored here and elsewhere; these include: 1) the use of spectrometers with higher magnetic fields, 2) use of the ^{1}H magnetization flip-back approach,[31] and 3) the use of dynamic polarization by irradiation of the electron spin resonances.

Several approaches are also being explored for enhancing the structural resolution that can be obtained on coal samples by ^{13}C CP/MAS, including the obvious (and, so far, not very promising) approach of higher static magnetic fields. It may be that the greatest advantage of large H_0 values in this regard is in improving the signal-to-noise ratio, so that various resolution

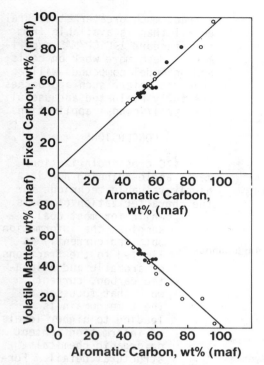

Fig. 15. Plots of wt.% aromatic carbon vs. wt.% fixed carbon and volatile matter.

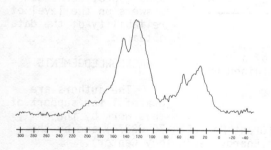

Fig. 16. ^{13}C CP/MAS spectrum of Exxon lignite no. 65-R9.

enhancement approaches become feasible. One approach that may prove fruitful is to employ temperature as a variable (e.g., from increased sensitivity at low temperature and possibly narrower lines due to motional averaging at higher temperatures).

Perhaps the most promising avenue at the present time for enhancing the amount of structural information to be obtained from ^{13}C CP/MAS is use of the time domain in the CP experiment. Several such approaches are under study. These approaches attempt to discriminate among structural types by means of differences in the various pertinent relaxation times (e.g., ^{13}C and ^1H T_1's; ^{13}C and ^1H $T_{1\rho}$'s; T_{CH} (the cross polarization relaxation time)). The effects of such differences are analyzed as a function of chemical shift region (structural type) in the spectrum, and can be manifested in difference and convolution difference spectra. Two preliminary examples of the use of the time domain in ^{13}C CP/MAS experiments on coal are seen in Figure 18, which compares the standard ^{13}C CP/MAS spectrum of a Wyoming subbituminous coal with what is obtained by employing a much longer cross polarization contact time (8 ms) and with a spectrum obtained by the interrupted decoupling approach.[32] By detailed studies of this type and by careful analysis of related spectra, one can

80

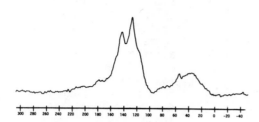

Fig. 17. ^{13}C CP/MAS spectrum of Exxon subbituminous C No. 36-IH.

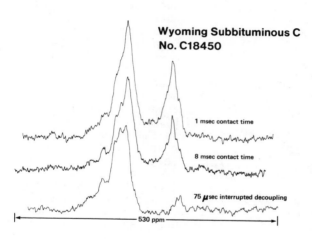

Wyoming Subbituminous C No. C18450

1 msec contact time

8 msec contact time

75 μsec interrupted decoupling

|← ——————— 530 ppm ——————— →|

Fig. 18. ^{13}C CP/MAS spectra of a Wyoming subbituminous coal obtained under three sets of conditions.

extract much greater structural detail than is available from the standard ^{13}C CP/MAS experiment. Much more work on coals and on model compounds will be required before such approaches are fully evaluated and useful for specific coal applications.

CONCLUSIONS

^{13}C cross polarization/magic angle spinning nmr is a highly valuable technique for the characterization of coals. While for most coal samples, the information obtained currently is limited to the fractions of aromatic and aliphatic carbon, current work that focusses on the time domain is leading to higher levels of information content on organic chemical structural detail. Further studies should make it possible to markedly increase the speed of the experiment and should provide needed answers on the level of reliability of the data obtained.

ACKNOWLEDGEMENTS

The authors are grateful for support of this work by the National Science Foundation (Grant No. GEAR 76-83374) and by the Department of Energy (Laramie Energy Technology Center).

REFERENCES

1. A. Pines, M.G. Gibby and J.S. Waugh, J. Chem. Phys., 56, 1779 (1972).
2. A. Pines, M.G. Gibby and J.S. Waugh, J. Chem. Phys., 59, 569 (1973).

3. J. Schaefer and E.O. Stejskal, J. Am. Chem. Soc., 98, 1031 (1976).
4. J. Schaefer and E.O. Stejskal in "Topics in Carbon-13 NMR Spectroscopy," Vol. 3, edited by G.C. Levy, Wiley-Interscience, New York, 1979, pp. 283-324.
5. E.R. Andrew, Progr. Nucl. Magn. Reson. Spectrosc., 8, 1 (1971).
6. a) I.J. Lowe, Phys. Rev. Letters, 2, 285 (1959).
   b) H. Kessemeier and R.E. Norberg, Phys. Rev., 155, 321 (1967).
7. V.J. Bartuska and G.E. Maciel, J. Magn. Resonance, submitted.
8. K.W. Zilm, D.W. Alderman and D.M. Grant, J. Magn. Resonance, 30, 563 (1978).
9. V.J. Bartuska, G.E. Maciel, J. Scahefer and E.O. Stejskal, Fuel, 56, 354 (1977).
10. G.E. Maciel, V.J. Bartuska and F.P. Miknis, Fuel, 58, 391 (1979).
11. F.P. Miknis, M.J. Sullivan, V.J. Bartuska and G.E. Maciel, Organ. Geochem., submitted.
12. H.L. Retcofsky and D.L. VanderHart, Fuel, 57, 421 (1978).
13. K.W. Zilm, R.J. Pugmire, D.M. Grant, R.E. Wood and W.H. Wiser, Fuel, 58, 11 (1979).
14. G.E. Maciel, V.J. Bartuska and F.P. Miknis, Fuel, 58, 155 (1979).
15. F.P. Miknis, G.E. Maciel and V.J. Bartuska, Organ. Geochem., 1, 169 (1979).
16. D. Vucelic, N. Juranic and D. Vitorovic, Fuel, 58, 759 (1979).
17. H.A. Resing, A.N. Garroway and R.N. Hazlett, Fuel, 57, 450 (1978).
18. F.P. Miknis, V.J. Bartuska and G.E. Maciel, Amer. Lab., 11, 19, (1979).
19. G.E. Maciel, L.W. Dennis and P.G. Hatcher, Science, submitted.
20. P.G. Hatcher, D.L. VanderHart and W.L. Earl, Organ. Geochem., in press.
21. S.J. Opella, M.H. Frey and T.A. Cross, J. Am. Chem. Soc., 101, 5856 (1979).
22. G.E. Maciel, M.P. Shatlock, R.A. Houtchens and W.S. Caughey, J. Am. Chem. Soc., submitted.
23. R.H. Atalla, J.C. Gast, D.W. Sindorf, V.J. Bartuska and G.E. Maciel, J. Am. Chem. Soc., 102, 3249 (1980).
24. W.L. Earl and D.L. VanderHart, J. Am. Chem. Soc., 102, 3251 (1980).
25. J. Schaefer, E.O. Stejskal and R. Buchdahl, Macromol., 10, 384 (1977).
26. E.O. Stejskal, J. Schaefer and T.R. Steger, Faraday Symp., 13, Chap. 5 (1979).
27. M.D. Sefcik, E.O. Stejskal, R.A. McKay and J. Schaefer, Macromol., 12, 423 (1979).
28. A.N. Garroway, W.B. Moniz and H.A. Resing, Organic Coatings and Plastics Preprints, 172nd ACS Meeting 36, 133 (1976).
29. Mention of a manufacturer's or product name does not imply endorsement by the U.S. Government.
30. G.E. Maciel and N. Szeverenyi, to be published.
31. J. Tegenfeldt and U. Haeberlen, J. Mag. Res., 36, 453 (1979).
32. S.J. Opella and M.H. Frey, J. Am. Chem. Soc., 101, 5854 (1979).

^{29}Si NMR: A New Tool for Coal Liquids Characterization

K. D. Rose* and C. G. Scouten[+]
*Analytical and Information Division
+Corporate Research Laboratories - Fuels Science
Exxon Research and Engineering Company
Linden, New Jersey 07036

ABSTRACT

Protonated heteroatom functionalities (COOH, OH, SH, NH) have a major impact on the chemical and physical properties of coal materials. Characterization of these functionalities will, therefore, be important to efficient development of new coal utilization technologies. Silicon-29 NMR spectroscopy of the trimethylsilyl derivatives of these functional groups is a powerful new tool for this characterization. Preparation of the trimethylsilyl derivatives is carried out in the NMR sample tube and the ^{29}Si NMR spectrum of the products is accumulated under conditions similar to those routinely used in ^{13}C NMR. Studies on derivatized model compounds show that ^{29}Si chemical shifts are generally segregated into three regions characteristic of COOH, OH and SH, and NH functionalities. The ^{29}Si resonances of aromatic OH derivatives are further differentiated so that the major oxygenated components of coal liquids can be monitored as a function of processing and distillation conditions. Quantitative ^{29}Si NMR results are used to calculate total OH and COOH concentrations in several solutions. Comparison of these results with the elemental oxygen content permits an estimate of the percentage oxygen present in non-derivatizable (e.g., ether) groups. The technique is illustrated using soluble coal liquid distillate fractions boiling in the range of initial boiling point to 1050°F (566°C).

INTRODUCTION

   Characterization of the heteroatom functionalities in coal
materials is important to provide a more detailed description of the
impact of these groups on the chemistry and physics of coal utilization. Protonated heteroatoms, such as -OH, -SH, >NH, and -COOH,
are prominent in most coal materials and are thought to contribute
to the secondary structure of coal through strong intermolecular
hydrogen bonding.[1] The influence of these groups is expected to
strongly affect those processes which depend upon the mechanical
and mass transport properties of coal including cleaning, grinding,
solvent extraction, and conversion. The presence of these functionalities in coal liquids contributes to important solution properties
such as viscosity, volatility, storage stability, and odor.[2] The
reactivity of these heteroatom functionalities is certain to have
a major impact on both the economic and technical aspects of coal
liquid upgrading and the utilization of coal liquids as synthetic
fuels. Thus, the identification of the heteroatom functional
groups in coal materials is critical to the development of a
satisfactory strategy for coal utilization.

   A variety of experimental techniques have now been used
both individually and collectively to characterize coal and its
products.[3] The ability of these techniques, however, to detect
and differentiate specific heteroatom functionalities is often
limited. The molecular complexity and physical properties of coal
materials contribute to the analytical difficulties. The limited
solubility and volatility of coal and many coal liquids as well as
their significant reactivity (especially toward oxygen) restrict
the kinds of techniques which can provide reliable results. The
reactivity of the protonated heteroatom groups, on the other hand,
has often been used to generate chemically modified products which
have substantially altered physical properties compared to the
underivatized material.

   Selective derivatization of the acidic heteroatoms is
advantageous for developing new analytical methods based upon mild
and specific chemical reactions. Conditions are now well documented,
for example, by which the acidic proton of particular heteroatom
groups can be replaced by an alkyl,[1] acyl,[4] trifluoroacyl,[5] or
substituted silyl group.[6] Replacement of the potentially hydrogen
bonding proton with a relatively non-polar group produces derivatives which are generally more volatile, less associated, and
reactively more stable than the starting material. In most cases,
the chemical derivatization also enhances the experimental sensitivity for detection of reactive heteroatoms. The groups
incorporated by chemical reaction are usually more amenable to
analytical techniques such as elemental analysis, gas chromatography (GC) coupled with element sensitive detection, mass
spectrometry (MS), and nuclear magnetic resonance (NMR). As a
result, important functional groups can be indirectly detected
and quantitatively measured by generally conventional methods. A

technique is described in this paper which combines the advantages of chemical derivatization and silicon-29 nuclear magnetic resonance spectroscopy to investigate the heteroatom functional groups in typical coal liquids.

Trimethylsilylation reactions, in particular, are well known for enhancing the volatility and stability of reactive molecules and have historically been used to facilitate GC and GC/MS analyses on complex materials.[7] In a recent NMR application using this derivatization technique, Schweighardt and co-workers[8] demonstrated that the total OH contents of coal liquid samples could be determined by ^{1}H NMR detection and integration of the incorporated trimethylsilyl group protons. The excellent NMR sensitivity to ^{1}H detection and the nine-fold trimethylsilyl protons per reacted heteroatom permitted rapid analysis by this technique. Since the trimethylsilyl protons are several bonds removed from the heteroatom attachment point, the observed dependence of the ^{1}H chemical shift on the substrate structure and neighboring substituents was slight requiring a high magnetic field spectrometer to minimize resonance overlaps. Similar difficulties are expected to exist in differentiating types of trimethylsilyl groups by ^{13}C NMR chemical shifts in addition to a significant sacrifice in detection sensitivity compared to ^{1}H NMR measurements. Thus, although quantitative measurements of the total OH content should be possible by both NMR techniques, molecular identification by ^{1}H or ^{13}C NMR chemical shift measurements on trimethylsilylated coal liquids appears more suited to high magnetic field spectrometers. Since removal of the silylating agent and its by-products is generally required for quantitative determinations of the total OH content using these methods, a simpler and more direct NMR method was sought.

Silicon-29 NMR capability is now routinely available on most commercial spectrometers and is a potentially more sensitive probe than ^{1}H and ^{13}C NMR for discrimination of trimethylsilyl derivatives. Since the silicon atom in a trimethylsilyl group is directly bonded to the heteroatom, the [29]Si chemical shift depends more strongly on the heteroatom type and substituent effects in the vicinity of the heteroatom than the corresponding ^{1}H and ^{13}C chemical shifts. Typical coal materials contain little silicon, which if present generally exists as silica and alumino-silicates. The difference in silicon environment between these inorganic silicon compounds and the covalently bonded trimethylsilyl groups results in a substantial [29]Si chemical shift difference for these structures. Discrimination of the incorporated groups from potentially existing compounds is straightforward by ^{29}Si NMR. Unlike conventional ^{1}H and ^{13}C NMR spectra of coal liquids which contain many individual resonances arising from all organic molecules in the sample, ^{29}Si NMR spectra characterizing the derivatized materials contain tolerably few resonances representing only those heteroatom functionalities bearing acidic protons.

An abundant literature on silylation techniques[6] and ^{29}Si NMR spectroscopy provides a firm basis for extending these procedures to fossil fuel research. The prevalence of silicon in organic and inorganic chemistry as silanes, silicones, silicates, and silyl derivatives has stimulated extensive research in ^{29}Si NMR over the past 25 years. Recent reviews of progress in this field[9] and common pitfalls encountered in ^{29}Si NMR detection[10] have appeared. Early experiments on this nucleus investigated the significance of substituent effects on [29]Si chemical shifts in a wide variety of organosilicon compounds.[11] Polymeric siloxanes and block copolymers containing silicon have more recently been studied by ^{29}Si NMR for monitoring end-group and tacticity effects.[12] The sensitivity of the [29]Si chemical shift to molecular structure and heteroatom types in silyl derivatized amino acids reported by Schraml and coworkers[13] has also been exploited in other biochemical areas including sugars[14] and steroids.[15] On the basis of this past research, the combination of ^{29}Si NMR and trimethylsilyl derivatization is expected to prove convenient and informative for simplifying complex coal materials using commercially available instrumentation. Results obtained thus far on model compound mixtures and coal liquid fractions illustrate the potential of this technique.

EXPERIMENTAL PROCEDURES

Samples were prepared for ^{29}Si NMR analysis by accurately weighing between 0.5 and 1.5g of coal liquid and 100 mg of benzoic acid into a vial. Higher boiling coal liquids sometimes required mild heating in order to become fluid and homogeneous. The weighed sample was diluted with 1.0 ml of N, O-bis(trimethylsilyl) trifluoroacetamide (Aldrich Chemical Company) and 1.5 ml of p-dioxane solvent. The solution was mixed thoroughly with approximately 30 mg of chromium (III) 2,4-pentanedionate and several drops of tetramethylsilane and transferred to a 10 mm NMR tube. Dioxane solvent used in all experiments was redistilled from lithium aluminum hydride and stored until use under nitrogen atmosphere. This pretreatment eliminated complications that could result from dissolved water and dioxane solvent stabilizer, 2,6-di(t-butyl)-4-methylphenol.

Silicon-29 NMR spectra were examined after several transients for evidence of residual BSTFA (δ = 24.8 and -2.4 ppm) or its mono-trimethylsilyl reaction product (δ = 10.5 ppm). If these signals were not observed in the initial spectrum, additional silylating agent was added to the sample. Since trimethylsilyl ethers and esters are known to be sensitive to hydrolysis,[6] experiments performed in the presence of excess silylating agent avoided potential sample instability and also eliminated all workup procedures.

The spin-lattice relaxation characteristics of ^{29}Si necessitated the addition of chromium (III) 2,4-pentanedionate relaxation reagent to all samples (about 0.04M) in order to avoid lengthy experiments and non-quantitative results.[10] Gated proton decoupling was used throughout to minimize intensity variations due to residual Nuclear Overhauser effects. Ninety degree pulses separated by 4.2 second delays were used for quantitative measurements. Although chemical shift determinations could be accomplished after only a few transients, 600 to 2400 accumulations depending on sample characteristics were generally required to obtain signal-to-noise ratios suitable for detailed interpretation. The integral of the ^{29}Si resonance for O-trimethylsilyl benzoate internal standard was compared to the total integral for all other O-TMS resonances in order to calculate the total hydroxyl content (including carboxylic acid) of each sample. Percentage compositions were determined from inflections in the integration trace. The calculated total OH values are estimated accurate to $\pm$10 percent.

Silicon-29 NMR chemical shifts due to solvent effects are known to be as large as one ppm.[11] Since the range of resonances for O-trimethylsilyl derivatives is only approximately ten ppm, accurate and reproducible chemical shift referencing is of considerable importance. Two approaches were used throughout in order to minimize solvent and acquisition induced ambiguities in resonance assignments. First, p-dioxane was consistently used as cosolvent because it is a satisfactory solvent for most coal liquids and it produces good resolution of the O-TMS resonances. Further, p-dioxane possesses a single ^{13}C resonance at +66.5 ppm which facilitates ^{13}C NMR spectroscopy on the same solution, if desired. Second, a concentric tube arrangement was used in which a sealed 4 mm capillary tube was positioned within the 10 mm NMR sample tube by Teflon spacers. The capillary tube contained d_8-dioxane, d_{12}-tetramethylsilane, tetramethylsilane and a trace of relaxation reagent, a combination of reagents which provided a reproducible deuterium signal for spectrometer field locking and a distinctive resonance from which both ^{29}Si and ^{13}C NMR spectra could be referenced. This tube arrangement permitted NMR measurements on non-deuterated samples eliminating costly reagents and sample dilution. Some sensitivity to ^{29}Si observation is, of course, compromised by this technique although at the concentrations examined this was not a serious handicap. Silicon-29 chemical shifts using these procedures were reproducible to $\pm$0.05 ppm.

RESULTS AND DISCUSSION

Trimethylsilylation of the acidic groups in coal liquids is an attractive derivatization technique for NMR investigations. Reaction conditions are known by which quantitative conversions of heteroatom functional groups can be accomplished upon mixing at ambient temperatures.[6] As a result, the derivatization for subsequent ^{29}Si NMR measurements is carried out in the NMR tube shortly before data accumulation; removal of excess reagents is not required. The non-specific silylating agent, BSTFA (N, O-bis-(trimethylsilyl)trifluoroacetamide), used in these experiments reacts readily with all acidic heteroatom groups.

$$- X - H + F_3C - \underset{O-TMS}{C} = N - TMS \longrightarrow - X - TMS + F_3C - \underset{\underset{H}{|}}{\overset{\overset{O}{\|}}{C}} - N - TMS$$

where X is O, S, N, and COO

Even hindered phenolic compounds such as 2,6-di(t-butyl)-4-methylphenol were found to react to completion within fifteen minutes using this highly reactive silylating agent.[16] BSTFA also quantitatively reacts with dissolved water to produce the mono-trimethylsilyl derivative of BSTFA and hexamethyldisiloxane. Since trimethylsilyl ethers and esters are known to be sensitive to hydrolysis, experiments performed in the presence of excess silylating agent avoids incomplete reactions. Although BSTFA readily attacks protonated heteroatoms, non-protonated heteroatoms such as ethers, pyridine derivatives and tertiary amines are generally inert; complications due to unexpected side reactions are therefore minimized. Finally, the comparative hydrophobicity of the incorporated trimethylsilyl group usually improves solubility in typical non-aqueous NMR solvents.

A principal heteroatom functional group observed in coal liquids is hydroxyl, predominantly phenolic hydroxyl. A variety of pure compounds were examined by ^{29}Si NMR after derivatization in order to establish the resolution and range of chemical shifts which might be expected for trimethylsilylated coal liquids. Although data on NH and SH derivatives are limited, preliminary chemical shift measurements corroborate the conclusions of Schraml and coworkers[13] that the chemical shift range for trimethylsilyl derivatives is approximately 30 ppm and is generally segregated into smaller ranges characteristic of NH, OH and SH, and COOH derivatives. As seen in Figure 1, trimethylsilyl derivatives of oxygen (including carboxylic acids) and sulfur derivatives resonate from approximately 14 to 26 ppm with respect to tetramethylsilane. In general, the aromatic hydroxyl and aromatic acid derivatives have ^{29}Si resonances at lower field (more positive chemical shift values) than their aliphatic counterparts. Phenol and its alkyl

88

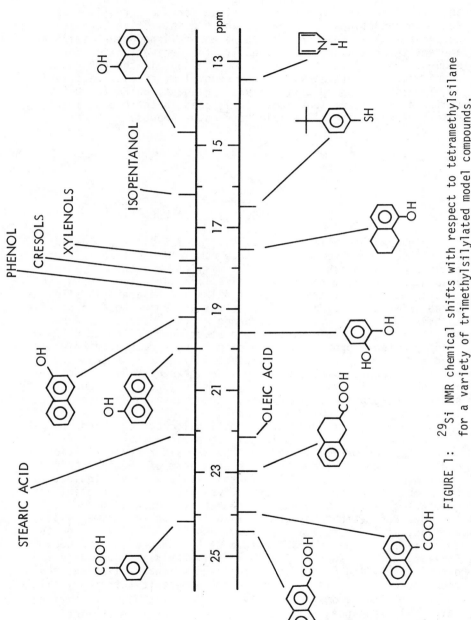

FIGURE 1: ^{29}Si NMR chemical shifts with respect to tetramethylsilane for a variety of trimethylsilylated model compounds.

substituted derivatives are resolved from the multi-ring OH and
dihydroxybenzene derivatives while additional differentiation is
observed among the single ring phenolics. The cresol isomers are
not differentiated by ^{29}Si chemical shift of the trimethylsilyl
group although the geometrical isomers of hydroxynaphthalene are
clearly resolved. Considering the inherently narrow linewidths
observed for ^{29}Si resonances, typically 0.05 ppm, this chemical
shift dispersion is entirely sufficient to resolve the major
oxygenated components of typical coal liquids.

The ^{29}Si NMR spectrum of a trimethylsilylated coal liquid
is shown in Figure 2. As noted previously, inorganic silicates
and the broad glass resonance sometimes observed in ^{29}Si NMR
spectra are far upfield from tetramethylsilane (δ = -110 ppm) and
consequently do not appear in the spectra reported here. Several
resonances in this NMR spectrum are not indigenous to the coal

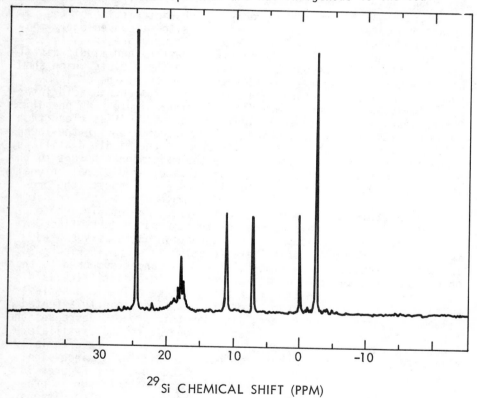

^{29}Si CHEMICAL SHIFT (PPM)

FIGURE 2: A typical ^{29}Si NMR spectrum of a
trimethylsilylated coal liquid.

liquid, namely those of internal tetramethylsilane (δ = 0.0 ppm) and BSTFA (δ = -2.4 and 24.8 ppm). The reaction products of BSTFA with trace amounts of water in the coal liquid sample are also observed, namely hexamethyldisiloxane (δ = 7.0 ppm) and the mono-trimethyl-silyl derivative of BSTFA (δ = 11.0 ppm). The O-trimethylsilyl chemical shift region (δ = 14 to 26 ppm) is unobstructed by most of these reagent resonances and contains several narrow peaks which can be assigned to coal liquid components. The trimethylsilyl deriva-tives of phenol, monoalkylphenol, and dialkylphenol are easily iden-tified as the three major resonances in this spectrum both on the basis of chemical shift assignments and by model compound addition. Additional intensity downfield from phenol is presumably due to multi-ring OH and dihydroxy aromatic molecules. Thus, identifi-cation of the major oxygenated components and their relative con-centrations in this coal liquid can be made by inspection of the ^{29}Si NMR spectrum.

The distillation and processing characteristics of the oxygenated molecules in coal liquids can also be conveniently mon-itored by this approach. Silicon-29 NMR spectra are shown in Figures 3A and 3B for trimethylsilylated naphtha and middle distil-late fractions of a second coal liquid. In these plots which show only the O-trimethylsilyl chemical shift region of the ^{29}Si NMR spectra, one BSTFA resonance and another due to benzoic acid inter-nal reference are observed downfield from the single ring phenolic resonances attributed to coal liquid components. It is clear from these results that phenol, which is a major component in the lower boiling naphtha fraction, is not detected in the middle distillate. Monoalkylphenol and dialkylphenol are the major contributors to the oxygenated molecules in the middle distillate. Additional intensity in Figure 3B which is not observed in Figure 3A suggests the pre-sence of other higher boiling oxygenated molecules which are not resolved in this spectrum by chemical shift. Presumably, these molecules would include a variety of multi-ring OH derivatives.

The influence of sample hydrogenation on the oxygenated molecules in the middle distillate fraction is shown in Figure 3C. The total intensity of O-trimethylsilyl resonances compared to that observed in the spectrum of the untreated middle distillate is sig-nificantly reduced. On the other hand, the relative intensities of single ring phenolic resonances are not significantly affected by sample hydrogenation. Thus, although this process decreases the concentration of oxygenated molecules compared to the distillate fraction feed, the hydrogenation does not appear to preferentially eliminate particular oxygenated molecules. Similarly since signi-ficant quantities of phenol are not observed in either Figures 3B or 3C, cleavage of alkyl substituents from aromatic rings is pro-bably not an important reaction under these processing conditions. From the analysis of qualitative ^{29}Si NMR measurements on coal liquid samples such as these, the consequences of distillation and sample processing on the oxygenated molecules can be inferred. Similar information is much more difficult to obtain from conven-tional ^{1}H or ^{13}C NMR spectra.

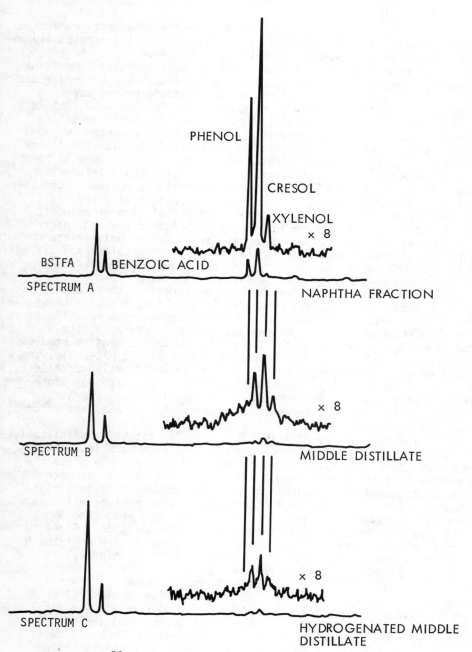

FIGURE 3: ^{29}Si NMR spectra of three trimethylsilylated coal liquid distillate fractions. The 0-trimethylsilyl chemical shift region of the ^{29}Si spectrum is shown in each plot.

QUANTITATIVE ASPECTS OF ^{29}Si NMR MEASUREMENTS

Intensity anomalies in ^{29}Si NMR spectra originate primarily from two effects: the extraordinarily long spin-lattice relaxation times for this nucleus and an unfavorable Nuclear Overhauser enhancement resulting from ^{29}Si-^{1}H dipole-dipole interactions.[10] Quantitative measurements in the presence of these effects are extremely difficult since the resonance intensity depends both on the spectrum acquisition conditions and on the method used to eliminate ^{29}Si-^{1}H scalar couplings. Fortunately, the addition of a small amount of paramagnetic relaxation agent to the NMR sample[17] produces two positive results: the ^{29}Si spin-lattice relaxation time is significantly shortened and the dominant dipolar interaction with protons is replaced by a more efficient electron-^{29}Si dipole-dipole interaction. The sensitivity of the ^{29}Si resonance intensity to nuclear properties and to spectrum accumulation conditions is dramatically reduced. After relaxation agent addition, reasonable real time experiments can be performed on this nucleus in which proton-decoupled spectra are quickly accumulated and co-added for signal-to-noise ratio improvement. With proper consideration of these problems associated with ^{29}Si NMR detection, measurements can be routinely performed with optimum sensitivity and quantitative accuracy.

Quantitative measurements were performed in these experiments by comparing the ^{29}Si resonance integrations for an unknown signal with that of an internal reference such as benzoic acid. From the known weights of unknown and internal reference, the OH concentration of unknown was calculated. Through extensive studies on pure compounds, unknown mixtures, and coal liquid materials, the total OH values determined by ^{29}Si NMR were consistently accurate to within ±10 percent with comparable repeatability. Although higher boiling coal liquid fractions sometimes benefited from overnight accumulations, most samples could be satisfactorily analyzed after less than one hour of acquisition.

The reliability of this NMR method for obtaining quantitative results was tested by comparison of ^{29}Si NMR data with independent measurements of the total OH concentration. One example of this comparison is shown in Figure 4 and Table I. A mixture of more than sixty model compounds was prepared by weighed additions of reagent grade chemicals. The components of this mixture which were susceptible to trimethylsilyl derivatization were less than 30 weight percent of the total mixture in toluene solution. Three samples of the model compound mixture were prepared and analyzed by ^{29}Si NMR without prior knowledge of the weighed addition values. A typical ^{29}Si NMR spectrum of the derivatized mixture expanded about the O-trimethylsilyl chemical shift region is shown in Figure 4. Each major resonance identified in this spectrum by chemical shift assignments was confirmed by model compound additions. The small resonance at approximately 21.9 ppm could not be positively identified by molecular structure but was assigned on the basis of its chemical shift position to an aliphatic acid. From the solution preparation, this component was later identified as decanoic acid.

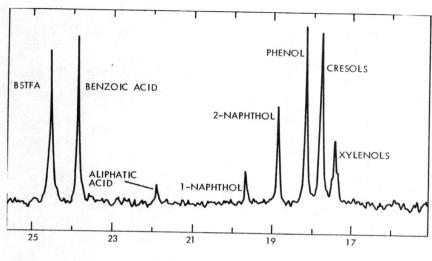

FIGURE 4: ^{29}Si NMR spectrum of a trimethylsilylated mixture of model compounds.

Comparison of the integral for the benzoic acid resonance
with that of the total O-trimethylsilyl chemical shift region produced the results summarized in Table I. Inflections in the integration trace were used to estimate the percentage of each identified component. With the exception of the monoalkyl and dialkylphenol concentrations, resonances for which there is some overlap in
the ^{29}Si spectrum, the comparison of the NMR data with the weighed
addition results is very good both in terms of total OH and fractional OH concentrations.

An advantage of the ^{29}Si NMR technique for quantitative
analysis of coal materials is its applicability to soluble samples
which may distill over a wide temperature range. Techniques other
than NMR which are suited to higher boiling components of coal
liquid samples include elemental analysis of derivatized samples[18]
and various non-aqueous titration procedures.[19] These methods,
however, provide little differentiation of the oxygenated molecules
present in the sample. Compared to the ^{29}Si NMR experiment, techniques such as GC or GC/MS measure more accurately the distribution
of oxygenated molecules but are limited to samples which are readily
volatilized and which do not decompose under the instrument operating conditions. Thus, the ^{29}Si NMR technique is particularly
useful for analysis of total coal liquids and for higher boiling
fractions which are difficult to analyze by other techniques. In
addition, since the NMR method uses between 0.5 and 1.0 grams of
sample per experiment, inconsistencies due to volatility or sample
heterogeneity are more easily controlled.

Table I

Comparison of Results Obtained by ^{29}Si NMR Spectroscopy
and by Weighed Additions on a Mixture of Model Compounds

	Mole Percent of Total OH	
Oxygenated Molecule	By ^{29}Si NMR*	By Weighed Addition
Aliphatic Acid	3	3
2-Naphthol	5	6
1-Naphthol	16	17
Phenol	25	26
Monoalkylphenol	30	33
Dialkylphenol	18	13
Unidentified	3	2
Total OH	2.4 mmoles/G	2.28 mmoles/G

* Average values of three trials

The distribution and distillation characteristics of oxygenated molecules in coal liquids provide a useful description of the starting materials. Two liquids products derived from coal were prepared under similar conditions and subsequently distilled into five fractions ranging from the initial boiling point to 1050 + °F. The ^{29}Si NMR spectra of the four lower-boiling trimethylsilylated fractions for one of these coal liquids are shown in Figure 5. Resonances due to BSTFA silylating agent and benzoic acid internal standard are observed in these expanded plots.

The distillation of the single ring phenolics and other oxygenated molecules can easily be monitored by inspection of the ^{29}Si NMR spectra. Only small amounts of phenol and monoalkylphenol are present in the lowest boiling fraction while phenol, monoalkylphenol, and dialkylphenol dominate Cut 2. In Cut 3 (450-675°F), phenol is no longer detected and the O-trimethylsilyl ^{29}Si intensity is almost evenly divided between substituted single ring phenolics and multi-ring OH molecules. The OH molecules in Cut 4 (675-1050°F) are probably quite diverse since little or no resolution is observed in the ^{29}Si NMR spectrum. The chemical shift range for the envelope of ^{29}Si resonances in this fraction is consistent, however, with aromatic OH derivatives, probably associated with multi-ring or highly substituted OH molecules. There is no evidence in these spectra for significant quantities of aliphatic acids or nitrogen derivatives. The ^{29}Si NMR spectra (not shown) for the distillate fractions of the second coal liquid were, in general, very similar to those shown in Figure 5. Some differences in relative intensities of ^{29}Si resonances were observed by comparing spectra from Sample 1 with those from Sample 2. These differences seemed more likely to be a consequence of the coal liquid distillation rather than of the procedure used to generate the liquid product from coal.

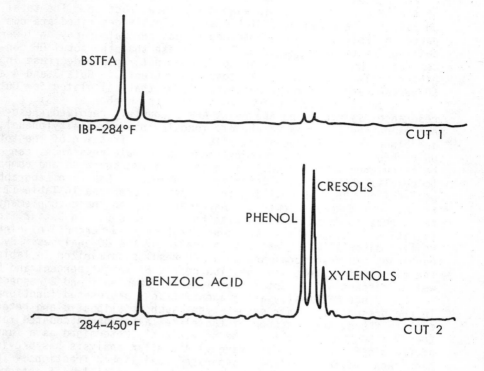

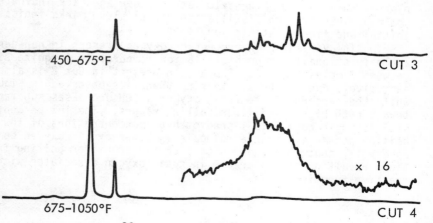

FIGURE 5: ^{29}Si NMR spectra of four trimethylsilylated distillate fractions derived from coal liquid Sample 1. The O-trimethylsilyl chemical shift region is shown in these expanded plots.

Integration of the ^{29}Si NMR spectra permitted a calculation of the OH concentrations in each distillate fraction. The total OH values determined by ^{29}Si NMR as previously described are compared in Table II to data obtained by gas chromatography on lower boiling fractions. These results indicate that the total OH concentration which maximizes in Cut 2 does not rapidly decrease in higher boiling fractions; the total OH contents in Cuts 3 and 4 are 60 percent and 30 percent respectively of that calculated for Cut 2.

In lower boiling fractions where the number of ^{29}Si resonances is small and the peaks are adequately resolved, resonances can be assigned to phenol, monoalkylphenol, dialkylphenol, and other types of oxygenated molecules. The fraction of the total OH concentration contributed by these assignable resonances can be calculated from inflections in the ^{29}Si NMR spectrum and compared to results obtained by alternative techniques. Gas chromatographic analyses on the lower-boiling fractions are compared in Table II to the ^{29}Si NMR measurements. Agreement between these complementary techniques is found once again to be satisfactory. In Cuts 2, integration distinguishes a low concentration (3-5 percent) of higher boiling material which was not indicated by the GC analyses. By combining the fractional and total OH results summarized in Table II, the single ring phenolics comprise nearly 67 weight percent and 61 weight percent of the Cut 2 fractions for samples 1 and 2 respectively.

Since BSTFA silylating agent attacks protonated functionalities in the presence of other oxygen types such as aliphatic and heterocyclic ethers, the fraction of the elemental oxygen accounted for by the ^{29}Si NMR measurement can be conveniently determined as a function of the distillation boiling range. A similar analysis has previously been reported for chemically separated coal liquid fractions.[8] Combining the elemental oxygen contents of each distillate fraction[20] and the total OH concentrations determined by ^{29}Si NMR (Table II), the percentage of total oxygen present as OH has been calculated and plotted in Figure 6. The calculated percentages are indicated at the midpoints of each distillation range; slash marks indicate the initial and final distillation temperatures

The variation in percentage oxygen present in non-derivatizable functionalities on distillation temperature is quite similar between Samples 1 and 2. The oxygen present in Cut 2 is almost exclusively due to the single ring phenolics observed in Figure 5 while the percent of elemental oxygen as OH decreases substantially toward both higher and lower boiling ranges. The low percentages of O as OH in both Cuts 1 predict high concentrations of low-boiling non-derivatizable molecules. Heterocyclic oxygen compounds apparently contribute more significantly in higher boiling fractions resulting in a decrease in total oxygen associated with reactive functionalities.

TABLE II

Comparison of ^{29}Si NMR and Gas Chromatographic Analyses
on the Distillate Fractions Derived from Two Similar Coal Liquids

Cut #	Percent by ^{29}Si NMR					Percent by GC			
	Phenol	Monoalkylphenol	Dialkylphenol	Other	mmoles OH/g	Phenol	C_1 Phenol	C_2 Phenol	mmoles OH/g
Sample 1:									
1	46	41	13	0	0.9	45	38	17	1.12
2	30	45	22	3	6.5	36	42	22	6.02
3	0	18	33	49	3.7	--	--	--	--
4	--	--	--	--	2.1	--	--	--	--
Sample 2:									
1	48	44	8	0	0.7	42	39	19	0.85
2	28	44	24	5	5.9	34	42	24	5.88
3	0	13	27	60	4.0	--	--	--	--
4	--	--	--	--	2.4	--	--	--	--

98

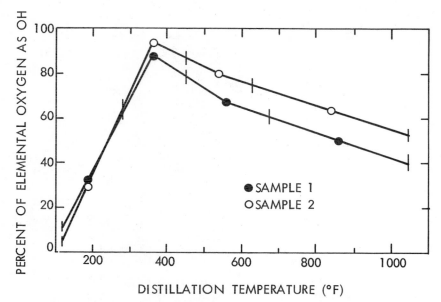

FIGURE 6: The fraction of total oxygen present as
OH as a function of sample distillation
temperature. Results obtained on two
similar coal liquids are shown.

CONCLUSION

Trimethylsilyl derivatization of acidic heteroatom functionalities in coal materials improves the sensitivity of many analytical
techniques for molecules containing these important groups. Silicon-
29 NMR measurements on derivatized samples are particularly useful
for simplifying the interpretation of rather complex mixtures. The
^{29}Si chemical shift of incorporated trimethylsilyl groups is sensitive to the type of covalently bonded heteroatom and to substituents in the vicinity of the heteroatom group. Sample derivatization in these experiments is conducted in a conventional NMR tube
at ambient temperatures and the NMR spectrum is accumulated without
additional sample workup. The acquisition techniques used to obtain
^{29}Si spectra with optimum sensitivity and quantitative accuracy are
quite similar to methods already in common use for high-resolution
^{13}C NMR. Commercially available NMR instrumentation is entirely
adequate for performing these ^{29}Si NMR measurements on trimethylsilyl derivatized samples. Both qualitative and quantitative
functional group distributions can be obtained on soluble materials
which span a wide range in distillation temperature. Finally, since
the chemical derivatization technique only attacks the reactive
heteroatom sites in coal materials, detailed information can be
obtained by ^{29}Si NMR spectroscopy in the presence of excess silylating agents, non-reactive NMR solvents and other coal liquid
diluents.

ACKNOWLEDGEMENTS

Samples discussed in this report were provided by R. P. Rhodes and R. H. Schlosberg of Exxon's Corporate Research Laboratories. Sample preparations and NMR measurements were made by H. J. Malone and C. F. Pictroski.

REFERENCES

1. R. Liotta, Fuel, 58, 724 (1979).

2a. K. A. Gould, M. L. Gorbaty, and J. D. Miller, Fuel, 57, 510 (1978).

 b. F. R. Brown, F. S. Karn, Fuel, 59, 431 (1980).

 c. J. R. Kershaw, D. Gray, Fuel, 59, 436 (1980).

3. T. Aczel, R. B. Williams, R. A. Brown, R. J. Pancirov, in "Analytical Methods for Coal and Coal Products, Volume 1". (C. Karr, Jr., Ed.) Chap. 17, Academic Press, New York, 1978.

4a. L. Blom, L. Edelhausen, D. W. van Krevelen, Fuel, 38, 537 (1957).

 b. Z. Abdel-Baset, P. H. Givens, R. F. Yarzab, Fuel, 57, 84 (1978).

5. P. Sleevi, T. E. Glass, H. C. Dorn, Anal. Chem., 51, 1931 (1979).

6. C. A. Roth, Ind. Eng. Chem. Prod. Res. Develop., 11, 134 (1972).

7a. J. F. Klebe, Advan. Org. Chem., 8, 97 (1972).

 b. C. W. Gehrke, A. B. Patel, J. Chromatogr., 130, 103 (1977).

8. F. K. Schweighardt, H. L. Retcofsky, S. Friedman, M. Hough, Anal. Chem., 50, 368 (1978).

9. E. A. Williams, J. D. Cargioli, Annu. Rep. NMR Spectroscopy, 9, 221 (1979).

10. G. C. Levy, J. D. Cargioli, in "Nuclear Magnetic Resonance Spectroscopy of Nuclei Other than Protons" (T. Axenrod and G. A. Webb, Eds.) Chap. 17, Wiley-Interscience, New York, 1974.

11. R. L. Scholl, G. E. Maciel, W. K. Musker, J. Amer. Chem. Soc., 94, 6376 (1972) and references therein.

12. R. K. Harris, B. J. Kimber, Appl. Spectroscopy Rev., 10, 117 (1975).

13. J. Schraml, J. Pola, V. Chvalovsky, H. C. Marsmann, K. Blaha, Coll. Czech. Chem. Comm., 42, 1165 (1977).

14a. A. H. Haines, R. K. Harris, R. C. Rao, Org. Magn. Reson., 9, 432 (1977).

   b. D. J. Gale, A. H. Haines, R. K. Harris, Org. Magn. Reson., 7, 635 (1975).

15. J. Schraml, J. Pola, H. Jancke, G. Englehardt, M. Cerny, V. Chvalovsky, Coll, Czech. Chem. Comm., 41, 360 (1976).

16. C. G. Scouten, unpublished results.

17a. R. Freeman, K. G. R. Pachler, G. N. La Mar, J. Chem. Phys., 55, 4586 (1971).

   b. O. A. Gansow, A. R. Burke, W. D. Vernon, J. Amer. Chem. Soc., 94, 2550 (1972).

18. S. Friedman, M. L. Kaufman, W. A. Steinger, I. Wender, Fuel, 40, 33 (1961).

19. D. W. van Krevelen, "Coal", Elsevier Press, Amsterdam, pp. 160-176, 1961.

20. The elemental oxygen values determined by neutron activation analyses were corrected for the measured water contents.

ELECTON SPIN RESONANCE (ESR) CHARACTERIZATION
OF COAL AND COAL CONVERSIONS

L. Petrakis and D. W. Grandy
Gulf Research & Development Company
Pittsburgh, Pennsylvania 15230

ABSTRACT

This paper provides an overview of the nature, behavior and
significance of free radicals in coals, coal components and
fractions during coal conversion processes. Special emphasis is
placed on the characterization of free radicals and their dependence on pyrolysis and liquefaction process conditions. Some
preliminary results are provided from the recently developed
technique for the in situ observation of free radicals under
preheater/reactor conditions.

INTRODUCTION

Much effort is being expended in the development of several
processes for the conversion of coal to environmentally clean
products. Many of these processes have several common elements
in that they involve depolymerization of coal, stabilization of
the lower molecular weight chemical moieties through hydrogen
abstraction, and the presence of mineral matter and/or added
catalyst in the coal/solvent slurry that is being reacted.

The overall conversion of coal to liquid products in a
direct hydroliquefaction process is subject to a complex set of
process and chemical or feed variables. The former include the
temperature of the reactor vessel, the pressure, heating rate,
and residence time. The latter include the coal/coal fractions,
solvents, gas, particle size, and maceral and mineral contents
of the coals. The interplay of these process and chemical or
feed variables determines coal conversion and affects the overall
process. The complexity of the phenomena involved in coal
hydroliquefaction has made it necessary to approach this problem
from a variety of viewpoints, including the role of petrographic
components in hydroliquefaction;[1,2] chemical functionalities;[3,4]
the question of analytical separations;[5-7] and the characterization of feeds and products.[8-10]

It is generally agreed that during dissolution of coal in
direct hydroliquefaction processes, complex and interrelated
physical and chemical phenomena occur that are not completely
understood. For example, in important papers, Neavell[11] and
Wiser[12] have described succinctly these complex and interrelated
events. The reaction of high volatile bituminous coal at temperatures of 400-500°C in a hydrogen donating slurry causes both
physical and chemical changes, as a result of which certain

components of the coal (vitrinite and exinite) are converted to low molecular weight benzene-soluble moieties. Initially, the coal particles swell forming a plastic mass with the pyrolysis products. The vitrinite particles disperse in the slurry, regardless whether the organic vehicle is hydrogen-donating or not. Thus, a pyrolytic breakdown of the "chicken-coop wire structure" of coal is effected that involves the formation of free radicals. These free radicals may be quite reactive species and are expected to behave differently in different environments. In the presence of hydrogen-donating solvents, the free radicals may abstract hydrogen from the solvents and thereby become stabilized as lower molecular weight benzene-soluble species. In the absence of hydrogen-donating solvents, the pyrolytically formed free radicals may recombine to form highly refractory benzene-insoluble materials that could cause reactor problems (i.e., coking) or affect adversely the product distribution to less desirable modes. Of course, the overall behavior of free radicals (their very formation, reactivity, stabilization, and recombination) will be determined not only by the chemical and physical make-up of the feed (coal/solvent/gas) but also by the process conditions at which the free radicals find themselves.

Coals have been known since 1954 to contain stable free radicals that were formed during coalification in geologic time under a set of conditions which are much milder than those encountered in coal conversion processes, e.g., the SRC-II preheater reactor.[13,14] These stable free radicals in coals have been studied quite extensively by numerous workers[15-17] using Electron Spin Resonance (ESR) techniques. Coal rank as well as other coal characteristics correlate well with the behavior and nature of these stable free radicals. The temperature and pressures in a direct hydroliquefaction preheater/reactor are such that they make the pyrolytic generation of additional and reactive free radicals quite likely. The presence of relatively large amounts of heteroatoms (S, N, 0) in coals would also tend to result in higher concentrations of free radicals, since the cleavage of bonds between heteroatoms and carbon would tend to proceed more readily than carbon-carbon bond cleavage. Loss of electrons from the coal molecules, additionally, may result in radical ions without bond rupture. These moieties also could be quite reactive. The propensity of the "coal molecules" to undergo bond rupture and/or free radical formation is likely to be further affected by the presence of heteroatoms in the "coal molecules" and also by minerals (especially minerals with transition metals) in the coal. The former would generally tend to lower the ionization potential of the molecule in which they find themselves, while the latter would be expected to affect electron transfer reactions from the "coal molecules" to the mineral substrate.

From batch experiments that have been reported, it is accepted that free radicals play a key role in the rate and mechanism of coal liquefaction.[11,12] The mode of stabilization

of the free radicals could indeed be critical in the product distribution whether liquid or solid refractory materials are obtained.
Yet, only recently have the concentration and nature of free radicals in coal liquefaction products[18-21] and liquefied coals been
measured. The baseline picture of the free radicals of coals and
coal fractions is quite complex; and it would be expected that the
overall behavior of coal free radicals would become even more
complex once these species are considered when solvents, gases,
and variable process conditions are introduced. Given the significance of free radicals in coal conversion processes, we
attempt in this paper a discussion of free radicals in coals.
More specifically, we summarize and discuss our findings of the
free radicals of whole coals, coal fractions and coal components.
We also investigate the effect of pyrolytic and liquefaction conditions and we conclude with some results of the in situ observation of free radicals under SRC-II preheater reactor conditions.

EXPERIMENTAL

The samples used in our ESR measurements were prepared by
placing 0.3 grams of material in a 2 mm I.D. Pyrex ESR tube. All
of the coals were dried overnight in air at 100°C before further
use. Samples were also sieved into various particle size fractions. All evacuation heat treatments were conducted by first
evacuating the sample to 10^{-4} torr for 2 hours at the treatment
temperature and then sealing. Duplicate samples were prepared at
a later date for many of the samples to check reproducibility.
The samples for ESR measurements for liquefaction experiments were prepared by placing approximately 0.3 grams of coal,
mixed in a one-to-one ratio by weight with the appropriate solvent, in a 2 mm I.D. ESR tube. The tube and sample were then put
in a 6 mm I.D. stainless steel bomb reactor, purged repeatedly
with either H_2 or N_2 gas and pressurized to about 9.3 MPa (1400
psig). A tube furnace was then placed around the reactor and
heating was begun. After heating for the specified period of
time, the reactor was cooled to ambient temperatures in about
3 minutes. The system was then depressurized and the sample tube
removed from the reactor and sealed in air. At least one duplicate sample was prepared for each set of experimental conditions.
All spectra (except for in situ studies) were recorded at
room temperature with Varian V-4500 single- and dual-cavity
spectrometers. Concentration measurements were done with the
dual-cavity instrument by the usual sample-standard interchange
method. Reproducibility between duplicate samples was usually
within 20%. Linewidth determinations were of variable quality,
depending on the signal-to-noise ratio and lineshape. Errors in
linewidths are 10% or less in most cases. Spectral lineshapes
were examined for many of the samples. In some cases, multiple
superimposed signals were observed.

TABLE I

CHEMICAL ANALYSIS OF WHOLE COALS

Coal	Weight % (MAF)					Ash (MF)	
	C	H	O	S_{ORG}	N	Wt %	Rank
1 Lignite, Hagel*	71.0	4.9	21.8	0.72**	1.6	9.7	Lignite
2 Wyodak, Gillette, WY	72.2	5.5	21.0	0.3	1.0	9.2	Sub C
3 Deitz, WY*	73.9	5.5	18.4	0.6**	1.7	5.9	Sub B
4 Kentucky #11*	76.4	5.4	10.0	5.9*	2.2	16.0	HVB
5 Lower Dekoven	80.6	5.6	4.8	7.7	1.8	21.9	HVB
6 Powhatan #5	80.8	5.6	8.4	1.9	1.2	10.3	HVA
7 Illinois #6	82.0	5.4	9.5	1.0	1.7	6.5	HVB
8 Pocahontas #3	86.4	3.9	6.8	0.9**	1.2	3.3	LV
9 Lower Kittaning*	89.1	4.8	3.9	1.2*	0.9	7.0	LV
10 Hiawatha Seam Utah							HVC

*Peter Given

TABLE II

BASELINE ESR MEASUREMENTS OF VARIOUS COALS

Coal	Spin Concentration (10^{18}/gram)		G*		Δ H (gauss)	
	25°C	100°C	25°C	100°C	25°C	100°C
1 Lignite Hagel	1.3±0.5	4.7	2.00297±2	2.00396±7	2.8±( )	5.67
2 Wyodak Gillette	-	5.8	-	2.0037	-	7.4
3 Deitz, WY	3.8±0.05	8±0.1	2.00347±1	2.00359±2	7.4±0.8	7.8±0.06
4 KY #11	12.8	19.1	2.00303	2.00291±4	2.4	1.9±0.6
5 Lower Dekoven	-	4.1	-	2.0028	-	0.9
6 Powhatan #5	7.5	8.6±0.3	2.00287	2.00276±1	1.2±0.01	1.2±0.01
7 Illinois #6	-	6.8	-	2.0029	-	1.3 and 6.0
8 Pocahontas #3 W VA	-	15.8	-	2.029	-	1.0 and 6.0

*Error is for last decimal

TABLE III

g-VALUES OF SOME POSSIBLE COMPOUND-TYPES IN COAL

RADICAL TYPE	g-VALUE	REFERENCES
1) AROMATIC HYDROCARBONS		
π RADICALS		
1-5 RINGS	2.0025 (CATION)	(23,24)
	2.0026 - 2.0028 (ANION)	
7 RINGS	2.0025 (CATION)	
(CORONENE)	2.00291 (ANION)	
2) ALIPHATIC HYDROCARBONS		
σ RADICALS	2.0025 TO 2.0026 (NEUTRAL)	(25)
3) OXYGEN CONTAINING		
FREE RADICALS		
σ TYPE	2.0008 - 2.0014 (NEUTRAL)	(26,27)
π TYPE		
QUINONES 1-3 RINGS	2.00469 - 2.00380	(23,28)
ETHERS		
1-RING, MONO, DI, TRI	2.00350 - 2.00398 (CATION)	(29)
METHOXYBENZENES		
4) N-CONTAINING RADICALS	2.0031	(30)
5) S-CONTAINING RADICALS	2.0080 - 2.0081	(30)
6) GRAPHITE	2.0025 - 2.015	(31)

The apparatus for the in situ observation of free radicals is desribed in Ref. 22. Samples are prepared for ESR measurements by placing 0.5 g of coal in a quartz 4 mm O.D., 2.5 mm I.D. ESR tube. When the coal is to be run with a solvent, a very thin capillary tube is placed in the bottom of the ESR tube and the coal is poured in around the capillary. The desired amount of the solvent is then forced through the capillary, which is later removed. After the coal and solvent have been placed in the tube, it is placed into the heater assembly in the cavity.

FREE RADICALS IN COAL, COAL FRACTIONS AND COMPONENTS

A series of coals have been investigated prior to any thermal or other treatments. This section reports these findings using a number of coals ranging from lignite to low volatile bituminous. Table I presents the analytical data of the coals investigated. Table II summarizes the baseline ESR measurements for these nine coals. Spin concentrations, g-values, and linewidths in gauss are shown for each coal after evacuation at both 25°C and of 100°C.

The results obtained in the ESR experiments on the total coals, Table II, are essentially in agreement with the results obtained by earlier workers.[16] Since the g-value of a radical is quite sensitive to the chemical environment of the unpaired electron, some information as to the types of radicals in coal can be obtained. In the case of the higher ranking coals, the g-values are typical of those found for π-type aromatic hydrocarbon radicals (see Table III). The subbituminous coal and lignite have much higher g-values which indicates that atoms other than carbon and hydrogen (probably oxygen) are important in the electronic structure of the radical.

Among the heteroatoms found in the hydrocarbon framework of coal, oxygen content varies in a regular fashion with rank, whereas nitrogen and sulfur content are more or less constant. Correlations between g-values of coals and oxygen content, which varies from 20% to 4% by weight going from low- to high-ranking coals, have been made and it has been suggested that the radicals may be of the quinone type.[16] Methoxy benzene radicals and related compounds are known to have g-values typical of those found in the lower rank coals, agound 2.0035 to 2.0037.[29] In view of the current opinions regarding the structure of coal, either or both may be possibilities, depending on the rank of the coal in question.

The spectral linewidth (ΔH_{p-p}) of all samples was measured as run in air or vacuum. With the exception of the subbituminous coal and lignite, many evacuated samples yielded spectra which consist of a sharp line (ΔH_{p-p} 1 gauss) superimposed on a broader signal. The majority of the samples have much broader linewidths in air or, as for those samples with two resolvable components, the narrow component often disappears. This is the expected result of dipolar broadening by paramagnetic molecular

TABLE IV

ESR PARAMETERS OF SELECTED MACERALS†

MACERAL	SPIN CONCENTRATION (10^{18}/G)	G*	ΔH (GAUSS)
FUSINITE (ILL #6)	229	2.00275 (±1)	0.7±0.04
FUSINITE (WAYNESBURG)	260	2.00276 (±0)	1.04±0.0
VITRINITE (WAYNESBURG)	3.8	2.00301 (±2)	7.20±0.04
VITRINITE (PITTSBURGH #8)	4.6	2.00292 (±2)	8.11±0.01
VITRINITE (ARMSTRONG)	7.7	2.00395 (±4)	8.08±0.05
RESINITE (ARMSTRONG)	0.17	2.00359 (±1)	7.46±0.10
RESINITE (HIAWATHA)	0.11	2.00359 (±3)	7.46±0.10

† SAMPLES EVACUATED AT 25°C

* UNCERTAINTY REFERS TO 5TH DECIMAL

FIGURE 1

FREE RADICAL CONCENTRATION vs TEMPERATURE— COALS AS RECEIVED

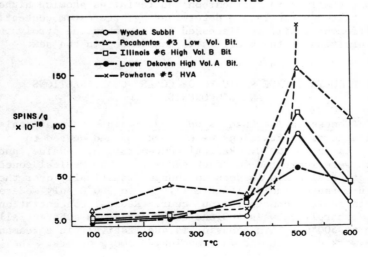

oxygen. As noted by Retcofsky,[16] spectra of some lithotypes found in coal show large differences when the samples are evacuated relative to those run in air.

Spin Concentration and Lineshapes - Concentration measurements were performed to find the number of unpaired electrons per gram of sample. The radical concentrations found for the coals in air follow the expected increase with rank.[16] With regards to particle size, the finer sieve fraction has a higher spin concentration than the coarse fraction. The difference may be attributed to the concentration of fusain in the fine fraction since fusain is known to have a spin concentration about five times greater than associated vitrains.[16]

The lineshapes of the spectra were estimated whenever possible. The two lineshape functions considered were the Lorentzian and the Gaussian function. Evacuated samples yielded spectra which range from some nearly equal linear combination of the two functions, as is the case with the subbituminous coal, to the very broadly sloping, multicomponent spectra of the higher rank coals. Multicomponent spectra were more common in total coal and coarse samples. Lineshapes are affected by unresolved hyperfine or anisotropic effects, or both. Of those samples run in air, only the subbituminous coal had appreciable Gaussian character. Signal asymmetry was far more prevalent in air-exposed samples than those which were run while evacuated.

In addition to coals and coals fractionated according to particle size, we have made measurements of the baseline ESR parameters of coal components or macerals .[32] These are summarized in Table IV.

The most spectacular difference is observed in the free radical concentration that varies by 4-5 orders of magnitude. The g-value also reflects the expected variation with fusinite being essentially pure hydrocarbon and resinites showing highest g-values that apparently are determined by heteroatom content. These differences will be discussed further when the pyrolysis and liquefaction of the macerals is considered in the next section.

PYROLYSIS FREE RADICALS OF COALS, COAL FRACTIONS AND COMPONENTS

Spin Concentration Changes upon Pyrolysis - Several coals and available sieve fractions were evacuated and heated to temperatures up to 600°C. Radical concentration, g-value, and linewidth were monitored for each sample.[34] The radical concentration of total coals is seen to change dramatically over the temperature range studied (Figure 1). Up to 300°C only moderate changes in radical concentration occur. Radical concentration increases sharply between 400 and 500°C and drops substantially on going to 600°C. This behavior is qualitatively in agreement with the work of Smidt and VanKrevelen,[33] among others. The change in free radical concentration as a function of temperature may be due to the rupture of bonds starting around 400°C and continuing to a point where the coal becomes fluid enough for the radical sites formed on bond-breaking to combine and the coal then resolidifies or polymerizes into a char.

Pyrolytic studies have also been carried out with pure macerals. A plot of the log of radical concentration as a function of treatment temperature is shown in Figure 2. Clearly, the free radicals from the various macerals, regardless of the coal from which they were picked, show great differences in their behavior. The initial concentrations of the macerals vary by some four to five orders of magnitude. The resinites have the lowest radical concentration, on the order of 10^{17} spins/g. Extreme difficulty was encountered in treating resinite samples beyond 400°C. The material, a brown solid not unlike brown sugar in appearance at room temperature, became a brown liquid at 400°C and vaporized at higher temperatures.

The radical concentations of vitrinites (Figure 2) are on the order of 3 to 8 x 10^{18} spins/g in their natural state, increase rapidly beyond 400°C and peak around 10^{20} spins/g at 400 to 550°C. Unheated fusinite radical concentrations are about two orders of magnitude greater than the vitrinites studied. Upon heating, the radical concentration in these materials increases by a factor of about 3, from 3 x 10^{20} to 9 x 10^{20} spins/g at 550°C. The very modest increase in radical concentration with temperature is consistent with that observed by others.

The relative order of spin concentrations in the unheated macerals, fusinite>vitrinite>resinite is expected in light of the decreasing aromaticity and mean size of the aromatic clusters in the above sequence.[35] Large aromatic structures would tend to stabilize free radicals to a greater degree than aliphatic chains or small aromatic units over geologic time.

The free radicals in resinite show very little dependence of their concentration of temperatures below 400°C, however, the change in g-value and the change in their physical appearance suggest that chemical reactions are taking place. It is quite likely that the material is losing heteroatoms in the form of CO_2, H_2O, etc., but that any radicals formed are satisfied by internal hydrogen transfer in this relatively hydrogen-rich material.[35]

In their natural state, the g-values of the free radicals in the resinites are quite high, indicating that the unpaired electron is somewhat delocalized on a heteroatom, probably oxygen. The g-values around 2.0036 are quite common among oxygen containing aromatic hydrocarbons such as quinones.[23] In the case of the subbituminous resinite, the g-value of the radicals were essentially the same as those in the associated vitrinite. With increasing treatment temperature, the g-values are observed to decrease. At 400°C and above, the g-values of

110

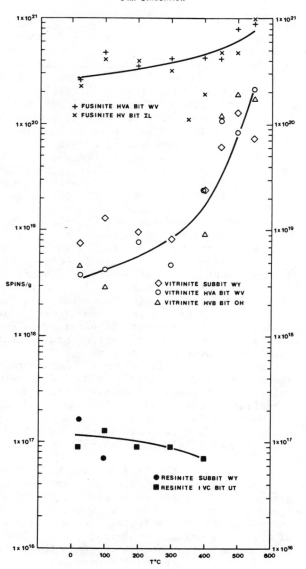

FIGURE 2
LOG RADICAL CONCENTRATION
VS
TREATMENT TEMPERATURE
2 HR EVACUATION

these materials are similar to those of pure aromatic hydro-carbon radicals.[23] It is probable, in view of our earlier work on a subbituminous coal,[34] that the decrease in g-values with increasing treatment temperature is due to oxygen-containing functional groups being split off in the form of CO_2, H_2O, etc. leaving the aromatic hydrocarbon skeleton behind.

At room temperature, the well-known rank dependence of g-values of radicals in vitrinites and total coals is ob-served.[16,34,36] Upon heating, the g-value of the radicals in subbituminous vitrinite decreases, becomes the same as that of the higher rank vitrinites.[32] This change is probably due to loss of oxygen-containing functional groups during pyrolysis. The overall behavior of the g-values of the radicals in heat-treated vitrinites is similar to that observed for total coals.[34] The g-values of fusinites, from 2.0026 to 2.0028, are typical of aromatic hydrocarbon radicals.[23]

We summarize this section by making the following observa-tions: 1. The initial concentrations of free radicals from various macerals vary by about 5 orders of magnitude (fusinite> vitrinte>resinite). 2. Free radical concentration of all macerals remain essentially unaffected by temperatures up to 400°C. At higher temperature, the free radical concentration of vitrinites change greatly, approaching the fusinite spin concen-tration. 3. The initial g-values of free radicals of fusinites and of HV bituminous vitrinites (i) are characteristic of aromatic hydrocarbon radicals, and (ii) they remain unaltered by high temperatures. The initial g-values of free radicals of resinites and subbituminous vitrinites (i) are characteristic of heteroatom containing aromatic radicals and (ii) at 400°C they also become aromatic hydrocarbon-like radicals.

LIQUEFACTION: FREE RADICALS OF COALS, COAL FRACTIONS AND COMPONENTS

Free radicals are believed to play an important role in the mechanism of coal liquefaction.[11,12] The mode of stabilization of these reactive speices may determine the type and value of the product obtained.[18,19,20] An attempt has been made to correlate the concentration or properties of the free radicals with process or chemical variables, such as residence times, heating rates and solvent and gas systems used.[37]

Free Radicals Quenching in Wyodak Subbituminous Coal - Wyodak subbituminous coal has been used to probe the sensitivity of residual free radicals to a range of experimental conditions: solvents (vacuum, naphthalene, tetralin), temperatures (425-480°C), residence times (10 minutes, 2 hours), heating rates (13°C/min), and different gases (H_2, N_2). Spin concentration, linewidths and g-values have been measured.

   Results of these experiments are shown in Figures 3 and 4.
It can be seen in both figures that a distinct difference between
the Wyodak coal samples treated in the two donor systems exists,
with the tetralin/H$_2$ treated coal having a lower residual radical
concentration. By comparing the two figures, a difference in
magnitude of this solvent effect is evident. The Wyodak coal
treated at the faster heating rate/shorter residence time (Figure
8) shows an enhanced solvent effect.

   In addition to these solvent and heating rate/residence time
effects, there is an effect of temperature upon radical concentration (Figure 4) where samples heated to higher temperatures
had higher concentrations of residual free radicals. The magnitude of this temperature effect appears to be larger in the
naphthalene/H$_2$ treated samples. A toluene soxhlet extraction of
the tetralin/H$_2$ and naphthalene/N$_2$ treated Wyodak coal run at
480°C with rapid heating (32°C/min) gave results of 48% vs 31%
conversion, respectively, on a DAF basis.

   As a follow-up to the investigation of the behavior of the
Wyodak coal, a variety of coals ranging in rank from lignite to
low volatile bituminous were reacted under the conditions discussed above.[37] The spin concentration of Kentucky #11 and of a
Lower Dekoven high volatile bituminous coal as a function of temperature and solvent show differences similar to the Wyodak.
The behavior of some of the other coals (e.g., Dietz subbituminous and Hagel lignite) upon temperature and solvent treatment
is distinctly different, however. These coals show a very strong,
steady temperature effect with no apparent differences due to
solvent treatment system. Lower Kittanning low volatile bituminous exhibits very little dependence of radical concentrations
on temperature and no noticeable solvent treatment effect.

   The g-values and in widths measured for these coals treated
in the two solvent system generally showed no solvent or temperature effects over the range of temperatures studied (400-480°C).
Most of the g-values fell in the range from 2.0026 to 2.0029,
typical of aromatic hydrocarbon radicals.[23,24]

   Dependence of Free Radicals on Solvents. Wyodak subbituminous coal was treated with a variety of solvents which should
have a wide range of hydrogen donating capabilities under conditions similar to those used above. A mixture of coal and
solvent, approximately 1:1 by weight, was heated at 32°C/min
under 1800 psig of H$_2$ or N$_2$ gas, held at either 425 or 480°C
for 10 minutes and rapidly cooled. After venting, the tube was
sealed in air and the ESR spectrum obtained. The solvents used
were tetrahydroquinoline (THQ), tetralin, N,N'-di-sec-butyl
paraphenylene diamine (D22), methanol, decane, 2-octanol, and
naphthalene.

   The residual spin concentration of the coal by solvent
treatment is shown in Figure 5. Each bar is 95% confidence
limit of all of the data for that solvent, regardless of temperature and gas used. The bars in some cases, then, include

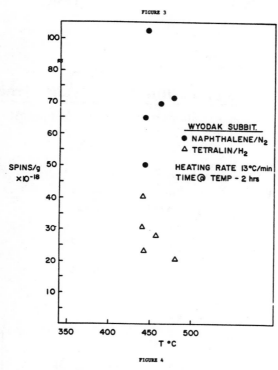

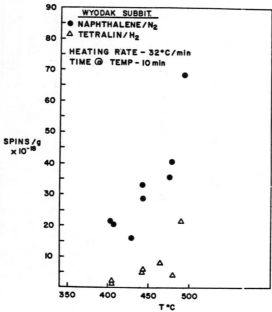

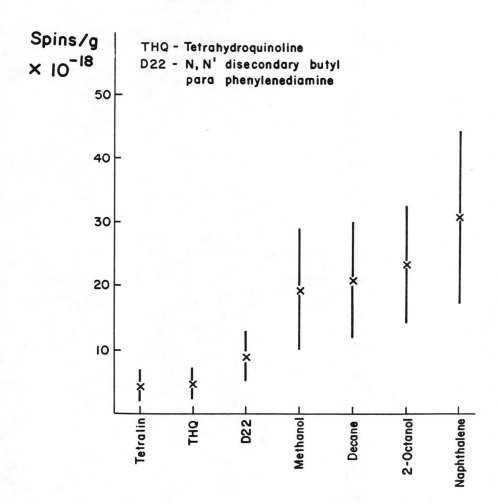

FIGURE 5

FIGURE 5

Radical Concentration by Solvent Treatment

Wyodak Subbituminous – All Temperatures

Spins/g
× 10^{-18}

THQ – Tetrahydroquinoline
D22 – N, N' disecondary butyl
     para phenylenediamine

Tetralin   THQ   D22   Methanol   Decane   2-Octanol   Naphthalene

some temperature and gas effects as well as experimental scatter. Tetralin and THQ were both very effective for quenching free radicals. The radical concentration in the coal heated with these solvents was about the same as that found for the unheated natural coal. Nearly all of the thermal radicals normally produced are quenched. Naphthalene, an aromatic, was visibly the poorest solvent for quenching free radicals, while all the other solvents were interemediate.

<u>Liquefaction of Macerals</u>. We report the results of room temperature ESR measurements of high purity macerals that have been liquefied in naphthalene or tetralin between 425 and 480°C under 12.4 MPa of hydrogen or nitrogen. The apparent reactivities of these macerals and the relationship between the conversion and free radical concentration are discussed.

Typical are the results of free radical concentration measurements of Pittsburgh #8 HVB vitrinite liquefied in tetralin or naphthalene are shown in Figure 6. Note the significantly lower concentration of free radicals in those samples where tetralin was the liquefaction solvent. The radical concentration appears to be independent of the pressurizing gas; however, it must be remembered that tetralin is in more than a 100-fold excess on a molecule/radical basis at 400°C, assuming the number of radicals measured in an evacuated heating experiment is the maximum number formed. Therefore, it is unlikely that hydrogen transfer from gas to solvent has an effect on the radical concentration. At 450° and higher, this excess of donor solvent molecules to radicals formed is about 10 to 1. This may explain the higher radical concentration of radicals in the 480° tetralin runs than those in the 425° tetralin runs, although no effect due to the gas used on the radical concentration is noticeable. The radical concentration in the vitrinite in the naphthalene runs is similar to that found in evacuated heating experiments.

The results of free radical concentration measurements on an Armstrong Seam WY subbituminous vitrinite liquefied in naphthalene and tetralin do not follow the same pattern observed for the high volatile vitrinites. In these experiments, an effect of temperature on the free radical concentration is present, however, there is no apparent solvent effect. A solvent effect on radical concentration has been observed in the high volatile bituminous coals and vitrinites but not in subbituminous or low volatile coals and vitrinites, with one exception,[21,37] out of about a dozen different coals.

There are no differences in radical concentrations due to solvent in the liquefaction runs done on Hiawatha seam HVC bituminous resinite. An increase in radical concentration due to temperatures is present.

The free radical concentration of Illinois #6 fusinite was measured after liquefaction with the four solvent/gas combinations. There are no clear gas or solvent effects on the free radical concentration. Temperature effects are small for the

116

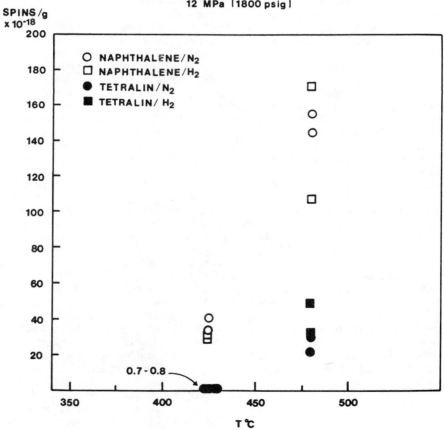

FIGURE 6

RADICAL CONCENTRATION
VS
TREATMENT TEMPERATURE
PITTSBURGH#8 VITRINITE HVB BITUMINOUS
SMITHFIELD OH
10 min. RESIDENCE TIME HR- 32°C/min.
12 MPa (1800 psig)

SPINS/g
x 10⁻¹⁸

○ NAPHTHALENE/N₂
□ NAPHTHALENE/H₂
● TETRALIN/N₂
■ TETRALIN/H₂

0.7 - 0.8

T °C

most part. Similar free radical concentration results were
obtained for a Waynesburg Seam HVA bituminous fusinite. The
radical concentration of both fusinites after liquefaction
appears to be insensitive to solvent or pressurizing gas and
shows only small temperature effects over this range.

Table V summarizes the significant changes that take place
in the various regiments under which the macerals were liquefied.
Resinites, not unexpectedly, show the least free radical concentration, while fusinites show the highest free radical concentrations and vitrinites are intermediates. Vitrinites have a
large solvent dependence both at 425 and 480°C as shown by the
ratio of the spin concentration in naphthalene to that in tetralin
(columns 4 and 5 in Table V). Columns 6 and 7 show that there is
no dependence of the free radical concentrations on the gas used
since the concentration ratios at both temperatures are essentially
constant. Columns 2 and 3 show the temperature coefficient for
the macerals arranged by solvent. Again, there is a temperature
coefficient for the vitrinite with the coefficient being higher
in naphthalene than in tetralin while both resinite and fusinite
show no temperature coefficients. These findings are consistent
with the fact that under all sets of conditions attempted here the
resinites liquefy completely while fusinites are inert and remain
uneffected. The vitrinites, which are the most abundant component
in coals and, therefore, more significant in coal liquefaction
processes, can be affected by the various conditions used.

Although the g-values of the free radicals in unheated
macerals may be quite different from each other,[32] they are
basically similar over the 425°C to 480°C range studied here.
The g-values of all of the macerals, regardless of treatment, fell
between 2.0026 and 2.0028, typical of aromatic hydrocarbon
radicals.[23,24]

Due to the wide range of free radical concentrations and the
markedly different appearance of the liquefaction products of the
different macerals in the ESR tubes, some soxhlet extractions,
using toluene as a solvent, were performed on a few select
samples. The product of several runs of Waynesburg vitrinite,
Hiawatha resinite and Illinois #6 fusinite in tetralin (50/50 by
weight) at 480°C under 12.4 MPa or H_2 with a 10-minute residence
time and 32°C/min heating rate were each combined in a cellulose
thimble for toluene extraction. The resinite was essentially
completely converted to toluene solubles, 98%. The Waynesburg
vitrinite had a conversion level of 36% and the fusinite, which
is apparently very unreactive had only a 5% conversion to toluene
solubles under these conditions.

The following conclusions can be drawn from these experiments.

1. Resinite and fusinite residual free radical concentrations appear to be insensitive to the liquefaction solvent used.

2. High volatile bituminous vitrinites have radical concentrations which are solvent sensitive.

118

3. The relationship between residual free radical concentration and the amount of toluene insolubles is non-linear. If suitable calibration procedures are used with the same feed coal, the free radical concentration of a product slurry might be a sensitive indicator of percent conversion.

IN SITU MEASUREMENTS OF FREE RADICALS DURING LIQUE-FACTION

We have recently designed and fabricated a high temperature, high pressure ESR cavity for the measurement of free radicals during coal liquefaction.[22] The temperature and pressure capabilities of this cavity are 500°C and 13.8 MPa (2000 psi) respectively. A small reference sample of $CuSO_4 \cdot 5H_2O$ is attached to the inside wall of the cylindrical cavity and is maintained at temperatures below 59°C. All spin concentration measurements on coal samples are made relative to this calibrated reference.

Upon thermal treatment (450°C) the free radical concentration of Powhatan #5 coal under 10.3 MPa of H_2 increases from 9 to 60 x 10^{18} spins per gram. This latter value has been corrected to account for the difference in the Boltzmann distribution in the electron spin states between 450°C and 20°C. Returning to 20°C, the residual free radical concentration is nearly equal to that observed at 450°C, being 53 x 10^{18} for this sample. Experiments are now underway to study the effect of solvent, residence time, heating rate and gas pressure on the in situ measurement of free radicals during coal liquefaction.

REFERENCES

1. P. H. Given, D. C. Cronauer, W. Spackman, H. L. Lovell, A. Davis, and B. Biswas, Fuel 54, 34 (1975).
2. Ibid, p. 40.
3. A. J. Szladow and P. H. Given, ACS Fuel Div. Preprints 23, No. 4, p. 161 (1978).
4. F. K. Schweighardt, C. M. White, S. Friedman, and J. L. Schultz, ACS Fuel Div. Preprints, Vol. 22, No. 5, p. 124 (1977).
5. M. Farcasiu, Fuel 56, 9 (1977).
6. R. G. Ruberto, D. M. Jewell, R. K. Jensen, and D. C. Cronauer "Characterization of Synthetic Liquid Fuels in Shale Oil, Tar Sands and Related Fuel Sources" (T. F. Yen, Ed.) ACS Advances in Chemistry, Series 151, ACS, Washington, DC (1976).
7. D. D. Whitehurst, et al, "The Nature and Origin of Asphaltenes in Processed Coals" NTIS PB-257, 569.
8. H. L. Retcofsky, F. K. Schweighardt, and M. Hough, Anal. Chem. 49, 585 (1977).
9. J. W. Larsen, P. Choudhury, and L. Urban, ACS Fuel Div. Preprints, Vol. 23, No. 4, 181 (1978).
10. I. Schwager, J. T. Kwan, J. G. Miller, and T. F. Yen, ACS Fuel Div. Preprints, Vol. 23, No. 1, 284 (1978).

11. R. C. Neavel, Fuel 55, 237 (1976).
12. W. H. Wiser, Scientific Problems of Coal Utilization", DOE Symposium Series 46, 1978, pp. 219-236.
13. J. Uebersfeld, A. Etienne, and J. Combrisson, Nature (London) 174, 615 (1954).
14. D. J. E. Ingram, J. G. Tapley, R. Jackson, R. L. Bond, and A. R. Murnaghan, Nature (London), 174, 797 (1954).
15. H. Tschamler and E. DeRuiter "Chemistry of Coal Utilization," Supplementary Vol., (H. H. Lowry, Ed.) John Wiley and Sons, Inc., New York, NY, 1963, pp. 78-85.
16. H. L. Retcofsky, J. M. Stark, and R. A. Friedel, Anal. Chem. 40, 1699 (1968).
17. D. E. G. Austen, D. J. E. Ingram, P. H. Given, C. R. Binder, and L. W. Hill, "Coal Science" Advances in Chem. Series 55, 344 (1966).
18. D. L. Wooten, H. C. Dorn, L. T. Taylor, and W. H. Coleman, Fuel 55, 224 (1976).
19. D. W. Grandy and L. Petrakis, Fuel 58, 239 (1979).
20. H. L. Retcofsky, M. Hough and R. B. Clarskon, ACS Fuel Div. Preprints, Vol. 24, No. 1, 1979, p. 83.
21. L. Petrakis and D. W. Grandy, ACS Fuel Div. Preprints, Vol. 23, No. 4, 1978, p. 147.
22. D. W. Grandy and L. Petrakis, accepted for publication in J. Mag. Res.
23. M. S. Blois, Jr., H. W. Brown, and J. E. Mailing, "Free Radicals in Biological Systems", Academic Press, New York, NY, 1961, p. 130.
24. B. G. Segal, M. Kaplan, and G. K. Fraenkel, J. Chem. Phys., 43, 4191 (1965).
25. R. W. Fessenden and R. H. Schuler, J. Chem. Phys., 39, 2147 (1963).
26. P. Smith, R. A. Kaba, L. M. Dorringuez, and S. M. Denning, J. Phys. Chem., 81, 162 (1977).
27. P. J. Krusic and T. A. Rettig, J. Am. Chem. Soc., 92, 722 (1970).
28. M. Adams, M. S. Blois, and R. H. Sands, J. Chem. Phys., 28, 774 (1958).
29. P. O'Neill, S. Steenken, and D. Schulte-Frohlinde, J. Phys. Chem., 79, 2173 (1975).
30. "Atlas of Electron Spin Resonance Spectra", Bielski and Gebicki, Ed., Academic Press, New York, NY, 1967, pp. 38, 606, 611.
31. Gerald C. Michael, Ph.D. Thesis, The Pennsylvania State University, 1969.
32. L. Petrakis and D. W. Grandy, in press (Fuel).
33. J. Smidt and D. W. VanKrevelen, Fuel, 38, 355 (1959); "Coal", Elsevier, Amsterdam, 1961, pp. 393-399.
34. L. Petrakis and D. W. Grandy, Anal. Chem. 50, 303 (1978).
35. H. Tschamler and E. deRuiter, "Coal Science" ACS Adv. Chem. Series, 55, 332 (1966).

120

36. H. L. Retcofsky, G. P. Thompson, M. Hough and R. A. Friedel, ACS Fuel Div. Preprints, Vol. 22, No. 5, 90 (1977).

37. L. Petrakis and D. W. Grandy, Fuel 59, 227 (1980).

UNDERSTANDING COAL USING THERMAL DECOMPOSITION
AND FOURIER TRANSFORM INFRARED SPECTROSCOPY*

P. R. Solomon and D. G. Hamblen
Advanced Fuel Research, Inc.
P. O. Box 18343, East Hartford, CT 06118

ABSTRACT

Fourier Transform Infrared Spectroscopy (FTIR) is being used to
provide understanding of the organic structure of coals and coal
thermal decomposition products. The research has developed a relationship between the coal organic structure and the products of
thermal decomposition. The work has also led to the discovery that
many of the coal structural elements are preserved in the heavy
molecular weight products (tar) released in thermal decomposition
and that careful analysis of these products in relation to the
parent coal can supply clues to the original structure.
Quantitative FTIR spectra for coals, tars and chars are used to
determine concentrations of the hydroxyl, aliphatic and aromatic
hydrogen. Concentrations of aliphatic carbon are computed using an
assumed aliphatic stoichiometry; aromatic carbon concentrations are
determined by difference. The values are in good agreement with
data determined by ^{13}C and proton NMR. Analysis of the solid products produced at successive stages in the thermal decomposition
provides information on the changes in the chemical bonds occurring
during the process. Time resolved infrared scans (120 msec/scan)
taken during the thermal decomposition provide data on the amount,
composition and rate of evolution of light gas species. The relationship between the evolved light species and their sources in
the coal is developed by comparing the rate of evolution with the
rate of change in the chemical bonds. With the application of these
techniques, a general kinetic model has been developed which relates
the products of thermal decomposition to the organic structure of
the parent coal.

INTRODUCTION

The increased use of complicated chemical processing of coal to
make alternative fuels or reduce pollutant formation has made it
imperative to obtain better understanding of coal structure and how
it comes apart during a chemical reaction. The study of thermal decomposition is important since all coal conversion processes (combustion, liquefaction and gasification) are initiated by this step.

*Work supported in part by the Department of Energy under Contract
ET-78-C-01-3167 with United Technologies Research Center

In addition, it apears that many of the coal structural elements are preserved in the heavy molecular weight products (tar) released in thermal decomposition so analysis of the products can supply important clues to the structure of the parent coal.

In a recent study, the thermal decomposition behavior of a large number of coals has been experimentally measured and theoretically modeled.[1-6] To provide quantitative organic structural information on the coals, and products of thermal decomposition (tar, char and light gases), techniques have been developed using a Nicolet 7199 Fourier Transform Infrared Spectrometer (FTIR). By measuring the absorption of infrared light the FTIR provides information on the nature of the chemical bonds. Analysis of the solid products produced at succesive stages of thermal decomposition provides information on the changes occurring in the chemical bonds. Rapid infrared scans taken during the thermal decomposition provide data on the amount, composition and rate of evolution of light gas species. A relationship can be developed between the evolved light species and their sources in the coal by comparing the rate of evolution with the rate of change in the chemical bonds.

With the application of these techniques, a general kinetic model has been developed which relates the products of thermal decomposition to the organic structure of the parent coal. The model uses a widely accepted view of coal structure which assume a polymer-like molecule consisting of groups of fused aromatic ring clusters ("monomers") linked by relatively weak aliphatic bridges. The ring clusters contain heteroatoms (oxygen, sulfur and nitrogen) and have a variety of attached functional groups. During thermal decomposition, the weak links are ruptured, releasing the clusters and attached functional groups. These large molecules comprise the coal tar. Simultaneous with the evolution of tar molecules is the competitive cracking of the bridge fragments, attached functional groups and ring clusters to form the light molecules of the gas. The evolution of each species is characterized by rate constants which do not vary with coal rank. The differences between coals are due to differences in the mix of sources in the coal for the evolved species. The sources are related to the functional group concentrations in the coal which may be determined by FTIR.

This paper reviews the relationship between functional group distribution and thermal decomposition behavior which has been developed in the papers referenced above and reports recent results obtained for a Pittsburgh seam coal and a Montana lignite at temperatures up to 1800°C. The temperature dependent evolution of corresponding products are similar for the two coals indicating that the use of coal independent kinetic rates is applicable for the higher temperatures. The dominant effect observed at higher temperatures is the trend toward increased yields of hydrogen gas and unsaturated compounds (olefins, acetylene and probably soot) at the expense of paraffins. These effects are modeled by including additional parallel reaction paths for the decomposition of the aliphatic content of the coal.

EXPERIMENTAL

The experimental apparatus is illustrated in Fig. 1. It consists of a small chamber in which the coal is pyrolized connected through a glass wool filter to a large gas cell for infrared

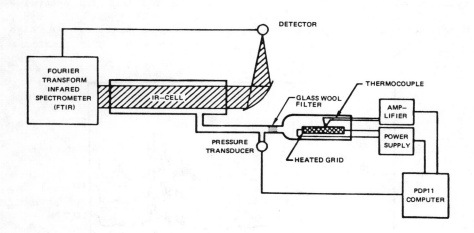

Figure 1   Pyrolysis Apparatus

analysis. The coal is evenly distributed between the folds of a stainless steel, molybdenum or tungsten screen and a current is passed through the screen to heat the coal. Coal temperatures of 1800°C and heating rates of 2000°C/sec were achieved using the tungsten screen.   The distribution of major products is determined by weighing the coal, char, gas and tar (which accumulates on the cold chamber walls).   Analysis of the solids is performed with a Perkin-Elmer 240 elemental analyzer, a scanning electron microprobe[7] and the FTIR. Gas analysis is performed with the FTIR which permits low resolution analysis at 120/msec intervals. Calibration of the FTIR has been made using pure gases or prepared gas mixtures. Unfortunately, most of the gases of interest show a marked increase in the integrated absorbance under a line if the pressure in the cell is increased by diluting with another gas. The explanation for this effect is that the absorption lines for these gases are extremely sharp and for moderate concentrations all the infrared energy is absorbed in the line center in a path shorter than the absorption cell. The instrument resolution is substantially broader than the line width so the lines do not appear to be truncated.

Dilution of the gas broadens the line, reducing the absorbance at line center so that a longer path contributes to the absorptivity, thus increasing the average absorbance. This effect makes calibration of these gases in the pyrolysis gas mixture difficult. The solution has been to dilute the mixture with nitrogen to a fixed pressure at which calibrations have been made. A typical set of spectra showing the evolution of gases during an 80 sec pyrolysis is illustrated in Fig. 2. The low resolution analysis can determine CO, CO_2, H_2O, CH_4, SO_2, CS_2, HCN, C_2H_2, C_2H_4, C_3H_6, benzene, COS, and heavy paraffins and olefins. A high

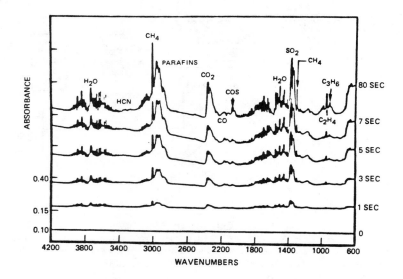

Figure 2   Infrared Spectra of Evolving Pyrolysis Gas

resolution analysis made at the completion of a run can determine all of the above plus C_2H_6, C_3H_8, C_4H_8, NH_3 and potentially many other species which have not yet been observed. H_2 is determined by difference. Other features of the apparatus are similar to those described previously.[1-6]

THE MODEL

The pyrolysis model has been based on the understanding of coal thermal decomposition developed by measuring and modeling vacuum pyrolysis for a large number of coals. The work is discussed in detail in Refs. 1-6. The model assumes a coal structure consisting of highly substituted aromatic ring clusters containing heteroatoms linked by relatively weak aliphatic bridges. Evidence suggests that during thermal decomposition these weak links break, releasing the clusters and attached bridge fragments which comprise the tar.

In vacuum devolatilization in a thin heated grid, the tar molecules may be removed quickly from the bed and undergo little secondary reactions. The evidence for this is the striking similarity between the tar and parent coal which has been observed in elemental analysis, FTIR spectra and NMR spectra[1-6, 8, 9]. The two materials are almost identical except for a higher concentration of aliphatic hydrogen (especially methyl) in the tar. This extra hydrogen is presumably abstracted from the char to stabilize the free radical sites formed when the bridges were broken. Similar arguments were given for pyrolysis of model compounds by Wolfs, van Krevelen and Waterman[10]. Since the abstracted hydrogen is most likely to come from the aliphatic portion of the coal, it is reasonable to expect the tar yield to depend on H_{al}. In Fig. 3, the tar yield in vacuum pyrolysis is plotted against H_{al} for a number of coals (circles). Also plotted are the yields of heavy

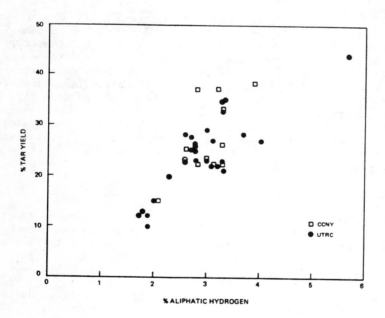

Figure 3   Variation of Tar Yield with Aliphatic Hydrogen

hydrocarbons, (i.e., oils and BTX) from hydropyrolysis (squares).[11] Indeed, there is a strong correlation between tar yield and H_{al}.

The observation that the tar is rich in methyl hydrogen but not in aromatic hydrogen when compared to the parent coal has implications concerning the nature of the aliphatic linkages. The result suggests that the bonds which were broken to free the ring clusters were predominantly between two aliphatic carbons not between an aromatic and an aliphatic. The increased concentration of methyl groups should relate directly to the density of ethylene or longer bridges in the parent coal.

Simultaneous with the evolution of tar molecules is the competitive cracking of bridge fragment, substituted groups and ring cluster to form the light molecular species of the gas. The quantity of each gas species depends on the functional group distribution in the original coal. At low temperatures there is very little rearrangement of the aromatic ring structure. There is decomposition of the substituted groups and aliphatic (or hydroaromatic) structures resulting in CO_2, H_2O, hydrocarbon gases and some CO release from the carboxyl, hydroxyl, aliphatic and weakly bound ether groups, respectively. At high temperature there is breaking and rearrangement of the aromatic rings. In this process, H_2 is released from the aromatic hydrogen and HCN and additional CO are released from ring nitrogen and tightly bound ether linkages.

There are therefore two distinct mechanisms for removal of a functional group from the coal; evolution as part of a "monomer" or evolution as a distinct species with cracking of the "monomer". To model these two paths with one path yielding a product which is similar in composition to the parent coal, the coal is represented as an area with X and Y dimensions. The Y dimension is divided into fractions according to the chemical composition of the coal Fig. 4a. $Y^o(i)$ represents the initial fraction of a particular component (carboxyl, aromatic hydrogen, etc.) and $Y^o(i)=1$. The evolution of each component into the gas (carboxyl into CO_2, aromatic hydrogen into H_2, etc.) is represented by the first order diminishing of the Y(i) dimension,

$$Y(i)=Y^o(i) \exp(-k_i t).$$

The X dimension is divided into a potential tar forming fraction X^o and a non-tar forming fraction $1-X^o$ with the evolution of the tar being represented by the first order diminishing of the X dimension

$$X=X^o \exp(-k_x t).$$

The amount of a component in each of the pyrolysis products is determined by integration yielding the following equations:[2]

$$W(i)_{char} = (1-X^o + X) Y(i) \tag{3}$$

$$W(i)_{tar} = (X^o Y^o(i)-XY(i)) k_x/(k(i)+k_x) \tag{4}$$

$$W(i)_{gas} = (1-X^o) (Y^o(i)-Y(i)) + W(i)_{tar}k(i)/k_x \tag{5}$$

Figure 4b illustrates the initial stage of thermal decomposition during which the volatile components H_2O and CO_2 evolve from the hydroxyl and carboxyl groups respectively along with aliphatics and tar. At a later stage (Fig. 4c) HCN, CO and H_2 are evolved from the nitrogen compounds, the ethers and aromatic H respectively. Some HCN and CO is released at low temperatures, so two kinetic rates are used for nitrogen and ether. Also a minor amount of NH_3 is evolved which has not been included in the model.

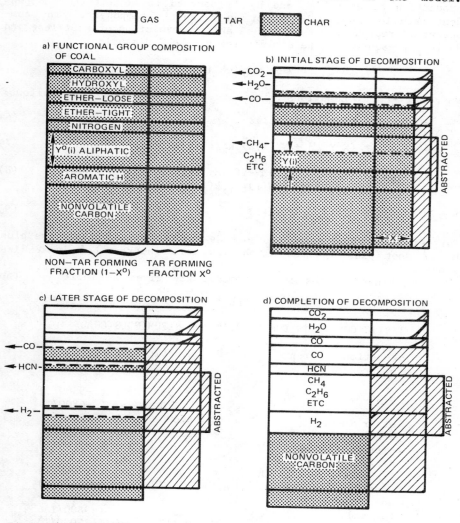

Figure 4   Progress of Thermal Decomposition according to Model
a) Functional Group Composition of Coal, b) Initial state of Decomposition, c) Later Stage of Decomposition, d) Completion of Decomposition

To simulate the abstraction of H by the tar, the aliphatic fraction in the tar is assumed to be retained together with some additional aliphatic material which may be added directly or may contribute its hydrogen. When hydrogen from the aliphatic material is contributed, its associated carbons remain with the non volatile carbon fraction which eventually forms the char (Fig. 4d).

Modifications of the model were made to include the high temperature production of unsaturated compounds. As an example, the production of acetylene and H_2 is assumed to be a third independent path for the evolution of an aliphatic component. The component is represented as a volume and the evolution of acetylene and H_2 is represented by the diminishing of the Z dimension,

$$Z = Z^O \exp(-k_z t),$$

where $Z^O = 1$. Equations 3-5 are then modified as follows:

$$W(i)_{char} = (1 - X^O + X) \, Y(i) Z \qquad (6)$$

$$W(i)_{tar} = (X^O Y^O Z^O - XY(i)Z) \, k_x / (k(i) + k_x + k_z) \qquad (7)$$

$$W(i)_{gas} = (1 - X^O) \, (Y^O(i) Z^O - Y(i)Z) \, k(i) / (k(i) + k_z) \qquad (8)$$
$$+ W(i)_{tar} \, k(i) / k_x$$

$$W(i)_z = W(i)_{gas} \, k_z / k(i) \qquad (9)$$

Further competitive processes such as the production of olefins plus H_2 or soot plus H_2 from aliphatics were incorporated in a like manner.

The kinetic rates for the model are given in Table I. The

TABLE I

KINETIC CONSTANTS FOR LIGNITE AND BITUMINOUS COALS

Functional Group or Product	Kinetic Rates		
carboxyl	$k_1 =$	75	exp ($-5800/T$)
hydroxyl	$k_2 =$	370	exp ($-6800/T$)
ether loose	$k_3 =$	87000	exp ($-12000/T$)
ether tight	$k_4 = 4.0 \times 10^9$		exp ($-35000/T$)
nitrogen loose	$k_5 =$	200	exp ($-7600/T$)
nitrogen tight	$k_6 =$	290	exp ($-12700/Y$)
light HC	$k_7 =$	33000	exp ($-12000/T$)
aromatic H	$k_8 =$	3600	exp ($-12700/T$)
non volatile C	$k_9 =$	0	
tar and heavy HC	$k_x =$	9300	exp ($-9800/T$)
olefins	$k_o = 5.0 \times 10^9$		exp ($-25000/T$)
acetylene	$k_a = 5.0 \times 10^{15}$		exp ($-50000/T$)
soot	$k_s = 4.0 \times 10^{19}$		exp ($-60000/T$)
sigma		5000	

determination of rate constants for the thermal decomposition model which will make it generally applicable to coal conversion processes requires input data to cover the temperature range from 400°C to 2000°C. This is regarded as a continuing project with updating of the constants as new data becomes available. The heated grid experiment has provided the model framework, the relation to coal properties and some kinetic information. The kinetic data from the heated grid experiment is reliable for all the species at low temperatures (400°- 700°C) and for the slowly evolved species, CO-tight, H_2 and HCN at higher temperatures. Above 700°C however, a substantial part of the reaction for the rapidly evolving species occurs during the heatup (roughly the first 1/2 sec of the experiment). High temperature kinetic data for these rates is therefore tenuous. Examination of the tar evolution rates presented previously[2] in comparison with rates for overall weight loss summarized in Ref. 12 suggests that the tar rate could be improved by being increased at high temperature. The rates presented in Table I have been chosen to fit; 1) the rates for all species derived from the heated grid experiment at low temperatures[2,6] 2) the rates derived from the heated grid experiment at high temperatures for the slowly evolved species.[2,6] 3) rates for tar and aliphatics which fit the weight loss data at high temperatures.[12] 4) rates for other fast species adjust to fit the heated grid data using the model. In performing the simulation computations the temperature history for the coal is assumed to follow a exponential heat up to 99% of the final temperature for the first 1/2 sec followed by an isothermal period. The heat up is approximated by 20 steps. Additional species evolution data at high temperatures is needed and is the subject of a continuing investigation.

COAL COMPOSITION

Quantitative information was determined for the coals, chars and tars from infrared spectra obtained on a Nicolet FTIR. The techniques for preparing KBr pellets and of correcting for scattering and mineral content were described in a previous publication[3]. Spectra for a lignite and a bituminous coal are shown in Fig. 5. Quantitative determinations of the hydroxyl H_{OH}, aliphatic H_{al}, and aromatic H_{ar} hydrogen concentrations are made from the indicated peaks as described previously[3]. If it is assumed that the average stoichiometry for the aliphatics is $CH_{1.8}$, then the aliphatic carbon concentration C_{al} is 6.7 H_{al}. The aromatic carbon concentration C_{ar} can be computed by difference from the total carbon concentration C_T,

$$C_{ar} = C_T - C_{al}.$$

Values of H_{ar}/H_T and C_{ar}/C_T determined by FTIR were compared to the corresponding data for 18 coals obtained by NMR by Bernard Gerstein et al.[13] The agreement is good.

130

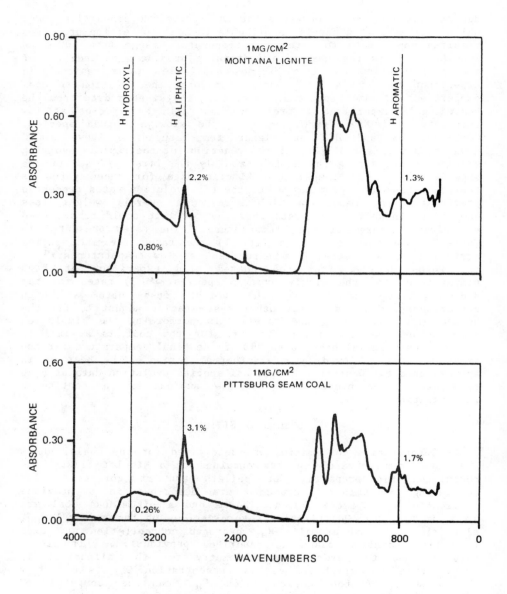

Figure 5.   Infrared Spectra for a Lignite and Bituminous Coal.

A summary of the coal functional group compositions and the source of data is given in Table II. The parameters for carboxyl, tar and the fraction of tightly bound nitrogen and ether were chosen to fit the pyrolysis data. All other parameters are derived from the ultimate analysis and infrared spectra. Water is assumed to be formed from the hydroxyl group according to

$$C-OH + C-OH \rightarrow C-O-C + H_2O$$

So the water component is 9/17 the hydroxyl concentration. The non-volatile carbon is given by:

$$C(non\ volatile) = C_{ar} - C(carboxyl) - C(ether)$$

TABLE II

Functional Group Composition for a Pittsburgh Seam Bituminous and a Montana Lignite

Composition Paramater (dmmf)	Source	Pittsburgh Seam Bituminous	Montana Lignite
C	Elemental Analysis	.8197	.6827
H	Elemental Analysis	.0536	.0457
N	Elemental Analysis	.0142	.0102
S(organic)	SEM[7]	.0189	.0067
O	By Difference	.0936	.2547
S(mineral)	SEM[7]	.0146	.0054
CO_2 – Carboxyl	Fit to Pyrolysis Data	.012	.050
H_2O – 9/17 Hydroxyl	Infrared	.024	.060
CO – Ether Loose	Fit to Pyrolysis Data	.013	.090
CO – Ether Tight	By Difference on O	.098	.199
N – Nitrogen Loose	Fit to Pyrolysis Data	.002	.002
N – Nitrogen Tight	By Difference	.012	.008
$CH_{1.8}$–Aliphatic	Infrared	.241	.170
H – Aromatic H	Infrared	.017	.013
C – Non Volatile	By Difference	.562	.401
S – Organic	SEM	.019	.007
Total		1.000	1.000
Tar	Fit to Pyrolysis Data	.43	.16

132

RESULTS

The comparison of theory and experiment is illustrated in Figs.6-9. Figures 6a and b show the distribution between char, tar, aliphatics and H_2, H_2O, CO and CO_2. The lignite has the smaller tar and aliphatic content but a much larger amount of oxygen containing gases. The high level of CO in the lignite produces the large slope in volatile yield at high temperatures not seen for the bituminous. While the shape of the volatile yields differ between the two coals, they are based on the same set of kinetic parameters. Only the mix of functional groups is different. The agreement between theory and experiment is good.

The composition of the char is shown in Figs. 6c and d. The temperature dependent variation in composition can be related to the evolution of individual products. For example, the initial sharp drop and subsequent gradual deminishing of H can be related to the initial tar and aliphatic evolution and subsequent H_2 evolution from aromatic species.

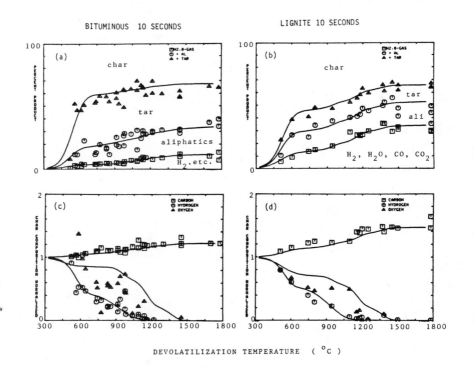

Figure 6. Pyrolysis Product Yields - Experiment and Theory

The yields of CO and CO_2 are illustrated in Figs. 7a and b. The shape of the CO yields show the two distinct sources for this product. While the yields are about a factor of 3 larger for the lignite, the shape of the temperature dependent evolutions are quite similar. The two sources for CO are evident.

Figures 7c and d show the evolution of H_2 and H_2O. The agreement between theory and experiment needs improvement. The observed reduction of H_2O at high temperature for both coals is not predicted. This is probably a steam-char reaction to produce H_2 and CO. Inclusion of this reaction in the model would also help the low H_2 predictions at high temperatures. Hydrogen production at low temperatures is higher than the model prediction and values observed in previous experiments.[1]

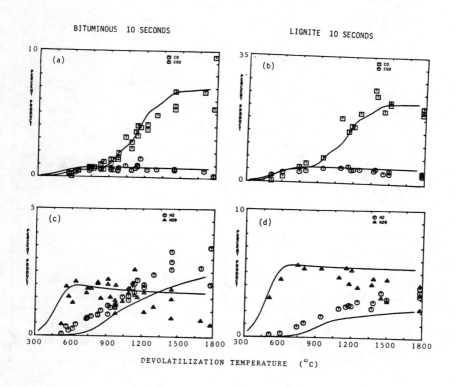

Figure 7. Pyrolysis Product Yields - Experiment and Theory

134

The evolution of methane is illustrated in Figs. 8a and b and of nitrogen in Figs. 8c and d. Volatile nitrogen is observed at low temperatures in the tar and as small amounts of NH_3 and HCN., Larger quantities of HCN are evolved at high temperature as the char nitrogen is reduced. The agreement between theory and experiment is good for temperatures up to 1200°C. Descrepancies are observed at high temperature. For the bituminous, the experimental results indicate that a subtantial amount of nitrogen is left in the char. For the lignite the nitrogen balance looks low indicating that some nitrogen form may not be detected.

Figure 9 shows results for several hydrocarbon species. The figure illustrates the tendency for high temperature pyrolysis to favor unsaturated compounds. As the temperature is raised there is a shift from paraffins to olefins to acetylene. The data show the yields of acetylene deminishing at the highest temperature, presumably due to the formation of soot. These processes also yield H_2 which would explain the continued increase in yield illustrated in Fig. 7.

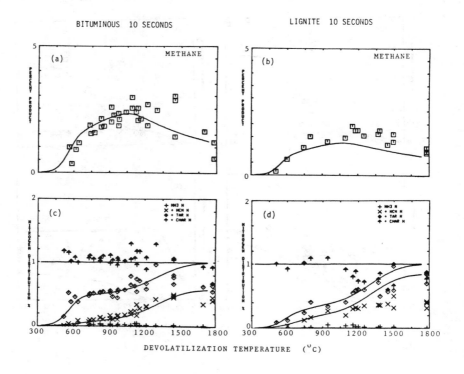

Figure 8.   Pyrolysis Product Yields - Experiment and Theory

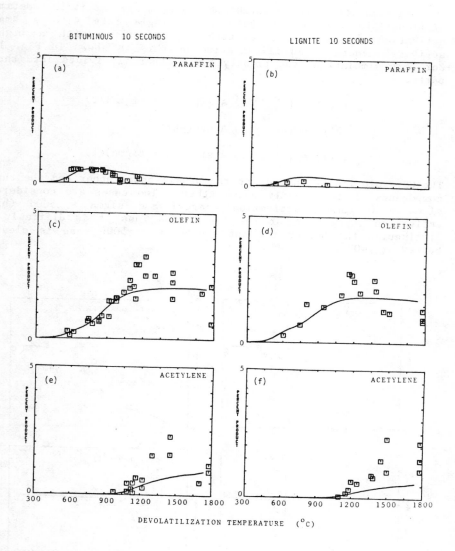

Figure 9   Pyrolysis Product Yield - Experiment and Theory

In simulating the evolution of a single species it is sometimes difficult to distinguish between a single rate with a small activation energy and a distribution of rates with a higher activation energy. An illustration of this is shown in Fig. 10. For this simulation, the tightly bound ether is divided into three parts.

$$Y1(i) = 1/4 \ Y^O(i) \ A \ exp((-B + sigma)/T)$$

$$Y2(i) = 1/2 \ Y^O(i) \ A \ exp(-B/T)$$

$$Y3(i) = 1/4 \ Y^O(i) \ A \ exp((-B - sigma)/T)$$

The CO evolution is the sum of the weakly bound ether and the three components of the tightly bound ether. Two cases are considered, sigma = 0 (small activation energy) and sigma = 5000 (high activation energy). For the ten second runs it is difficult to distinguish the better fit but the sigma = 5000 case is clearly better at 400 sec.

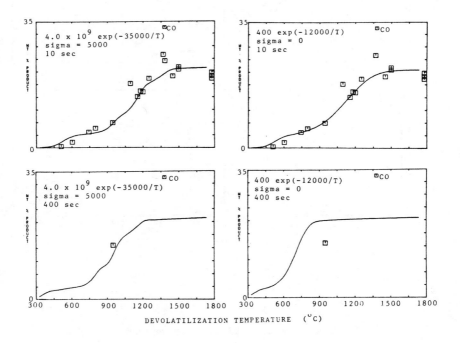

Figure 10. Comparison of Single and Distributed Kinetic Rates.

The pyrolysis process may also be followed by examining the infrared spectra of chars and tars. The spectra for a series of chars is illustrated in Fig. 11. The chars were produced by devolatilizing coal PSOC 212 for 80 sec at the indicated temperature. The results are similar to those observed by Brown[14] and Oelert[15] The rapid disappearance of the aliphatic and hydroxyl peaks is apparent. The aromatic peaks remain to high temperature and the ether peaks are observed to increase in intensity possibly from the creation of new ether linkages occuring as a result of the H_2O formation. For high temperature chars whose carbon content exceeds 92 percent, a broad absorption begins to dominate the spectrum. This is similar to the effect observed in high rank coals above 92% carbon which has been attributed to electronic absorbtion.[16,17]

The spectra for tars produced at various temperatures is illustrated in Fig. 12. The spectra indicate pyrolysis of the tar "monomers" at high temperatures. There is a decrease in hydroxyl and aliphatic groups and an increase in aromatic hydrogen, especially the peak associated with 3 and 4 adjacent hydrogens at 750 cm^{-1}.

CONCLUSIONS

1. Infrared and NMR spectra of a series of coals and tars demonstrate the very close similarity of tars to their parent coals, providing evidence that the tar consists of hydrogen stabilized "monomers" derived from decomposition of the coal "polymer".
2. The time and temperature dependent evolution of pyrolysis products may be predicted from a knowledge of the coals functional group distribution using a general kinetic model with rate constants independent of coal rank.
3. Most of the coal parameters used in the model may be obtained directly from an infrared and ultimate analysis of the coal.

ACKNOWLEDGEMENT

The author wishes to acknowledge the able technical assistance of David Santos and Gerald Wagner and helpful discussions with Med Colket, Daniel Seery and Jim Freihaut.

138

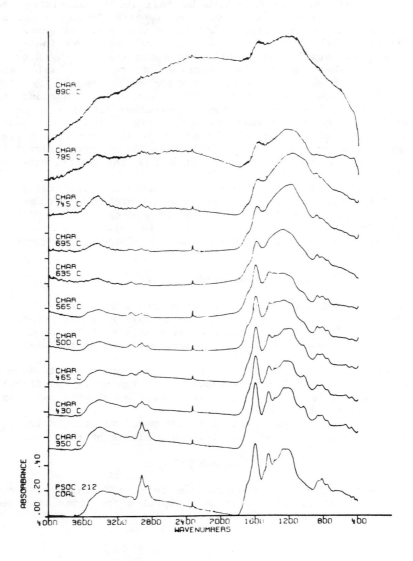

Figure 11   Spectra of Chars of PSOC 212 Coal-Chars from 80 Sec
Vacuum Pyrolysis at the Indicated Temperature

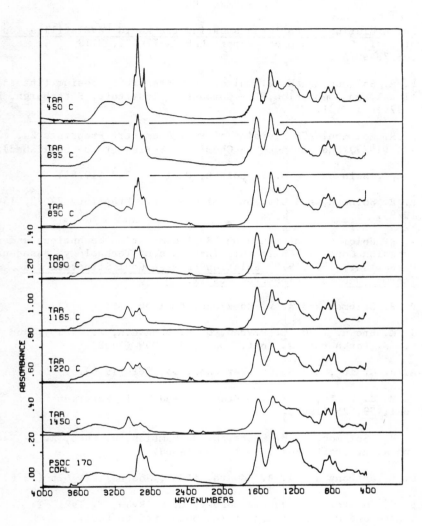

Figure 12   IR Spectra of Tars from PSOC 170 Coal from 10 sec Vacuum
            Pyrolysis at the Indicated Temperature

140

REFERENCES

1. The Evolution of Pollutants During the Rapid Devolatilization
   of Coal, P. R. Solomon Report NSF/RA-770422, NTIS
   #Pb278496/AS.

2. P. R. Solomon and M. B. Colket, Seventeenth Symposium (International) on Combustion, The Combustion Institute, Pittsburgh, PA
   (1979) Pg. 131.

3. P. R. Solomon, ACS Division of Fuel Chemistry Preprints 24, #2,
   184 (1979) and Advances in Chemistry Series (to be published).

4. P. R. Solomon and M. B. Colket, Fuel 57, 749 (1978).

5. P. R. Solomon, ACS Div. of Fuel Chemistry Preprints 24 #3, 154
   (1979).

6. P. R. Solomon, Conference on "Coal Combustion Technology and
   Emission Control" California Institute of Technology, Pasadena,
   CA February 5-7 (1979) Relation Between Coal Structure and
   Thermal Decomposition Products II.

7. P. R. Solomon and A. V. Manzione, Fuel 56, 393 (1977).

8. J. K. Brown, I. G. C. Dryden, D. H. Dunevein, W. K. Joy, and
   K. S. Pankhurst, J. Inst. Fuels, 31, 259 (1958).

9. A. A. Orning, and B. Greifer Fuel, 35, 381 (1956).

10. P. M. J. Wolfs, D. W. van Krevelen and H. I. Waterman
    Fuel, 39, 25 (1960).

11. P. R. Solomon, R. H. Hobbs, D. G. Hamblen, W. Chen, A. La Cava
    and R. A. Graff, Fuel, to be published.

12. D. B. Anthony and J. B. Howard, AIChE, Journal 22, 625 (1976)

13. B. C. Gerstein, P. Dubois Murphy, L. M. Ryan, T. Taki, T.
    Sogabe and P. R. Solomon, to be submitted to Fuel.

14. J. K. Brown, Chem. Soc., 752 (1955).

15. H. H. Olert, Fuel, 47, 433 (1968).

16. H. H. Lowry,Chemistry of Coal Utilization (Supplementary
    Volume), Wiley, NY, (1962).

17. D. W. van Krevelen and J. Schuyer, Coal Science,
    Elsevier, Amsterdam (1957).

ACOUSTIC MICROSCOPY FOR THE CHARACTERIZATION OF COAL

C. F. Quate
Edward L. Ginzton Laboratory
Stanford University, Stanford, CA 94305

I. INTRODUCTION

In characterizing coal it is now common to use the reflectance of optical waves as a measure of the carbon content. The reflectance is weak and light scattering from other surfaces can be a problem which is overcome with oil immersion lenses. The technique is fast and reliable and it has found widespread use in the coking industry.[1] With high rank coals as used there, the reflectance varies between 0.5 and 4.0%.[2] For low rank coals with carbon content less than 60% and with shales the reflectance is so small that the technique cannot be used. Fluorescent microscopy[3] is coming into use in these areas and there is much more to be done in that area.

It is our thesis that these characterization techniques — so successful with high rank coal — could be carried over to the lower rank coals and to the shales if we could find a way to increase the reflectance. The advent of acoustic microscopy[4] may provide us such a system. In one sample of coal where the optical reflectance was measured at 0.81% (in the laboratory of Ralph Gray and Sandra Todd at U.S. Steel) we have determined that the acoustic reflectance for acoustic waves is near 30%. It is this large increase in reflectivity that makes the acoustic microscope an interesting new tool for coal characterization.

In the acoustic microscope sound waves are generated in the instrument and these waves are used to illuminate the sample. In a second technique which we want to propose for coal characterization, we will generate the sound waves in the sample itself. This will be done through optical absorption of light waves impinging on the sample surface. The absorption of light will generate heat and this, in turn, will cause the sample to expand. When the intensity of the impinging light is modulated, the sample surface will expand periodically in response to the modulated light beam. If the surface of the sample is covered with a liquid the vibration surface will generate a propagating sound wave in the liquid which can be picked up and detected by the acoustic microscope. As it turns out, the acoustic wave can provide us with a direct measure of the optical absorption on the sample surface. This technique is now extensively used for the measurement of optical absorption and it comes under the term photoacoustics[5] or optoacoustics. It has proven to be a most sensitive technique that is competitive with other systems. In coal samples where the optical reflectance is near 1% it should be far easier to monitor the larger numbers associated with optical absorption. Furthermore, the wavelength of the impinging lights can be

varied to provide us with spectroscopic information on the sample.
The combination of these probes, all based on scanning a focused
beam of radiation across the surface of the coal, should provide us
with much more information than we now have in the instruments based
solely on optical reflectance.

II. THE ACOUSTIC MICROSCOPE

The scanning acoustic microscope centers on a cylindrical sapphire pellet with a thin film piezoelectric transducer on the back
side. This serves to convert electromagnetic energy into acoustic
energy in the form of a collimated beam which propagates through the
pellet to the front surface. There a spherical cavity filled with
the coupling fluid forms an acoustic lens. This lens is ideal — it
does not suffer from aberrations — and it brings the acoustic beam
into a diffracted limited focus near the center of curvature of the
spherical surface. The object under examination is placed near the
focus. It is then scanned with mechanical techniques point by point
and line by line in a raster pattern familiar to television display.
The cathode ray tube used to monitor the image has a scan synchro-
nized with the mechanical motion of the object. Thus each point on
the monitor screen corresponds to a given point on the object. An
electronic signal proportional to the acoustic power as reflected
from the object is used to modulate the intensity of the electron
beam in the monitor. The image on screen which reproduces the
"image of acoustic reflectivity" is recorded photographically.

If acoustic imaging systems suffer from the absence of photo-
graphic film (sensitive to acoustic radiation) they have an enormous
advantage (over optics) in that the wave velocity in solids can be
much larger than that in liquids. The sketch of Fig. 1 is useful
for explaining this advantage. The large velocity ratio means that
a wave is traveling through the solid-liquid interface undergoes a
large angle of refraction [Fig. 1(a)] and travels in the liquid at
an angle that is nearly normal to the interface. This has important
consequences in the microscope. At a spherical interface the rays
approaching from the solid will leave in a direction that is nearly
radial [Fig. 1(b)]. This permits us to construct a simple, single
surface, lens that is free from aberrations. It will focus the beam
to a diffracted limited waist.

The essential parts of the reflection mode microscope are shown
in Fig. 2. The electrical input signal is converted to a sound wave
at the piezoelectric transducer. It is a sputtered film of zinc
oxide sandwiched between the two films of gold. This transducer is
extensively used in a variety of microwave acoustic devices. It has
proved to be efficient (50% conversion efficiency) and it operates at
frequencies in excess of 10 GHz. In our instrument it generates a
plane wave that travels through the sapphire crystal to the spherical lens. We work in the Fresnel zone of the radiating transducer
and some care is required in the overall design to avoid nonuniform
illuminations of the lens. At the lens surface the impedance ratio
between sapphire and water is nearly 30:1 and this must be overcome
with a quarter-wave matching layer. Various combinations have been

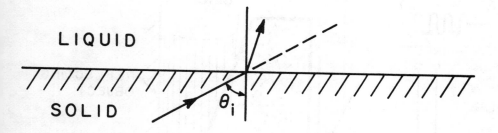

LIQUID

SOLID θ_i

(a) SNELL'S LAW

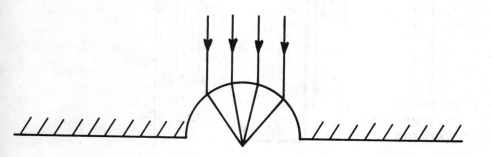

(b) LENS

FIG. 1--Illustration of strong refraction of acoustic waves at a
liquid-solid interface. (TIR - Total Internal Reflection)

144

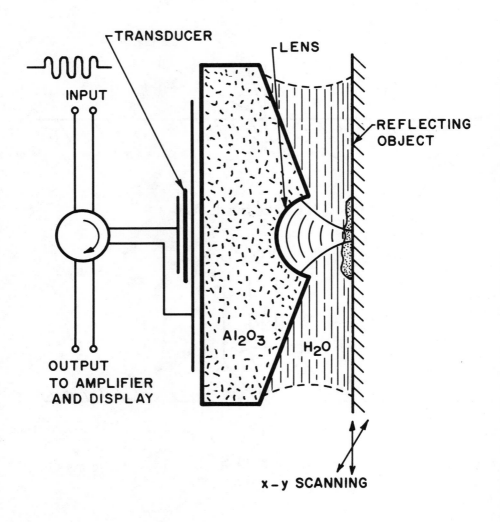

FIG. 2--Sketch of microscope components for reflection.

suggested — gold-quartz, glass, Arsenic-Tri-Selenide, and they all
are effective. Carbon films with an acoustic impedance of
9×10^6 kg/m^2-sec may be ideal for this purpose but they have not
yet been used.

With the sapphire water combination the beam converges to a
focus with a focal length that is 15% greater than the radius of
curvature.

The object itself is mechanically translated through the beam
waist by a loudspeaker. This unit moves the object along a line at
a rate of 60 Hz over a distance of 0.2 mm. A micrometer drive is
used to lift the loudspeaker through this same distance in a time
equal to several seconds. This slow scan speed (5 - 10 seconds per
frame) makes it necessary to store the image by some method before
display.

In the reflection mode (Fig. 2) the radiation is pulsed so that
the reflection from the object can be distinguished in time from
spurious pulses that come from other reflecting points within the
system. A block diagram of the reflection mode system is shown in
Fig. 3. We use short rf pulses in this system so as to separate
the pulse from the object from spurious pulses which come from
other reflecting pulses. After being amplified this pulse is detected and its amplitude is used to modulate the intensity (Z-axis)
of a television monitor. The photographs included in the later section are recorded by photographing the face of this monitor.

Some representative images are included in order to give the
reader an idea of what is now possible with the present state of
instrumentation.

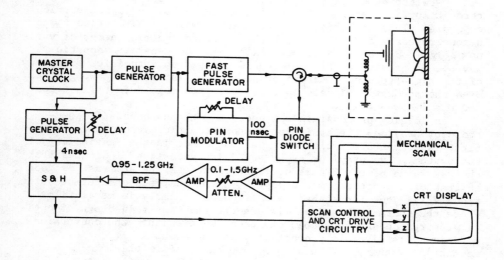

FIG. 3--Block diagram of the acoustic reflection microscope centered
at 1100 MHz.

In Fig. 4 we compare the optical and acoustic images of the polished surface of polycrystalline silicon. The optical index of refraction is uniform over the cross section but we see that the acoustic reflection shows the variations in the elastic properties of this surface. In Fig. 5 we can see the same effect for a polished surface of α-brass. Figure 6 illustrates the appearance of coal samples in the acoustic microscope and the comparison with the optical image. It is a high volatile bituminous "A" rank coal with 35% volatile matter and 58% fixed carbon content.

III. THE PHOTOACOUSTIC MICROSCOPE

Materials with optical absorption bands can be used to transfer energy from an optical beam to acoustic radiation — the photoacoustic effect. The change in temperature and the thermal expansion which accompanies the absorption of optical energy generates acoustic waves at frequencies corresponding to the amplitude modulation of the optical beam. A definitive experiment was carried out in 1971 by Kreuzer and Patel.[6] They used the acoustic energy generated by the photoacoustic effect to monitor trace impurities in gases. Since 1973 Rosencwaig[7] has carried out extensive photoacoustic work with solids, liquids, and biological materials. Several others[8] since that time have used photoacoustic spectroscopy (PAS) to study a variety of problems and the results of all this has been summarized in two reviews by Maugh[9] and Farrow.[10]

Here we will discuss the feasibility of a photoacoustic microscope which follows more or less directly from a report by Brienza and De Maria.[11] In 1967 they demonstrated that mode-locked lasers with Q-switching could be used to generate intense sound beams through surface heating of metal films deposited on piezoelectric crystals. This was preceded by the early work of R.M. White[12] who demonstrated in a classic paper that electromagnetic energy of various forms could be used to heat materials and generate acoustic waves. He predicted that this could be used as a tool for "...study of the high-frequency thermal properties of materials." Wong[13] has used photoacoustic signals in a gas cell to image defects in ceramics of silicon nitride.

An imaging proposal in conjunction with photoacoustic spectroscopy provides the basis for a photoacoustic microscope. The first experimental results give us confidence in the feasibility of the system. Our previous experience with acoustics[5] indicates that the resolving power of this new instrument will be at least as good as that of the optical microscope. In the analysis of thermoelastic signals[12,14] one finds that in some cases — namely in metal films where the thermal skin depth is greater than the skin depth of optical absorption and less than the sound wavelength — the conversion efficiency is directly proportional to the acoustic frequency. Cachier's experimental results at 800 MHz confirm this.[14]

We do have some initial results on this form of imaging which were obtained with a modified acoustic microscope. The input acoustic lens has been replaced with an optical objective lens (NA = 0.25) and in this way the optical input beam is focused to a diameter of

A

B

FIG. 4--Comparison of the optical A and the acoustic B micrograph of
a polished surface of polycrystalline silicon.

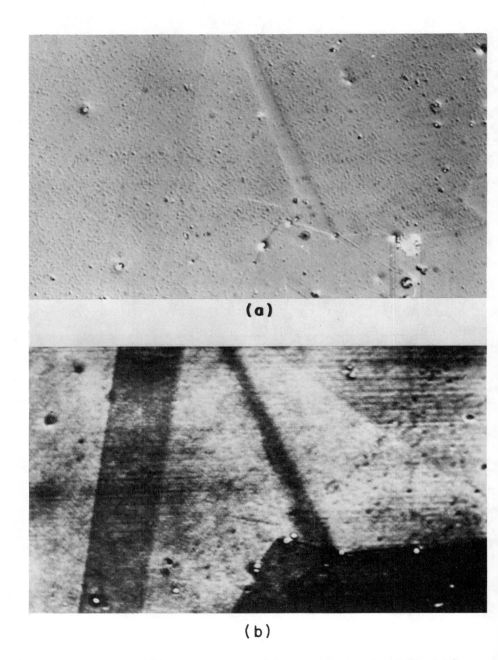

(a)

(b)

FIG. 5--Optical (a) and acoustic (b) comparison of polished brass
surface. Field of view is 55 × 90 μm.

149

(a) ACOUSTIC

(b) OPTICAL x250

(c) ACOUSTIC

FIG. 6—Comparison of optical (b) and acoustic (a) and (c) images of coal.
It is "A" rank bituminous coal.

two micrometers. It is this focused optical beam, properly modulated, that is the source of our photoacoustic signal. The output lens and the transducer from the original acoustic microscope were used to detect the acoustic energy as radiated from the heated specimen. The schematic for the overall system is shown in Fig. 7.

In the photoacoustic image of Fig. 8 we used a chromium pattern on a glass cover slip. The chrome was a thickness of 150 nm and it was overcoated by 200 nm of aluminum. The signal-to-noise ratio at the output for the acoustic signal was 20 dB. The acoustic loss in the water path was estimated to be 32 dB.

The theory for the configuration of Fig. 7 has not been worked out but we do know something about the parameters that will come from this theory. The coupling coefficient relating the generated acoustic power to the square of the absorbed optical power is given by White.[12] We know that the photoelastic coupling constants are large in those materials with a small heat capacity, a large expansion coefficient and a high value for the thermal conductivity. Furthermore, we know that the thermoelastic effect in many metals is the major source of acoustic attenuation. Because of this similarity we expect high values for the photoacoustic coupling in those materials such as gold or silver where the acoustic attenuation is large and small photoacoustic coupling constants in materials such as nickel or acoustic power. A proper coating for matching the acoustic energy into the liquid would correct this. We estimate the conversion efficiency from light-to-sound to be near 10^{-4}. This is consistent with the calculations of Cachier[14] if we consider the metal-glass interface to be clamped.

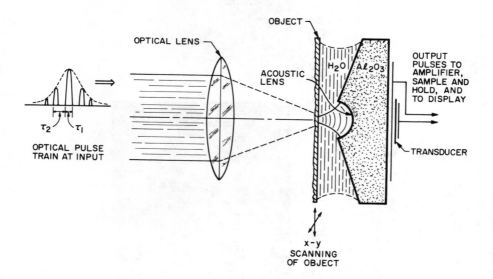

FIG. 7--The photoacoustic set-up. The optical pulse train is from a mode-locked Q-switched laser.

151

GLASS/Au/GLASS

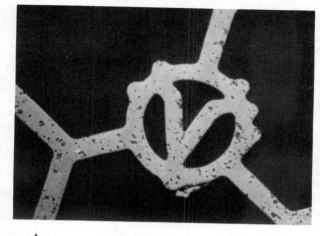

|← 100μ →|

OPTICAL REFLECTION

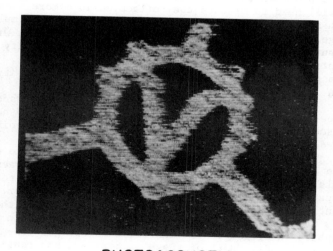

PHOTOACOUSTIC

OPTICAL $\lambda = 1.06\mu$
ACOUSTIC f = 840 MHz
$\lambda = 1.83\mu$

FIG. 8--Photoacoustic image of chrome on glass.

152

In our projected work on coal we believe the large absorption of the vitrinite components will give us strong photoacoustic signals with moderate levels of optical power. We do know that graphite coating is often used to enhance the strength of the photoacoustic signals.

IV. CONCLUDING REMARKS

The prime purpose in our presentation is to argue that we should use other forms of radiation to examine coal and expand the use of scanning microscopy in the field of coal petrology. In addition to the optical reflectance and fluorescence that are now in use we argue that acoustic microscopy and photoacoustic microscopy can be used with equal ease to characterize coal. The combination of the four different techniques should give us more, much more, information on coal samples than we now have.
We now have an acoustic microscope operating at 2500 MHz with a spot diameter of 0.4 microns and this instrument can be used as a guide for the work in coal petrology. A spot diameter of 2 microns should be adequate and for this a lower operating frequency can be used. The range of acoustic frequencies from 500 to 2500 MHz should suffice. The work to date has been related to layered structures which comprise integrated circuits and to materials with subsurface defects. The theories of imagery have been developed with these problems in mind. In coal petrology the emphasis will shift. We will need to measure the reflection in a more quantitative manner with a goal of identifying macerals rather than simply recording the image. The elastic properties of the coal maceral will have to be measured and we will have to find the sound velocity and sound absorption for the vitrinites and for the liptinites. The background data will be essential for understanding and quantifying the reflection images.
The photoacoustic instrument still in its infancy may well prove to be the most versatile and most useful. I would not be surprised to see the use of this instrument growing to the point where it plays the dominant role in the characterization of coal.
With the optical microscope it is necessary to use oil immersion lenses in order to increase the contrast between different macerals. With oil, the reflectance varies from 0.5 to 4.0%. In the acoustic microscope we believe it to be much higher. We have little quantitative data on coal, but we do know some of the elastic properties of graphite. The intensity of the reflectance at a water-graphite interface is 25% and for a liquid-gallium-graphite interface it is 38%. Both of these numbers are substantially higher than the maximum optical reflectance. It is the basis for our optimism in regard to coal petrology and acoustic microscopy.

Partial support was provided by the Institute for Energy Studies at Stanford University.

REFERENCES

1. <u>Coal Petrology</u>, ed. E. Stach (Gebrüder Borntraeger, Berlin, 1975).
   Also see J. Microscopy, Part 1, <u>109</u> (January 1977), special issue on Microscopy of Organic Sediments.
2. A. Davis and F.J. Vaslola, J. Microscopy <u>109</u>, Part 1, 3-12 (January 1977).
3. M. Teichmuller and M. Wolf, J. Microscopy <u>109</u>, Part 1, 49-74 (January 1977).
4. C.F. Quate, Scientific American <u>241</u>, 62-70 (October 1979).
5. H.K. Wickramasinghe, R.C. Bray, V. Jipson, C. Quate, and J.R. Salcedo, Appl. Phys. Lett. <u>33</u>, 923-925 (1 December 1978).
6. L.B. Kreuzer and C.K.N. Patel, Science <u>173</u>, 43 (1971). Also L.B. Kreuzer, J. Appl. Phys. <u>42</u>, 2935 (June 1971).
7. A. Rosencwaig, Optics Communications <u>7</u>, 305 (1973). Also A. Rosencwaig, Physics Today <u>28</u>, 23 (1975).
8. L.C. Aamodt, J.C. Murphy, and J.G. Parker, J. Appl. Phys. <u>48</u>, 927 (1977). Also F.A. McDonald and G.C. Wetsel, Jr., J. Appl. Phys. <u>49</u>, 2313 (1978).
9. T.H. Maugh, II, Science <u>188</u>, 38 (1975).
10. M.K. Farrow, R.K. Burnham, M. Auzanneau, S.L. Olsen, N. Purdue, and E.M. Eyring, Appl. Optics <u>17</u>, 1093 (1 April 1978).
11. M.J. Brienza and A.J. DeMaria, Appl. Phys. Lett. <u>11</u>, 44 (1967). Also see Proc. IEEE <u>57</u>, 5 (1969).
12. R.M. White, J. Appl. Phys. <u>35</u>, 3559 (1963).
13. W.H. Wong, R.L. Thomas, and G.F. Hawkins, Appl. Phys. Lett. <u>32</u>, 538 (1 May 1978).
14. G. Cachier, J. Acoust. Soc. Am. <u>49</u>, 974 (1971).

LOW TEMPERATURE CHEMICAL FRAGMENTATION OF COAL

N. C. Deno, K. Curry, A. D. Jones, R. Minard,
T. Potter, W. Rakitsky, K. Wagner, and R. J. Yevak
Pennsylvania State University, University Park, PA 16802

ABSTRACT

Trifluoroperoxyacetic acid chemically fragments coals at 25-80°.
The reagent is so selective for oxidizing aromatic rings and is so
inert towards benzylic hydrogen that isopropylbenzene, tetralin,
indan, and dihydrophenanthrene have their benzene ring(s) totally
destroyed without any evidence of oxidation at the benzylic hydrogen. The fragments contain more and different structural information
relative to conventional oxidations with Mn (7+), Cr (6+), HNO_3, and O_2.
These latter selectively oxidize benzylic hydrogen.

Products from 27 selected coals show that 9,10-dihydroanthracene
and 9,10-dihydrophenanthrene units are prominent features in most
bituminous coals in accord with Given's structure of coal. Their
frequency accounts for the fact that no continuous fused ring
structure or graphite structure exists. The only simple alkyl substituent is methyl.

Liquefaction (solvent refining) sharply increases the amount of
arylmethyl and aromatic structure and causes the appearance of unsubstituted phenyl groups.

Nitric acid degradations have been reinvestigated and are
developed into a premier method for analyzing the amounts and lengths
of linear alkane chains in coals. Generally the amounts of such
chains represent 0-2% of the carbon in coal, but in one Utah coal
they account for about 42% of the carbons.

INTRODUCTION

If a chemical reaction could be developed which would cleave
coals to identifiable fragments without loss of carbon or loss of
structural information, the problem of the chemical structure of
coal would be largely solved. Such an achievement would also provide
a basis for producing marketable organic chemicals without loss of
carbon. Although present methods of cleavage fall far short of the
above ideal, much has been learned, and it is timely to review two
low temperature degradative methods which have been recently
developed.

OXIDATIONS WITH TRIFLUOROPEROXYACETIC ACID (CF_3CO_3H)

This oxidizing agent has the unusual property of oxidizing
alkylaromatics to form alkyl carboxylic acids instead of aromatic
acids.[1-4] For example, propylbenzene forms butanoic acid and cyclo-
hexylbenzene forms cyclohexanecarboxylic acid in 70-80% yields. With
isopropylbenzene, the combined yields of isobutyric acid and

isopropylmaleic acid are 70-80% so again the alkyl group is preserved in high yield. Hydroaromatic structures and alkane bridges are oxidized to aliphatic diacids, again with preservation of aliphatic structure and destruction of aromatic structure. Examples are the formation of succinic acid from 1,2-diphenylethane and 9,10-dihydrophenanthrene, succinic and glutaric acids from indan, and glutaric and adipic acids from tetralin. Yields are in the 50-80% range.[2,3]

The above results suggest that coals will dissolve in CF_3CO_3H, and the products will reveal the aliphatic structures. In fact lignites and bituminous coals (but not graphite) dissolve to give colorless or near colorless solutions. However, the interpretations proved to be more subtle than first envisioned. This was due to complexities of the reaction and particularly complexities associated with dihydrobenzene components of the coal structure. These will be discussed before turning to the results with coals.

The first complexity involves lower homologous products. With propylbenzene, although the main product is butanoic acid, it is accompanied by lesser amounts of propanoic and acetic acids. This formation of lower homologs is not due to degradation of the butanoic acid, but arises from multiple pathways in the oxidation of intermediates. The formation of minor amounts of lower homologs is also found in the oxidation of cyclic aliphatic structures fused to aromatic rings.

A second complexity is the formation of acetic acid from structures in which no methyl group is present. For example, indan and dihydrophenanthrene produce acetic acid as a major product when oxidized by CF_3CO_3H alone. Fortunately, in the standard procedure[3] with CF_3CO_3H-H_2SO_4, this spurious acetic acid is not produced.

A third complexity is a consequence of the need for added H_2SO_4. With CF_3CO_3H alone, small secondary alcohols are stable. Even large secondary alcohols simply hydroxylate at remote positions until all carbons are brought within about four carbons of an alcohol group.[5] With the addition of H_2SO_4 and reaction temperatures of 70°, secondary alcohols undergo oxidation and cleavage of the carbon chain. Primary alcohols are still stable, and 1-butanol was inert to CF_3CO_3H-H_2SO_4 for 3 hours at 70°. Primary alcohols with longer alkyl chains can be expected to hydroxylate to secondary alcohols which will cleave to diols and hydroxy acids.

A fourth complexity was the formation of alkylmaleic acids. With isopropylbenzene the yield of isopropylmaleic acid was comparable to the yield of isobutyric acid.[2] This ratio did not change on addition of H_2SO_4. In contrast, the yield of cyclohexene-1,2-dicarboxylic acid from tetralin dropped from 36% to below 10% on addition of H_2SO_4.[3] Hydroxy substituents in the ring favor the production of alkylmaleic acids and this phenomenon needs more study.

A fifth complexity was the decarboxylation of benzenepoly-carboxylic acids. Fortunately, the added H_2SO_4 strongly inhibits these decarboxylations so that they are not significant in the usual CF_3CO_3H-H_2SO_4 procedure.[3] In the absence of H_2SO_4, the decarboxylations are much faster and benzenepentacarboxylic acid degrades to benzene-1,2,3,4-tetracarboxylic acid and on to benzene-1,3,5-tricarboxylic acid.

A sixth complexity arose in connection with methylene (CH_2) groups attached to two aryl rings as in diphenylmethane or dihydroanthracene. While some malonic acid was produced, yields were low and erratic. To compound the problem, malonic acid was produced from many polyaromatic compounds. The formation in this latter case is undoubtedly via hydroxylation, ketonization to the cyclohexadienone, and excising out the methylene unit as malonic acid.[3] The result was to render malonic acid yields of doubtful value in determining coal structure.

With the above background, the products from coals can be interpreted. Acetic acid is always produced. Its amount is determined from NMR (nuclear magnetic resonance) spectra of the reaction mixture. Providing H_2SO_4 is used in the oxidation, the yield of acetic acid is a good measure of the amount of arylmethyl groups in the coal. In general, about 1% of the carbon in the coals were present as arylmethyl. The yield of acetic acid increases to 2-4% of the carbon in solvent refined (liquified) coals (Table I).[4,6] This increase in arylmethyl is a result of thermal cleavage to benzyl radicals and hydrogen abstraction by these benzyl radicals to form arylmethyl.

Propionic, butyric, and other simple aliphatic monoacids were not observed. This question is being further investigated by trying to reduce the lower homolog problem and by developing better isolation of products. However, despite an early misinterpretation, the best data show no evidence for these acids from which it is concluded that arylmethyl is the only small arylalkyl group in coals.

Arylalkyl groups larger than propyl are also absent. These would form linear acids longer than butanoic. Such acids would undergo remote hydroxylation[5] and oxidative cleavage to form diacids and ω-hydroxy acids. The latter would be distinctive, but they largely polymerize to non-volatile polyesters under the present isolation procedures. Fortunately the 40% HNO_3 method, described in the next section, clearly shows the absence of simple arylalkyl groups where the alkyl contains four or more carbons.

Closely related to the existence of arylmethyl groups is the formation of minor amounts of methylmalonic, methylmaleic, and methylsuccinic acids. The first two are attributed to arylmethyl and the last to 9-methyl-9,10-dihydrophenanthrene units. The formation of these and the absence of other alkyl substituents reinforces the view that arylmethyl is the only arylalkyl in coals.

Succinic acid is a significant product from all coals. In Illinois no. 6 Monterey coal the four carbons of succinic acid account for 19.6% of the carbon of the coal.[3] At present there is no simple basis for deciding whether the succinic acid comes from dihydrophenanthrene or from 1,2-diarylethane components of coal structure.

The most intriguing product is benzene-1,2,4,5-tetracarboxylic acid (I). Although this is a major product from treatment of polyaromatics and polyalkyl aromatics with HNO_3, Mn (7+), Cr (6+), and O_2, it is not a product from these substrates when CF_3CO_3H or CF_3CO_3H-H_2SO_4 is the oxidant. Specifically it is not formed from a wide

Table I Yields of acetic acid from oxidative
degradations with $CF_3CO_3H-H_2SO_4$

Name	Penn State coal base number[a]				
	372 Kentucky Imboden	330 Penn. Middle Kittaning	256 Penn. Lower Freeport	312 Arizona Red	405 Oklahoma Lower Hartshorne
% liquified[a] (3 min)	79	70	65	51	25
% C (maf)	85.9	83.5	88.2	78.4	89.8
rank	HVA	HVB	med. vol.	HVC	low vol.
Yield of acetic acid (meq. per g of maf)[b]					
parent coal	0.44	0.33	0.31	0.42	0.31
SRC (3 min)	0.77	0.74	1.05	1.46	1.51
SRC (90 min)	1.45	1.15	1.06	1.60	1.69
residue (3 min)	0.33	0.41	0.27	0.27	0.24

[a]Samples and liquefaction data were supplied by D. D. Whitehurst, Mobil Research and Development Corp.

[b]The percentages of moisture (m) and ash (a) are available from the Penn State Coal Base computer printouts. The SRC samples had negligible moisture or ash. In calculating the yields of acetic acid from the residue, it was assumed that all of the ash in the coal was retained in the residue in correcting to a maf (moisture and ash free) basis.

variety of polyaromatics including anthracene and benzanthracenes nor is it formed from 9,10-dihydroanthracene or tetraalkylbenzenes. The key to this mystery was that naphthalene-2,3-dicarboxylic acid (II) produced I as the major product and 9,10-dihydroanthracene produced phthalic acid. This suggested that if an additional benzene ring were fused onto dihydroanthracene, the dihydroanthracene would degrade to II which would form I. Not only was this realized, but oxidation of 5,12-dihydronaphthacene also gave many of the minor products observed from coals such as oxiranetricarboxylic acid, oxiranetetracarboxylic acid, ethanetricarboxylic acid, and benzene-1,2,4-tricarboxylic acid. Table II illustrates these parallels.

Methanol is a product from lignites. Its amount is an accurate determination of the amount of arylmethoxy groups. In North Dakota lignite, 16% of the hydrogen of the dry lignite was present in arylmethoxy groups,[2] and this is comparable to the amounts in lignin as determined either by CF_3CO_3H oxidative degradations or by the Zeisel method. With lignin and lignites, the high volatility of methanol and methyl trifluoroacetate are not a problem because the oxidations are complete at 25°. With bituminous coals where temperatures of 60-100° are required, special precautions would be required to trap these volatile products. Such precautions have not been taken because Zeisel determinations had indicated the absence of arylmethoxy in bituminous coals. Finally, the CF_3CO_3H oxidations of lignite and lignin showed the absence of any higher homologs of methanol.

Polyaromatic clusters of three or more rings are rare or absent. Model studies (reported in part[3]) have shown that these produce predominantly phthalic (benzene-1,2-diCOOH) acid and minor amounts of certain lactones. In oxidation of coals, phthalic acid was absent or minor product and the distinctive lactones were not observed. These results support the many lines of evidence which indicate that polyaromatic clusters are largely limited to 2-4 fused rings. After liquefaction (solvent refining), phthalic acid becomes a significant product (Table III) indicating the formation of poly-aromatic clusters and the appearance of the lactones (Table III) provides added evidence.

A phenyl substituent attached to a more oxidizable aromatic system reveals itself by formation of benzoic acid. Benzoic acid was not a product from the original coals, but it was a product from the liquified coals (Table III). This generation of unsubstituted phenyl in the process of liquefaction is interpreted as a result of removal of the heteroatom in benzofurans, benzothiophenes, and possibly benzopyrroles. It shows that at least some of these benz rings are unsubstituted in the heterocycle.

Anthraquinone was a product from a few coals. The structures responsible for its formation have not been fully defined. It does not form anthrone and is only a minor product from dihydroanthracene. It is the major product from 9-formylanthracene. We also have data indicating that the yield of anthraquinone increases after exposure of the coal to air. In one coal sample (PSOC-155), 22% of the carbon of the coal was accounted for by anthraquinone. Although the structural precursors are not yet completely defined, production of anthraquinone shows the presence of some kind of anthracene structure with the A or C ring unsub-stituted.

The CF_3CO_3H oxidative degradations of coals provide evidence for arylmethyl as the only arylalkyl and an analytical method for its estimation.[3] They provide direct evidence for dihydroanthracene and dihydrophenanthrene units as the principal aliphatic components as proposed by Given in his famous structure for coal.[8] A remarkable

159

Table II A comparison of products from per TFA oxidation of
5,12-dihydronaphthacene (I) and Monterey coal

Product	Relative area of g.c. peak of methyl ester			
	I[b]	I[c]	I[d]	coal
malonic acid	23	7	8	45
benzoic acid	48	22	12	9
ethanetriCOOH	3	2	0	3
oxiranetriCOOH	11	13	15	16
propane-1,2,3-triCOOH	6	6	18	11
benzene-1,2-diCOOH[a]	60	60	60	60
benzene-1,4-diCOOH	18	8	11	16
oxiranetetraCOOH	13	9	3	12
benzene-1,2,4-triCOOH	41	34	35	60
benzene-1,2,4,5-tetraCOOH	25	22	61	54

[a]Arbitrarily set at 60.

[b]Peak heights from a capillary column.

[c]Areas from a 0.25 inch packed column.

[d]In this experiment, the initial CH_2Cl_2 extraction was aided by
saturating the aqueous layer with NaCl. The relative increase
in the tetramethyl ester of benzene-1,2,4,5-tetracarboxylic
acid is evident. Areas were obtained on a 0.25 inch packed
column.

Table III Absolute yields (mg per g maf) of products[a] from oxidative degradations with $CF_3CO_3H-H_2SO_4$

MW of methyl ester	132	146	136	204	218a	218b	194a	194b	194c	276	252a	252b	310a	310b
PSOC-372														
coal	17.7	4.9	---	1.2	0.7	4.3	---	---	---	1.3	3.3	---	2.1	---
SRC (3 m)	10.5	3.7	2.3	0.5	0.7	5.6	---	0.9	0.7	2.0	2.0	0.8	0.9	0.1
SRC (90 m)	13.7	6.8	3.5	0.6	0.5	7.2	---	2.4	1.3	2.0	6.7	2.4	2.4	0.4
res (3 m)	7.6	0.7	0.6	0.3	0.7	1.6	1.0	0.4	0.4	0.2	2.8	---	1.7	0.3
PSOC-330														
coal	21.0	6.1	---	1.2	0.9	2.3	---	0.2	0.2	1.5	1.5	---	1.6	---
SRC (3 m)	18.4	6.8	6.1	1.0	0.4	10.0	---	2.4	---	2.6	6.0	---	1.7	0.2
SRC (90 m)	2.5	7.1	7.1	0.1	0.7	2.8	7.7	4.9	3.2	3.0	9.3	0.1	4.4	1.0
res (3 m)	5.1	4.2	1.5	0.1	0.3	0.6	0.5	0.4	0.4	0.6	2.2	---	1.9	0.7
PSOC-256														
coal	18.7	3.9	---	1.1	0.6	1.6	0.2	0.5	0.4	0.9	3.2	---	2.2	---
SRC (3 m)	6.8	6.2	5.9	0.1	0.8	0.7	2.0	1.9	1.9	2.7	9.3	0.2	5.3	0.7
SRC (90 m)	5.8	8.4	6.6	0.4	1.0	10.2	3.2	3.3	2.3	3.5	12.0	0.1	0.5	1.1
res (3 m)	12.5	3.1	2.3	1.0	0.2	2.4	1.7	0.8	0.6	1.6	7.1	---	2.9	0.5
PSOC-312														
coal	13.8	4.9	---	2.0	1.8	10.7	---	---	0.5	2.8	2.5	0.2	1.1	0.2
SRC (3 m)	6.7	12.8	5.9	1.6	0.8	10.4	---	3.3	---	2.8	3.9	1.3	0.9	0.2
SRC (90 m)	11.7	14.3	4.8	0.3	1.8	6.7	---	2.6	3.1	3.3	8.2	0.1	2.8	0.4
res (3 m)	14.8	3.4	1.6	1.2	0.6	4.9	0.7	0.5	0.7	1.6	3.3	0.2	2.3	0.5
PSOC-405														
coal	7.9	2.4	---	0.6	0.1	2.5	1.3	0.9	0.5	0.8	2.5	0.2	1.0	0.3
SRC (3 m)	3.5	10.7	5.8	0.2	1.4	5.2	2.8	3.6	2.7	2.6	7.2	0.2	4.5	0.6
SRC (90 m)	0.9	9.1	5.3	0.3	1.8	1.3	4.0	6.9	---	3.6	9.1	0.2	2.9	0.4
res (3 m)	10.9	3.9	2.4	0.2	0.5	2.8	2.5	1.6	1.8	3.6	10.0	---	6.1	1.1

[a]Identified in Table IV.

Table IV Identification of products in Table III

MW of methyl ester	Relative GC ret. time	Name of corresponding acid (X is COOH)
132	3.20	malonic acid (XCH_2X)
146	4.60	succinic acid (XCH_2CH_2X)
136	5.15	benzoic acid
204	9.23	1,1,2-ethanetricarboxylic acid
218a	10.54	1,2,3-propanetricarboxylic acid
218b	10.97	oxiranetricarboxylic acid
194a	11.10	benzene-1,2-dicarboxylic acid
194b	11.33	benzene-1,4-dicarboxylic acid
194c	11.58	benzene-1,3-dicarboxylic acid
276	14.33	oxiranetetracarboxylic acid
252a	15.79	benzene-1,2,4-tricarboxylic acid
252b	16.14	benzene-1,3,5-tricarboxylic acid
310a	19.18	benzene-1,2,4,5-tetracarboxylic acid
310b	19.50	benzene-1,2,3,5-tetracarboxylic acid

observation is that of some 27 coals studied, generally over 90% of the peak areas in the gas chromatograms are due to the same 10-15 compounds. In contrast, many pure substrates such as anthracene, chrysene, and adamantane produce more complex mixtures of products. Perhaps coals are simpler and more regular than their black, amorphous, heterogeneous nature would suggest.

Many minor products have been identified. A few of these are shown in Table V.

The CF_3CO_3H technique is recently discovered[1-4] and is not fully developed. Esterifications have been conducted with BF_3-CH_3OH rather than CH_2N_2 because of the desire to avoid the hazardous and toxic CH_2N_2 and because of the large amounts of CH_2N_2 that would be required to destroy all the H_2SO_4. The BF_3-CH_3OH procedure may give incomplete esterification with certain polyacids. There are also problems in quantitatively isolating the methyl

Table V Absolute yields (mg per g maf) of selected minor products from oxidative degradations with CF_3CO_3H-H_2SO_4

	MW of products containing N[a]					MW of lactones[a]								
	209	239	253	297	311	192	250b	250c	250d	250e	264	308a	308b	322
PSOC-372														
coal	---	---	0.4	0.4	0.4	---	---	---	---	0.1	---	0.4	0.3	---
SRC (3 m)	0.2	---	---	0.4	---	0.2	---	0.2	0.2	---	---	0.3	tr	---
SRC (90 m)	0.3	---	---	---	---	0.8	0.5	0.5	0.7	0.1	---	0.6	0.3	---
res (3 m)	0.1	0.5	---	0.4	---	---	---	0.1	---	---	---	---	---	---
PSOC-330														
coal	---	0.1	0.2	0.4	0.5	---	---	---	---	---	---	0.3	0.4	---
SRC (3 m)	---	---	---	---	---	0.2	0.2	0.2	0.2	---	---	0.5	0.3	---
SRC (90 m)	0.3	0.3	---	0.7	---	0.2	0.5	0.5	0.3	0.7	0.3	0.7	1.0	0.4
res (3 m)	---	0.2	---	---	---	---	---	---	0.2	0.5	---	---	0.7	---
PSOC-256														
coal	---	0.1	0.3	0.6	0.2	---	---	---	---	0.2	---	0.1	0.3	---
SRC (3 m)	---	0.4	---	1.1	---	---	0.7	0.3	0.1	0.6	---	0.2	0.7	---
SRC (90 m)	0.2	---	---	---	---	0.3	0.5	0.4	0.5	0.3	---	1.3	1.0	---
res (3 m)	---	0.2	---	0.3	---	---	---	---	---	---	---	0.3	---	---
PSOC-312														
coal	---	---	---	0.3	---	---	---	---	---	0.1	---	0.4	0.4	---
SRC (3 m)	---	---	---	---	---	0.4	0.5	0.8	0.3	0.2	---	0.3	---	---
SRC (90 m)	0.3	---	---	---	---	0.3	0.4	0.2	0.4	0.2	0.2	0.6	0.4	0.2
res (3 m)	0.2	0.1	---	0.7	---	---	---	0.1	---	---	---	---	0.4	---
PSOC-405														
coal	---	---	---	0.1	---	---	tr	tr	---	tr	---	0.1	tr	---
SRC (3 m)	---	0.6	---	1.8	---	0.2	0.4	0.2	---	0.6	0.2	0.4	0.2	0.2
SRC (90 m)	0.6	0.5	---	3.4	---	0.8	0.9	0.2	0.3	1.1	---	0.2	0.2	0.4
res (3 m)	0.4	0.9	---	3.1	---	---	---	0.9	---	---	---	---	0.4	---

Table V (continued)

The nitrogen containing products can be formulated as substituted pyridines. Using X for $COOCH_3$, they are 209 (CH_3pyX_2), 239 ($X_2pyCOOH$), 253 (pyX_3), 297 ($X_3pyCOOH$), and 311 (pyX_4).

The lactones appear to be derived from phthalide, 1 (3H)-isobenzo-furanone. Using the symbols P for phthalide and X for $COOCH_3$, viable structures are PX for 192, PX_2 for 250, $XP-3-CH_2X$ for 264, PX_3 for 308, and $X_2P-3-CH_2X$ for 322.

The above speculations are based on GC retention times, MW, and most particularly on the electron impact mass spectra. The details of these analyses will be published at a future date when more model spectra are available.

sters, and the data in Table II show the beneficial effect of salting out.

The five nitrogen containing products in Table V appear to be pyridinepolycarboxylic acids. Their formation is direct evidence for aromatic clusters containing a pyridine ring and a measure of the amount of pyridine rings in the coal. Pyridine is inert towards $F_3CO_3H-H_2SO_4$ whereas benzene is rapidly oxidized.[2,3] Thus, oxidation of an aromatic cluster containing pyridine can be expected to give pyridine polyacids in high yield. The absence of pyrrole polyacids may simply reflect the reactivity of pyrroles. For example, carbazole reacted but did not produce pyrrole polyacids.

OXIDATIONS WITH NITRIC ACID

Nitric acid acts on alkylaromatic systems in two distinct ways. First is to nitrate aromatic rings. This is favored by high acidities which generate the active agent (NO_2^+).[9-11] Second is the oxidizing action of nitric acid. Oxidation is favored by lower acidities which inhibit the competing nitration, addition of nitrites to generate the active agent (NO_2), and higher temperatures.[12] Addition of urea and other agents that scavenge NO_2 block oxidation.[12]

It has been found that 40% aqueous nitric acid at 60° oxidatively cleaves alcohols, ketones, and alkylbenzenes to form carboxylic acids. The aliphatic and aromatic carboxylic acids do not react further in general, and specifically, alkane chains do not react appreciably below 80°. The reaction is an excellent method for determining the amounts and lengths of saturated linear alkane chains in coals.[12] Table VI summarized data on six coals. To illustrate the interpretations, a 1,8-diphenyloctane would produce

Table VI GC peak areas of $(CH_2)_n(COOCH_3)_2$ products as percentage of total peak areas

n =	Coal[a]					
	Monterey	Wyodak	PSOC-110	PSOC-405	PSOC-124	PSOC-155
2	10.9	9.2	8.8	1.8	8.3	10.5
3	7.1	5.8	7.5	1.6	15.6	14.9
4	5.1	5.3	6.9	1.7	17.4	17.3
5	4.5	4.6	4.7	1.5	15.9	18.0
6	4.5	4.7	3.7	2.0	13.0	13.5
7	5.1	3.7	0.9	0	9.4	9.9
8	4.5	2.9	0	0	6.2	5.2
9	1.7	1.3	0	0	2.9	2.0
10-	0	0	0	0	7.7[b]	5.7[b]
total	43.4	37.5	32.5	8.6	96.4	97.0
% C (dry)	69.7	71.8	79.5	81.49	75.78	56.87
% H (dry)	4.98	5.20	5.20	4.29	6.33	5.36
% moisture			1.46		0.58	9.36
wt. of coal sample	0.201	0.203	0.216	0.211	0.522	0.524
wt. of methyl esters	0.0121	0.0104	0.0093	0.0072	0.017	0.208
fraction of carbon[c]	1.86 (2.14)	1.34 (1.50)	1.51 (1.63)	0.39 (0.42)	2.33 (2.34)	38.3 (42.2)

[a]The Monterey and Wyodak coals are the Illinois no. 6 Monterey mine coals and the Belle Ayre mine (Amax Co.) of the Wyodak-Anderson seam in Wyoming that were used before.[1,2] The PSOC numbers refer to the

Table VI (continued)

Penn State/DOE Coal Base collection of coals. These coals can be further characterized as follows: PSOC-110 is Pittsburgh seam, PSOC-405 is Oklahoma Lower Hartshorne, PSOC-124 is West Virginia Lower Kittaning, and PSOC-155 is Utah Cannel coal. The PSOC 110, 124, and 155 coals were used as received. The other three were used as dried samples.

[b]The homologous series continued in a declining manner. With PSOC-155, members of the series could be detected out to $(CH_2)_{38}(COOCH_3)_2$.

[c]This is the fraction of carbon in the coal which appeared in the diacids of four or more carbons. The number in parentheses is an estimate of the fraction of carbon in alkane chains assuming that each chain contained two more carbons that appeared in each diacid as discussed in the text.

adipic acid so that the yield of the six-carbon adipic acid measures the amount of eight-carbon alkane chains in coals.

The fraction of carbon in such chains varied from 0.4–2.3% in five of the six coals. In the sixth coal, it accounted for 42% of the carbon, and diacids out to C_{40} were observed. The yields of diacids formed a smooth continuous curve. The question is whether this reflects a smooth distribution of amounts and lengths of such chains or whether the chains are of a single length or a few selected lengths with an oxidizable and cleavable group (such as C=C) randomly distributed.

The absence of linear monoacids shows the absence of aryl-alkyl groups larger than propyl.

OTHER OXIDANTS

A superb review (Molecular Structure of Coal)[7] has appeared which summarizes other methods of chemically fragmenting coal at low temperature. The results are all in reasonable accord with our work on $CF_3CO_3H-H_2SO_4$ and 40% HNO_3 oxidative degradations.

ACKNOWLEDGMENTS

Grateful acknowledgment is made to the Department of Energy (Contract ET-78-S-01-3159) for support of this work. We are grateful to R. M. Davidson of IRA, R. E. Winans of Argonne National Labs., and Peter Given of Penn State for providing preprints of their work.

REFERENCES

1. N. Deno, B. A. Greigger, L. A. Messer, M. D. Meyer, and S. G. Stroud, Tetrahedron Lett., 1703 (1977).

2. N. Deno, B. A. Greigger, and S. G. Stroud, Fuel 57, 455 (1978).

3. N. Deno, B. A. Greigger, A. D. Jones, W. G. Rakitsky, K. A. Smith, and R. D. Minard, Fuel, two papers in press.

4. N. Deno, B. A. Greigger, A. D. Jones, and W. G. Rakitsky (Penn State) and D. D. Whitehurst and T. O. Mitchell (Mobil Oil Co.), Fuel, in press.

5. N. Deno, E. J. Jedziniak, L. A. Messer, M. D. Meyer, S. G. Stroud, and E. S. Tomezsko, Tetrahedron 33, 2503 (1977).

6. N. Deno, K. W. Curry, J. E. Cwynar, A. D. Jones, R. D. Minard, T. Potter, W. G. Rakitsky, and K. Wagner, paper to be presented at American Chemical Society Symposium on "Structure and Reactivitiy of Coal and Char," San Francisco, August 1980.

7. R. M. Davidson, Molecular Structure of Coal, International Energy Agency, London, January 1980.

8. P. H. Given, Fuel 39, 147 (1960); 40, 427 (1961).

9. R. J. Gillespie and D. J. Millen, Quant. Rev. 2, 277 (1948).

10. N. Deno and R. Stein, J. Am. Chem. Soc. 78, 578 (1956).

11. N. Deno, H. J. Peterson, and E. Sacher, J. Phys. Chem. 65, 199 (1961).

12. N. Deno, K. W. Curry, A. D. Jones, K. R. Keegan, W. G. Rakitsky, C. A. Richter, and R. D. Minard, Fuel, submitted.

PANEL DISCUSSION: A CRITIQUE OF DETERMINATIONS OF COAL STRUCTURE

H. L. Retcofsky
Pittsburgh Energy, Technology Center, U.S. Department of Energy,
Pittsburgh, Pa.

P. H. Given
The Pennsylvania State University, University Park, Pa.

R. H. Schlosberg
Exxon Research and Engineering Company, Linden, N.J.

Stephen E. Stein
West Virginia University, Morgantown, W.Va.

The original intent of the panel discussion was twofold:
(1) to present a critique of the ten papers that made up the formal
sessions on Coal Structure and the twenty eight related poster
papers, and (2) to discuss other approaches to studies of coal
structure deemed important by the panel members but not included in
any of the presented or poster papers. Clearly, the available time
prevented such an ambitious approach. The actual panel discussion
reflected those topics of special interest to panel members and
addressed several topics at the request of the audience. To aid the
reader, specific papers appearing elsewhere in this volume are
referenced by providing the name of the author in parentheses.

The panel discussion was opened with the following quotation
from the proceedings of the 1977 meeting:

"An understanding of the fundamental chemistry
of coal liquefaction is essential if we are to
optimize processes for industrial development."
- W. H. Wiser -

The panel members unanimously agreed that elucidation of the chemical and, in some cases, the physical structure of coal is a major
step in the development of this understanding (Figure 1).

While there continue to be real challenges and needs in terms
of our information gaps relating to the organic, inorganic and
physical structures of coals, considerable progress has and is
being made. It appears that we now seem to have at hand (or nearly
at hand) the tools to determine carbon (Maciel et al.) and, to a
lesser extent, hydrogen (Solomon, Deno et al.) types in coals. We
are, however, in very much poorer shape in terms of O, N, and S
determinations, including functionalities. Much more attention
should be directed toward a qualitative and quantitative determination of O (especially), N, and S functionalities in coal. In terms
of inorganic structural determinations, a variety of analytical
tools singly and in combination, are being used to elucidate some
inorganic structural features. Needs include more quantitative
analytical tools, characterization of chemical compounds in ash and
in slag, and identification of species responsible for catalytic
behavior in coal conversions. While such physical measurements as

168

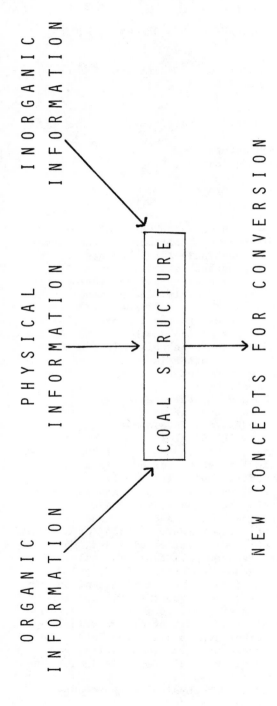

FIGURE 1. THE USEFULNESS OF COAL STRUCTURE DETERMINATIONS.

CO_2, He and N_2 surface areas, density measurements, mercury porosi-
metry measurements and diffusion measurements are available (Grimes),
more meaning needs to emerge from such determinations. Finally, in
the area of physical structure is an issue which may be overriding,
i.e., the issue of diffusional limitations or mass transfer limitations. A brief summary of needs in coal structure research is given
in Table 1.

It was pointed out that much of the major progress in understanding coal structure and coal chemistry have followed improvements
in analytical techniques and instrumentation. What is needed now is
more reliable information and the extrapolation of that information
to better formulate the actual chemical mechanisms involved in coal
conversion. In situ studies, such as the high temperature, high
pressure electron spin resonance studies of Petrakis and Grandy, to
observe fleeting intermediates during coal liquefaction are especially fertile areas for research.

A few comments regarding the utility of kinetic studies of
well-defined organic compounds for gaining insight into coal chemistry are in order at this point. Present state-of-the-art methods
in chemical kinetics allow unambiguous interpretation of many
classes of thermal reactions, often in a very quantitative manner.
Many aspects of the chemistry of gaseous hydrocarbon pyrolysis, air
pollution, oxidation and combustion are well understood, and viable
working models are in common use despite the highly complex nature
of these reactions. The present level of understanding of the
chemistry of these reactions would be in a primitive state without
the information obtained from kinetic and thermodynamic studies of
well-defined chemical systems. In view of significant recent advances and probably future advances in the analysis of coal and
coal products, there is reason to believe that a similar approach to
numerous aspects of coal chemistry may provide equally useful kinetic understanding.

One may cite the well-know chemical and physical heterogeneity
and variability of coal and coal products to support the pessimistic
view that true molecular understanding of aspects of coal chemistry
is not possible. While this problem is certainly a major obstacle
in carrying out coal reactions and makes generalization difficult in
this area, there is a considerable body of evidence supporting the
view that many major chemical and physical changes of coal occur by
means of a very limited number of pathways. Fundamental chemical
processes responsible for aromatization, hydrogen transfer and gas
evolutions, for instance, can now be hypothesized and tested in the
laboratory. The number of elementary processes that constitute the
whole of chemical kinetics is remarkably small and many of the
classes of reactions that dominate thermal organic reactions are
adequately understood in theory.

At present kinetic analysis cannot lead to unambiguous determination of reaction mechanisms in highly complex chemical systems,
such as are found in coal conversion, on the basis of studies of
these systems alone. Details of reaction pathways that control the
chemistry in these systems must be determined in carefully defined

Table 1. Research Needs in Coal Structure Determinations

ORGANIC STRUCTURE OF COALS

  ° Improved resolution in magnetic resonance of solids.

  ° Material balance data from coal oxidations.

  ° Determination of acidic OH groups in solid coal.

  ° Reliable methods for total organic oxygen and organic sulfur determinations.

  ° Quantification of oxygen, nitrogen, and sulfur functionalities.

INORGANIC STRUCTURE OF COALS

  ° More quantitative analytical tools.

  ° Characterization of chemical compounds in ash.

  ° Characterization of slag compounds.

  ° Identification of species responsible for catalytic behavior in coal conversions.

PHYSICAL STRUCTURE OF COALS

  ° Method for determining meaningful surface areas and pore volumes.

  ° Better understanding of pore size distribution.

    - Effect of grinding.
    - Effect of drying.

  ° Better understanding of diffusional limitations.

systems designed for exact study. Certainly studies of coal and
"model substances" are essential for achieving a better understanding of coal chemistry, but these experiments must be interpreted and
generalized in the light of results of fundamental studies. Fundamental reaction pathways, rate constants and ideas may be often
applied, with care, from well-understood systems to complex, incompletely undertstood systems. It is not nearly so easy to apply the
results of one incompletely understood, complex system to another.

Finally, the phrase "model compound studies" to describe
studies of well-defined organic systems is not without its critics.
Fundamental kinetic studies of well-defined chemical systems are
generally not meant to (nor can they realistically hope to) "model"
or simulate coal reactions any more than studies of O-atom reactions
with hydrocarbons are meant to "model" combustion. It is believed
that objections from workers outside the area of chemical kinetics
to "model compound reactions" are often due to the implication inherent in this unfortunate phrase that such studies are meant to
"model" coal reactions.

It is important to point out that the user of sophisticated
structural-determining techniques, in particular spectral techniques,
is confronted with many possible perils and pitfalls. The so-called
"86% carbon anomaly" noticed in many graphical presentations of
chemical and physical properties of coals as a function of carbon
content (Larsen) is frequently not the only inflection point. For
many coal properties, an inflection point representing a change in
bulk properties of coals occurs ~92-94% C. At this point, coals
begin to exhibit graphite-like properties. The transition from pure
insulator to weak semiconductor can invalidate conclusions drawn
from spectral data. Magnetic resonance measurements are particularly susceptible to errors if proper sample preparation techniques
are not used. Heat-treated coals, as well as very high rank coals,
require special attention in this regard.

In planning and interpreting work on coal structure, it is
important to remember that the original source of materials in
coals was biological (higher plants, algae, bacteria, etc.). Thus,
the input to coalification consisted, no doubt in somewhat altered
form, of lignin, cellulose, cutin and suberin (condensation polymers of long-chain hydroxy acids), sporopollenin (alleged to be a
co-polymer of carotenoids and fatty acids), lipids, bacterial amino
acids, and many complex phenolic substances.

It has often been assumed in the past that coals are such
complex substances, and that their constituents have become so
scrambled in structure over the years, that it was not worthwhile
to look for biological markers or chemical fossils. The situation
is certainly changed by the sophisticated analytical methods now
available. There is now evidence that solvent extracts of coal are
analyzable by GC/MS and contain many identifiable biological markers,
such as hydrocarbons and phenols related to abietic acid and also
terpenoid hydrocarbons such as the hopanes. Moreover, there are
reasons to thank that the involatile and insoluble part of coals
contains structures related to lignin, as is seen in the work of

Bimer and Raj, and also the work of Hayatsu and his co-workers.
Thus anyone proposing structural moieties for coals should be prepared at least to consider the question, "how did this structure
arise from what biological precursors?"

COMMENTS ON SPECIFIC PAPERS

In presenting the conclusions of his studies of coals using
FTIR, Peter Solomon faced up to the finding that the frequency of the
OH vibration indicates that the group is hydrogen bonded. He
showed a partial structure in which hydroxyl groups in different
molecules are hydrogen bonded. It seems unlikely that in any
(non-crystalline) structure of coals, groups will be close enough
to each other and with the right orientation to form inter-molecular
hydrogen bonds. It is a characteristic feature of the structure of
lignin that the aromatic part of the structure usually contains two
oxygen substituents in the ortho position to one another, so that
intra-molecular hydrogen bonding will usually be possible. This
explanation of the phenomena found in coal spectra is preferred not
only because it is inherently more feasible, but because it is in
line with what is known of the biological origins of coals.

The information about coal structure that is generated from the
products of Norman Deno's oxidation reaction is potentially so
valuable that further exploitation to assist in answering a number
of questions about coal chemistry is in order. Work performed at
the U. S. Bureau of Mines a number of years ago showed that the
content of hydroaromatic hydrogen varied widely with rank, and to
some extent also with geological province. This should be further
documented by application of the Deno reaction. How do other
aspects of coal structure, as revealed by the products of the reaction, vary with rank and geological history? How do the structures
of coal and of the fractions of their liquefaction products differ?
To answer these questions, extension of Deno's work to include good
quantitative analyses is necessary.

Dr. Peppas emphasized in his discussion of coal as a polymer-
like system that he had not based his ideas upon any particular
model concept, but surely, once one uses the words "chains", and
talks about cross-links between them, one has already made assumptions about a model. It would appear reasonable that the structure
of a vitrinite could bear some resemblance to that of a fully cured
phenol-formaldehyde resin or a glyptal resin (condensation co-polymer of phthalic acid with glycol or glycerol). Both these types of
polymer are highly cross-linked but can one really consider them
cross-linked chains?
(Response from N. A. Peppas: "Only more experimental results
will show whether the hypothesis of cross-linked chains is correct.
I agree that this structure is applicable mostly to vitrinite.")

Maciel et al. presented an excellent overview of applications
of cross-polarization ^{13}C NMR to studies of coal structure. They
appeared somewhat reluctant, however, to label the data quantitative.
Their conservatism is quite understandable in light of reports that

all the carbons in the coal are not being observed during the measurement. Recent cooperative work between the Pittsburgh Energy Technology Center and the National Bureau of Standards shows excellent agreement between carbon aromaticities obtained by the cross-polarization technique and those obtained by conventional high-resolution ^{13}C methods. The materials examined were all derived from coal and were soluble in solvents suitable for high-resolution studies. Cross-polarization measurements were made after removal of the solvents. The soluble materials included coal extracts, coal-derived asphaltenes, and coal-derived oils. Agreement, which was generally within ±.05 f_a units, increases confidence in the aromaticity values obtained by the cross-polarization technique.

Davies and Raymond discussed the direct determination of organic sulfur using the electron probe microanalyzer. The poor counting statistics, a problem in the early stages of this work, seems now to have been taken care of. The method may prove very useful, but it does warrant checking against the standard chemical method. Both types of measurement should be made on the samples, and any iron in the high temperature ash of the residue from the nitric acid treatment should be determined.

Dyrkacz has developed a new method for separating the macerals from a coal. Application of the method, however, requires grinding the coal particles to a top size of 3 μ. Surely it must be difficult to recognize the macerals after separation when the particle size is so small?

(Response from Dyrkacz: It is certainly difficult to distinguish vitrinite and micrinite in the fine particle size, but the use of fluorescence microscopy makes it considerably easier than might be supposed to identify other macerals; the technique of fluorescence microscopy was not available to earlier workers in this field.)

Schmidt et al. have used small-angle x-ray scattering to investigate pores in coals, including closed pores. It is interesting that the results they have obtained so far seem to indicate that transitional or intermediate pores are more abundant in coals from the Interior province; coals from this province were never deeply buried and so experienced a smaller pressure of overburden than other coals, and one might predict this should lead to a more open porosity.

OTHER EXPLORATORY TECHNIQUES

The relatively new technique of photoacoustic spectroscopy has recently been applied to coals. One advantage of the method is that spectra in the infrared and ultraviolet-visible regions free of light scattering effects can be obtained. Recent studies of coals gave infrared spectra comparable in quality to those obtained by conventional absorption techniques. Ultraviolet-visible photoacoustic spectra of coals show only a general monotonous increase in intensity with decreasing wavelength; no distinct spectral bands assignable to specific organic structures were observed.

Improvements in infrared reflectance techniques, such as those developed at Ohio University, show considerable promise in coal research. Fusains, which exhibit essentially total absorption in the infrared, yield reflectance spectra rich in detail and comparable in quality to those of vitrains.

CLOSING REMARKS

The material summarized above represents the input of the panel members as well as comments made by members of the audience. No attempt was made to obtain a consensus of opinion; it is possible that some of the viewpoints expressed are not shared by all. We extend our apologies to those members of the audience who made many excellent contributions to the informal discussions that unfortunately were not incorporated into this panel summary.

It is appropriate to conclude with an observation expressed by one of the panel members as follows: "One of the most encouraging and rewarding signs at this meeting is the appearance of a growing cadre of quality scientists committed to working on the problems of coal. There appears to be an increase in the amount of hard data and consequently more room for constructive controversy. This if true, argues well for the future of scientific endeavors in coal characterization and utilization."

ELECTRICAL PROPERTIES OF COAL AT MICROWAVE FREQUENCIES
FOR MONITORING

Constantine A. Balanis
Department of Electrical Engineering
West Virginia University, Morgantown, West Virginia 26506

ABSTRACT

The development of electromagnetic systems to detect and
monitor coal-related processes requires a thorough knowledge of the
electrical properties (dielectric constant and conductivity) of
coal. Using a two-path interferometer at microwave frequencies
($\approx$9 GHz), samples of solid eastern bituminous and eastern anthracite coals were tested as a function of polarization and direction
of travel of the electromagnetic wave. Temperature, moisture,
and pyrite distribution tests were performed on eastern bituminous
coal samples.

In general there were slight decreases in the values of the
dielectric constant and conductivity as a function of temperature,
from ambient to 700° F, and as the moisture content decreased
through drying. Pyrite layers along the bedding planes cause a
general increase in the conductivity. The anthracites (higher
rank coals) have larger values of conductivity (about a factor of
10) and permittivity (about a factor of 2 and 3) than bituminous
(lower rank) coals. No distinct relationship was found between
the direction of propagation and the electrical properties.

Recommendations are made for orienting electromagnetic
radiators to improve efficiencies of systems which are designed
for coal-related detection, monitoring and mapping applications.

INTRODUCTION

With coal assuming a more important role in this country's
energy budget, the need for a thorough understanding of the properties of coal is increasing. Many new techniques for energy
extraction from coal are being studied; and at the same time conventional underground mining is undergoing considerable changes,
such as the use of remotely controlled automated mining equipment,
remote detection, monitoring and communication systems. Parameters
monitored will include the burn-front of underground coal gasification processes, coal thickness and continuity, mine coal-roof
and coal-floor interfaces, roof rock characteristics, sulfur
content in coal, and other mine environmental factors. In addition,
new exploration and logging techniques are being developed. Many
of these will employ advanced electromagnetic techniques and will
require thorough knowledge of the electrical properties of coal
which can be obtained by laboratory measurements.

In the VHF range a capacitance measurement technique,
utilizing a paralled plate electrode configuration as a test cell

or as a sample holder is employed[1]. Several investigators have performed measurements on rocks and minerals to compute their electrical properties with this method[2, 3]. At microwave frequencies resonant cavity, waveguide and coaxial, techniques are employed which result in the use of a specimen that is physically small but electrically large[4]. Two laboratory methods, utilizing waveguide techniques, were used to determine the electrical properties of fine and granular Pittsburgh seam coal at two different moisture levels[5]. One method used two different lengths of a shorted waveguide and another employed a slotted waveguide filled with a very long sample. Both methods yielded representative values of the electrical properties.

This paper reviews the work performed at West Virginia University on the electrical properties of solid coal samples in the microwave frequency range ($\simeq$9 GHz). A laboratory method, using a two-path interferometer, is discussed which has been used to determine the dielectric constant (relative permittivity) and the conductivity of solid coal samples as a function of temperature, moisture, pyrite distribution, rank of coal, and polarization and direction of travel of the electromagnetic wave. This microwave system permits the measurements of phase and amplitude variations of the signal through a coal sample. The system utilizes a variable phase shifter and a variable attenuator to compensate for the phase and amplitude changes in the system due to the insertion of the sample. Orientations of electromagnetic radiators are recommended for improving the efficiencies of systems which are designed for coal-related detection, monitoring, and mapping applications.

TWO-PATH INTERFEROMETER

The two-path interferometer, shown in a block diagram form in Figure 1, consists of a reference and a working path. The reference path is composed of a variable phase shifter, a variable attenuator and a directional coupler, whereas the working path is composed of a variable attenuator, a slotted line, two horn antennas and a directional coupler. By proper adjustment of the variable attenuator and phase shifter in the working and reference paths, respectively, a null indication is produced at the phase detector. The change of amplitude and phase, which contains the information needed to compute the constitutive parameters, are obtained from the initial and final measurements of the system. The initial measurements are made without the presence of the dielectric medium in the system and with the reference and working paths having the same amplitude and phase which produce a null at the detector. The final measurements are made with the dielectric medium inserted in the system. With proper adjustments of the variable attenuator and phase shifter in the proper paths, a null indication at the detector is reproduced. The dielectric constant

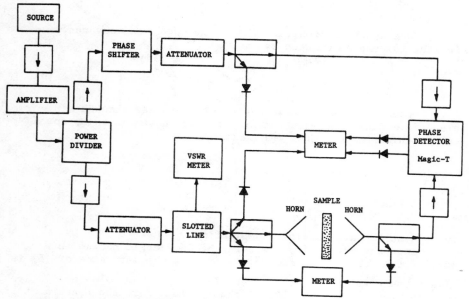

Fig. 1. Block diagram of a microwave two-path interferometer.

of the sample is computed from the changes in phase whereas the conductivity is computed from the changes in the amplitude.

A. Dielectric Constant

The dielectric constant is analytically determined from the initial and final measurements obtained from the phase shifter in the reference path. The initial phase measurement ϕ_0 is related to the phase constant β_0 and the thickness d of the free space medium through which the electromagnetic waves must travel by

$$\phi_0 = \beta_0 d \qquad (1)$$

assuming normal incidence. The final phase measurement, in the presence of the sample, is also related to the phase constant β and the thickness d by

$$\phi_1 = \beta d \qquad (2)$$

Thus from (1) and (2) we can write the change in phase $\Delta\phi$ as

$$\Delta\phi = \phi_1 - \phi_0 = (\beta - \beta_0)d \qquad (3)$$

Since the phase constant is defined as 2π radians per wavelength, (3) can be written as

$$\Delta\phi = (\frac{2\pi}{\lambda} - \frac{2\pi}{\lambda_0})d = (\frac{1}{\lambda} - \frac{1}{\lambda_0})2\pi d \qquad (4)$$

Because of the periodic nature of the phase through a dielectric medium, multiples of $2\pi n$ or π in phase can exist for a sample thickness d. Thus (4) becomes

$$(\frac{1}{\lambda} - \frac{1}{\lambda_0})2\pi d = \begin{cases} \pm 2\pi n \pm \Delta\phi \\ \\ \pm 2\pi n \pm (\pi - \Delta\phi) \end{cases} \qquad (5)$$

to account for all possible combinations.

Since the wavelength λ within the dielectric medium is smaller than the free space wavelength λ_0, the left side of (5) is always positive, hence, the right side of (5) must also be positive for any n multiples of phase. Therefore the constraint on (5) is

$$(\frac{1}{\lambda} - \frac{1}{\lambda_0})2\pi d = \begin{cases} \left.\begin{array}{l} |\Delta\phi| \\ |\pi - \Delta\phi| \end{array}\right\} \quad n=0 \\ \\ \left.\begin{array}{l} 2\pi n \pm \Delta\phi \\ 2\pi n \pm (\pi - \Delta\phi) \end{array}\right\} \quad n=1,2,3 \dots \end{cases} \qquad (6)$$

Assuming that $\mu = \mu_0$, (6) can be written as

$$\frac{2\pi d}{\lambda_0}(\sqrt{\epsilon_r} - 1) = K \qquad (7)$$

where

$$\epsilon_r = [\frac{\lambda_0}{\lambda}]^2 \qquad (7a)$$

$$K = \begin{cases} \left.\begin{array}{l} |\Delta\phi| \\ |\pi - \Delta\phi| \end{array}\right\} \quad n=0 \\ \\ \left.\begin{array}{l} 2\pi n \pm \Delta\phi \\ 2\pi n \pm (\pi - \Delta\phi) \end{array}\right\} \quad n=1,2,3,4 \dots \end{cases} \qquad (7b)$$

Solving (7) for the dielectric constant ε_r yields

$$\varepsilon_r = \left[1.0 + \frac{K\lambda_0}{2\pi d} \right]^2 \tag{8}$$

This equation will be used to compute the dielectric constant from measurements of the phase changes. Since there will be numerous solutions to this equation, many samples of different thickness will be used to resolve the ambiguities.

B. Conductivity

The initial and final amplitude measurements used to null the interferometer will be utilized to compute the conductivity. It is well known from basic electromagnetic theory that as the power propagates through a material it decreases according to the factor $e^{-2\alpha x}$. If the power at x=0 is P_o, then at x=d the power is given, assuming no reflections, by

$$P = P_o e^{-2\alpha d} \tag{9}$$

where α is the attenuation coefficient of the wave through the sample. The total decrease in power (in dB) is given by

$$\Delta P = -10\log_{10}\left(\frac{P}{P_o}\right) = -10\log_{10}(e^{-2\alpha d}) \tag{10}$$

The attenuation coefficient α in (10) accounts for the losses due to the effective conductivity of coal, which will be referred to as ohmic losses. Therefore the change in power ΔP in (10) must represent the change due to ohmic losses. Solving for the attenuation coefficient yields

$$\alpha = \frac{\Delta P_{ohmic}}{20\, d\, \log_{10}e} = \frac{\Delta P_{ohmic}}{8.686\, d} \qquad \text{Nepers/unit length} \tag{11}$$

In our system, in addition to ohmic losses, there are also reflection and diffraction losses. Because the samples are large, we can assume that diffraction losses are negligible[6]. Therefore

$$\Delta P_{ohmic} = \Delta P_T - \Delta P_R \tag{12}$$

where ΔP_T is the power difference determined by the final (P_2)

180

and the initial (P_1) attenuator readings in the working path. Substitution of (12) into (11) yields

$$\alpha = \frac{\Delta P_T - \Delta P_R}{8.686 \, d} \tag{13}$$

where

$$\Delta P_T = P_2 - P_1 \tag{13a}$$

The power loss due to reflections (ΔP_R) is computed from the application of basic electromagnetic theory. The transmitted power P_t through a medium is related to the incident power P_i by

$$P_{tk} = P_i(1 - |\Gamma_k|^2) \tag{14}$$

$$|\Gamma_k| = \frac{V_{sk} - 1}{V_{sk} + 1} \qquad k=1,2 \tag{15}$$

where V_{sk} is the voltage standing wave ratio (VSWR).

The initial transmitted power P_{t1} is obtained without the presence of the sample in the system. Thus (14) becomes

$$P_{t1} = P_i(1 - |\Gamma_1|^2) \tag{16}$$

where Γ_1 is given by (15) when k=1.

The final transmitted power P_{t2} is written from (14), assuming the same amount of incident power P_i, as

$$P_{t2} = P_i(1 - |\Gamma_2|^2) \tag{17}$$

where Γ_2 is given by (15) where k=2.

The ratio of the initial to the final transmitted power yields

$$\frac{P_{t1}}{P_{t2}} = \frac{(1 - |\Gamma_1|^2)}{(1 - |\Gamma_2|^2)} \tag{18}$$

Hence the power loss due to reflections is

$$\Delta P_R = 10\log_{10}(\frac{P_{t1}}{P_{t2}}) = 10\log_{10}\left[\frac{1 - |\Gamma_1|^2}{1 - |\Gamma_2|^2}\right] \qquad (19)$$

where Γ_1 and Γ_2 are given by (15) when k=1,2.

Once the attenuation coefficient is determined using (13), we need to find the conductivity. Again from well known electromagnetic theory, the complex propagation constant (γ) of a lossy medium is given by[7]

$$\gamma = (\alpha + j\beta) = [j\omega\mu(\sigma + j\omega\varepsilon)]^{\frac{1}{2}} \qquad (20)$$

where

γ = propagation constant

α = attenuation coefficient (Nepers/unit length)

β = phase constant (radians/unit length)

Solving (20) for α and β leads to

$$\alpha = \omega\sqrt{\mu\varepsilon}\left\{\frac{1}{2}\left[\sqrt{1 + (\frac{\sigma}{\omega\varepsilon})^2} - 1\right]\right\}^{\frac{1}{2}} \qquad (21)$$

and

$$\beta = \omega\sqrt{\mu\varepsilon}\left\{\frac{1}{2}\left[\sqrt{1 + (\frac{\sigma}{\omega\varepsilon})^2} + 1\right]\right\}^{\frac{1}{2}} \qquad (22)$$

From (21), the conductivity σ can be written as

$$\sigma = \omega\varepsilon\left\{\left[\frac{2\alpha^2}{\omega^2\mu\varepsilon} + 1\right]^2 - 1\right\}^{\frac{1}{2}} \quad \text{(S/m)} \qquad (23)$$

where $\varepsilon = \varepsilon_r\varepsilon_0$.

The above equations are used to compute the conductivity of the medium where α is obtained from (13), (13a) and (19).

In summary, the dielectric constant is computed by (8) from the knowledge of the phase change obtained by the phase shifter in the reference path. The conductivity is computed by (23) using the dielectric constant given by (8) and the attenuation coefficient given by (13).

SYSTEM DESCRIPTION AND MEASURING PROCEDURE

A block diagram of the microwave two-path interferometer is shown in Figure 1. An Ultra Stable Oscillator and a microwave amplifier constitute the electromagnetic wave source. The system operates at X-band frequencies and has two principle paths which permit measurements of amplitude and phase.

The reference path is composed of a variable phase shifter, a variable attenuator and a 10-dB directional coupler. The variable phase shifter provides a phase shift that compensates for phase changes in the working path due to the insertion of a sample. When the amplitude and phase of both paths are equal, a null is produced at the output of the phase detector. The difference in the final (ϕ_1) and the initial (ϕ_0) phase measurements results in $\Delta\phi$ of (3). By proper adjustment of the variable attenuator, the amplitude of the signal in the reference path can be made equal to that of the working path. This is monitored by an oscilloscope through the use of a 10-dB directional coupler and a crystal detector.

The working path is composed of a variable attenuator, a slotted line, a 20-dB directional coupler with a frequency meter at one port, two expontentially tapered horn antennas shown in Figure 2, and a 10-dB directional coupler. By adjusting the

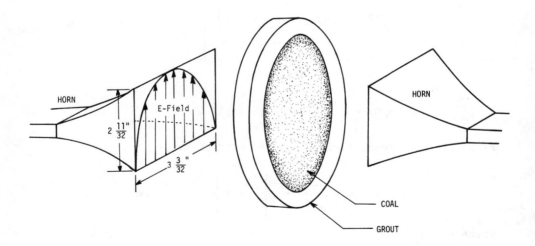

Fig. 2. Geometrical arrangement of coal sample between electromagnetic horns.

variable attenuator, the amplitude of the signal in the working
path is made equal to that of the reference path, for the initial
as well as the final measurement readings. This amplitude is also
monitored by an oscilloscope through a 10-dB directional coupler
and a crystal detector. The difference in the final (P_2) and the
initial (P_1) attenuation readings yields the quantity ΔP_T of (13a).
The voltage standing wave ratio (VSWR) is measured by the slotted
line with and without the sample in the system. These measurements yield the power loss due to reflections ΔP_R of (19). The
expontentially-tapered horn antennas are aligned so to minimize
the power loss and to provide the best match for a given spacing
between them. The separation of the horns must be large enough
to allow the thickest of the samples to be inserted between the
horns without difficulty. To avoid erroneous results, precautions
must be taken to insert the sample normal to both horns.

The phase difference of the electromagnetic waves in the
reference and working paths are detected at the output of the
phase detector. The phase detector is composed of a magic-T with
a pair of matched crystal detectors. The outputs of the E and H
arms of the magic-T are subtracted to give a net change in the dc
output voltage. When the amplitude in both paths is equal, the
phase shifter in the reference path is adjusted to produce a null
(zero volts). The normalized output characteristics of the phase
detector, with free space between the horns, has a sine curve
variation, shown in Figure 3. For one cycle of the output, two

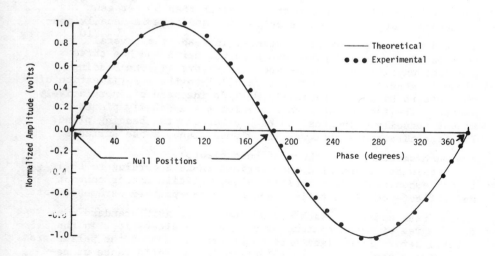

Fig. 3. Normalized output characteristics of the phase
detector with free space between the horns.

null positions exist. When a dielectric medium is inserted between the horns, the null positions will change from zero volts
to a different voltage level. If the magnitude of the voltage is
considered, an infinite number of possible phase values exist,
four of which are in one cycle[8]. Thus, an infinite number of
dielectric constant values. In principle, to remove the ambiguities of (8), at least two samples, each having a different
thickness, are needed. For each sample, a range of dielectric
constant values is computed but only one value is the same for
both samples. Thus, the ambiguous dielectric constant values can
be resolved. In this investigation 2-6 samples from each coal,
each having a different thickness, were measured because the
dielectric constant values of the samples are not exact due to
the nonuniform composition of the coal and the measuring errors.

COAL

Coal is a heterogeneous aggregate composed of organic and
inorganic materials, which is derived from variously decomposed,
physically and chemically altered plant remains which were deposited in sedimentary basins hundreds of million years ago. The
optically homogeneous organic materials are called macerals, which
are grouped as vitrinites, liptinites (exinites) and inertinites.
Vitrinites are composed of coalified woody tissues and usually
account for more than 70 to 80 per cent of any given coal seam.
The inorganic materials include clay minerals, carbonates,
sulfates and sulfides. For the medium to be classified as coal,
the total organic content must be greater than 50 per cent[9].
Fossil organic matter such as coal and coke is usually considered to be a colloidal substance, instead of a crystal[10].
Coal occurs in sedimentary layers in the earth and has three
distinct physical planes, as shown in Figure 4. The bedding
plane is generally more or less parallel to the stratification of
the host rocks and is typically well defined and often relatively
smooth. The face cleats and butt cleats are usually planes
almost mutually orthogonal to each other and the bedding plane.
The face cleats are more well developed planes whose surfaces
are much smoother than those of butt cleats[8].
Coals are classified into various ranks according to their
fixed carbon content, calorific value, volatile matter content
and reflectance, with reflectance being the primary parameter for
higher rank coals[9]. "Rank" is defined by an ASTM standard[11] as
the "degree of metamorphism, or progressive alteration, in the
natural series from lignite to anthracite". One of the parameters
most widely used by coal petrologists is the reflectance or reflectivity which is defined as the ratio of the reflected to the
incident light upon the substance[12]. Reflectance of vitrinite is

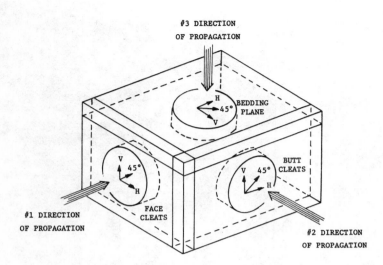

Fig. 4. Coal medium with coal sample
orientations.

a measure of the degree of metamorphism or coalification of coals
and is a good indicator of rank, increasing with higher ranks.
Reflectance is a function of the indices of refraction of the
sample being measured and the medium in which it is being
measured, along with the absorption index of the material.

In general, the reflectance is anisotropic[10]. For most
bituminous coals the optical axis (direction of minimum reflectivity), which also corresponds to the minimum conductivity
and dielectric constant, is oriented perpendicular to the coal bed
stratification. However high rank anthracites seem to have the
direction of minimum reflectivity (optical axis) at an angle
which is not necessarily perpendicular to the bedding plane. This
may be due to a combined vector sum of the lithostatic load forces
and the compressional forces acting within the highly folded and
contorted anthracite beds[13]. If the theory of optical anisotropy
holds for these high rank coals, there should be a cyclical relationship of the electrical parameters with polarization
direction; that is, the maximum and minimum values of both reflectivity and dielectric constant should lie at 90° to each other.
The difference between maximum and minimum reflectivity is known
as bireflectance and is a good indicator of coal rank[9].

To conduct our investigation, samples of eastern bituminous
coal were initially obtained from the Freeport and Upper
Kittanning coal seam regions. The samples ranged in thickness
from 0.25 to 1.5 inches and were bored out of solid blocks of coal
into cores of about 6.5 inches in diameter. These samples were
reinforced with grout as shown in Figure 5.

Fig. 5. Photograph of coal sample showing
cleatings.

Later, coal samples from three mines (Lewis, King Knob, and
Hansford) of medium-high and one mine (Allegheny) of low volatile
bituminous coal in West Virginia, and one mine (Reading) of
anthracite coal in Pennsylvania were obtained. These samples
ranged in thickness from 0.155 to 1.031 inches with diameters
from 6 to 7 inches. Reinforcing grout was not used with these
samples. Between two to six samples were cut for each of the
Figure 4 orientations (face cleats, butt cleats, and bedding
plane).

MEASUREMENTS, COMPUTATIONS, AND RESULTS

The coal samples were tested at microwave frequencies ($\approx$9 GHz)
using the two-path interferometer as a function of the temperature
of the sample, and direction and polarization of the electromagnetic wave. For the bedding plane samples, horizontal polarization was arbitrarily chosen to be perpendicular to the face
cleats.

In order to determine the constitutive parameters as a
function of direction and polarization, each of the samples was

placed within the horns as shown in Figure 2 and rotated for each measurement to obtain the electric field orientation indicated in Figure 4, to be referred to as vertical polarization (VP), 45° polarization (45°P), and horizontal polarization (HP). To gain information on the optical anisotropy, the samples were rotated every 15° and measurements of reflectance, dielectric constant, and conductivity were performed over the full 180° polarization direction. To obtain the constitutive parameters as a function of temperature, the samples were heated in an oven over a temperature range of 65° to 700° Fahrenheit and the measurements were taken at every 100° F intervals.

The computed, from measurements, data [dielectric constant ε_r (dimensionless) and conductivity σ (S/m)] for eastern bituminous coal[8], from the Freeport and Upper Kittanning coal seams, are shown listed in Tables 1 and 2. At the time the temperature measurements were made face and butt cleats samples were only available, and the results are shown in Table 1. Due to the decomposition (swelling) of the eastern bituminous coal samples at temperatures above about 650–700° F, the measurements were limited to an upper temperature of 700° F. Listed in Table 2 are the data for the wet and dry samples.

TABLE 1

Electrical Properties of Eastern Bituminous Coal as a Function of Temperature and Polarization

Frequency ≃ 9 GHz [ε_r (dimensionless), σ (S/m)]

Temperature (°F)			65	200	300	400	500	600	700
Face Cleats	Vertical	ε_r	3.89	3.70	3.79	3.68	3.65	3.78	3.53
	Polarization	σ	0.26	0.17	0.17	0.17	0.14	0.12	0.10
	45°	ε_r	3.84	3.72	3.93	3.60	3.79	3.72	3.58
	Polarization	σ	0.43	0.39	0.40	0.36	0.33	0.24	0.29
	Horizontal	ε_r	3.74	3.68	3.92	3.58	3.67	3.89	3.71
	Polarization	σ	0.73	0.69	0.77	0.74	0.73	0.66	0.61
Butt Cleats	Vertical	ε_r	3.88	3.38	3.39	3.41	3.43	3.48	3.48
	Polarization	σ	0.18	0.12	0.11	0.10	0.10	0.09	0.11
	45°	ε_r	3.85	3.40	3.36	3.38	3.44	3.50	3.54
	Polarization	σ	0.32	0.28	0.29	0.28	0.27	0.27	0.23
	Horizontal	ε_r	3.93	3.41	3.40	3.36	3.36	3.49	3.56
	Polarization	σ	0.51	0.44	0.48	0.47	0.49	0.53	0.54

TABLE 2

Electrical Properties of Eastern Bituminous Coal as a Function

of Direction and Polarization [ϵ_r(dimensionless), σ (S/m)]

f ≃ 9 GHz Ambient Temperature		FACE CLEATS			BUTT CLEATS		
		V P	45° P	H P	V P	45° P	H P
WET	ϵ_r	3.89	3.84	3.74	3.88	3.85	3.93
	σ	0.26	0.43	0.73	0.18	0.32	0.51
DRY	ϵ_r	3.71	3.72	3.68	3.38	3.40	3.41
	σ	0.17	0.39	0.69	0.12	0.28	0.44

It is evident that there are slight decreases in the values of dielectric constant and conductivity as the temperature is increased from ambient to about 700° F and as the samples become dry. Within measuring accuracies, there are no noticeable variations in the dielectric constant as a function of polarization and direction of travel of the electromagnetic wave.

It has been established, however, that there are unique polarization effects in the conductivity of an eastern bituminous solid coal sample. Conductivities lower by a factor of 3 to 6, depending upon the temperature and moisture content of the sample, for each of the following, by referring to Figure 4, have been measured:

    a. the electromagnetic wave travels perpendicular to the face cleats and its electric field is parallel to the face and butt cleats but perpendicular to the bedding plane.

    b. the electromagnetic wave travels perpendicular to the butt cleats and its electric field is parallel to the butt and face cleats but perpendicular to the bedding plane.

The conductivity showed a more-or-less random variation when the propagation was perpendicular to the bedding plane. These effects are probably due to the orientations of the pyrite layers which most tend to lie parallel to the bedding planes. However, the bedding plane samples are cored across the faces of these pyrite layers which give the pyrite a random distribution in the bedding plane view.

Physical analysis was conducted for the individual planes
on the Lewis, King Knob, Hansford, and Allegheny bituminous coals,
and Reading anthracite with a breakdown for each plane orientation[13]. Table 3 gives the percentages of fixed carbon, volatile

Table 3

Proximate Analysis by Sample Orientation

Sample Orientation	Volatile Matter (M.F.)%	High Temperature Ash (M.F.)%	Moisture Content (A.R.)%	Fixed Carbon (M.A.F.)%
Lewis Face Cleat	29.43	8.31	1.78	68.12
Lewis Butt Cleat	29.89	14.85	1.21	65.67
Lewis Bed Plane	22.98	17.49	0.48	73.00
King Knob Face Cleat	30.10	9.98	0.48	66.90
King Knob Butt Cleat	27.91	11.12	2.96	68.96
King Knob Bed Plane	28.52	12.71	0.19	63.75
Hansford Face Cleat	31.82	5.71	1.67	66.36
Hansford Butt Cleat	29.84	3.19	2.62	69.21
Hansford Bed Plane	29.43	8.31	1.78	68.12
Allegheny Face Cleat	10.67	13.75	0.99	87.86
Allegheny Butt Cleat	13.56	7.16	0.30	85.47
Allegheny Bed Plane	10.12	8.88	3.16	88.98
Reading Face Cleat	0.00	5.33	2.18	100.00
Reading Butt Cleat	0.69	6.96	2.09	99.26
Reading Bed Plane	2.36	3.15	0.78	97.57

matter and high temperature ash as well as the moisture content
for the coals. All of the coal samples were air-dried at standard
room temperature and humidity conditions. The Reading anthracite
shows the high fixed carbon (98.9 per cent average), and the
corresponding low volatile matter of a high rank coal. The
Allegheny bituminous is a low volatile bituminous coal with a
fixed carbon content of 87.4 per cent (average). The other three
bituminous coals are all high volatile coals of about 66-73 per
cent fixed carbon. The Lewis coal differs from the other two in
that it has somewhat more ash and less volatile matter. All of
the bituminous coals are reasonably low in moisture content. All
of the samples were strongly air-dried at 55° C for approximately

190

five hours. This resulted in their overall low moisture contents
with the bituminous samples being slightly lower than the
anthracites.

Table 4 gives the sulfur content of the coals with a breakdown by sulfur type[13]. The Lewis bituminous is very high in the
conductive mineral, pyrite (FeS_2), with 6.30 per cent (average).

The other coals are all reasonably low in pyrite with the Allegheny
bituminous being the next highest with 1.68 per cent (average).
The gross amount of pyrite is probably not as important as its
distribution within the coal, a point which will be discussed later.

Table 4

Sulfur Type Analysis by Sample Orientation

Sample Orientation	Organic Sulfur (wgt. %)	Sulfatic Sulfur (wgt. %)	Pyritic Sulfur (wgt. %)	Total (wgt. %)
Lewis Face Cleat	0.00	0.04	6.31	6.35
Lewis Butt Cleat	1.85	0.20	4.26	6.31
Lewis Bed Plane	0.06	0.87	8.32	9.24
King Knob Face Cleat	0.63	0.00	0.65	1.28
King Knob Butt Cleat	0.83	0.20	1.42	2.45
King Knob Bed Plane	0.57	0.06	0.54	1.17
Hansford Face Cleat	0.47	0.08	0.29	0.84
Hansford Butt Cleat	0.45	0.01	0.23	0.69
Hansford Bed Plane	0.42	0.04	0.23	0.69
Allegheny Face Cleat	0.37	0.30	1.47	2.15
Allegheny Butt Cleat	0.69	0.40	2.76	3.85
Allegheny Bed Plane	0.73	0.09	0.82	1.64
Reading Face Cleat	0.22	0.01	0.30	0.53
Reading Butt Cleat	0.29	0.01	0.21	0.52
Reading Bed Plane	0.29	0.01	0.23	0.53

Table 5 gives the averages (across sample size) of the dielectric constant and conductivity for each mine as a function of
polarization and direction of incidence[13]. Several relationships
between the chemical data and the electrical data stand out.

Table 5

Average Electrical Properties at 9 GHz as a Function
of Polarization and Direction of Incidence

Sample Company	Direction of Incidence	Vertical Polarization		45° Polarization		Horizontal Polarization	
		ϵ_r	σ (S/m)	ϵ_r	σ (S/m)	ϵ_r	σ (S/m)
Lewis Coal Company	Face Cleats	3.928	.298	4.051	.651	4.140	1.033
	Butt Cleats	4.193	.349	4.220	.685	4.045	1.165
	Bedding Planes	4.049	.581	4.173	.661	4.091	.621
Hansford Coal Company	Face Cleats	3.824	.134	3.818	.213	3.988	.261
	Butt Cleats	3.906	.203	4.303	.281	4.503	.408
	Bedding Planes	4.034	.684	4.170	.622	4.200	.576
King Knob Coal Company	Face Cleats	3.755	.200	3.769	.265	3.959	.373
	Butt Cleats	3.767	.230	3.943	.301	4.028	.341
	Bedding Planes	4.415	.357	4.495	.339	4.457	.349
Allegheny Coal Company	Face Cleats	4.087	.090	4.180	.137	4.281	.207
	Butt Cleats	3.827	.075	3.931	.126	4.218	.258
	Bedding Planes	3.740	.174	3.822	.171	3.863	.163
Reading Coal Company	Face Cleats	9.788	2.456	10.096	2.562	10.385	2.712
	Butt Cleats	9.068	1.979	9.435	2.142	9.420	2.171
	Bedding Planes	9.893	2.634	10.281	2.728	11.006	3.079

First, the high conductivity of the Lewis Bituminous, 2 to 4 times greater than that of the other bituminous coals, is probably related to the high pyrite concentrations of that coal. Second, the Reading anthracite has very high conductivity and permittivity which is probably related to its rank. The ash content of the coals does not seem to be related to the electrical properties, in keeping with former studies[14]. Within each company's coals, the moisture differences do not seem to be related to the electrical properties, probably because all of the moisture levels are reasonably low and any effects are masked by the spread of experimental error. It is difficult, however, to explain the low values of conductivity of the Allegheny bituminous coals, which are fairly high rank coals.

Several anomalous samples were found. The King Knob butt cleat sample 01 and the Lewis face cleat sample 01 have somewhat high values of conductivity; but as the radiograph of Figure 6 shows these samples had large concentrations of pyrite (dark horizontal bands or dark distributed areas). This same trend was also observed in the Hansford bedding plane samples. The samples containing large areas of pyrite had much higher conductivities than those with little pyrite.

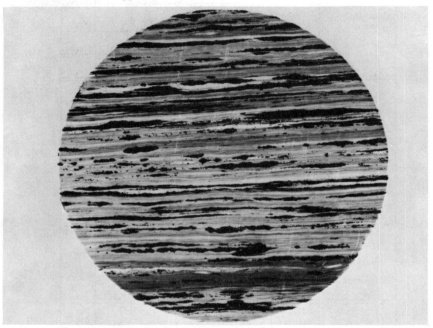

Fig. 6.   Radiograph of the Lewis coal face cleat sample 01.

Probably the most important observations made were the spatial relationships of the electrical properties of the coals. As Table 5 shows, there seem to be no clear cut relationships between propagation direction within the coal and the electrical properties; but this is not the case for polarization effects. There are probably at least two factors involved in these polarization effects. One is the effect of pyrite bands in the samples, and the other is the optical anisotropy.

For the bituminous coals, with few exceptions, conductivity is lowest for the vertical direction and highest for the horizontal direction on the face and butt cleat samples, but shows a more-or-less random orientation on the bedding plane samples. The amount of conductivity change with polarization varies with the coal type; but the coal that shows the most change is the Lewis

which also has the highest pyrite content. These effects are probably due to the orientations of the pyrite layers, which tend to
lie along the horizontal bedding planes, as can be seen in Figure
6. However, the bedding plane samples are cored across the faces
of these pyrite layers which give the pyrite a random distribution in the bedding plane view. The dielectric constants of the
bituminous coals do not show the degree of change with polarization direction that the conductivities show, suggesting that
the primary cause of anisotropy in the lower rank coals is the
presence of the pyrite layers.

To explain the directional effects of these samples such as
the Reading anthracite bedding plane sample 02, another cause
must be found. The radiograph of this sample showed only small
amounts of pyrite, particularly in the central region which affects
the measurements most radically. If the theory of optical
biaxility[10] holds for these high rank coals, there should be a
cyclical relationship of the electrical parameters with polarization direction; that is, the maximum and minimum values of both
reflectivity and dielectric constant should lie at 90° to each
other, and the degree of anisotropy as measured by bireflectance
should match the difference between maximum and minimum permittivity. Further, these maxima do not have to lie along the
bedding planes. To test this, reflectance, dielectric constant
and conductivity were measured at 15° intervals on some of the
anthracites samples[13].

The reflectance, dielectric constant and conductivity for the
Reading anthracite face cleat sample 02, butt cleat sample 02 and
bedding plane sample 02 are plotted in Figures 7, 8, 9, respectively. Zero degrees corresponds to horizontal polarization.

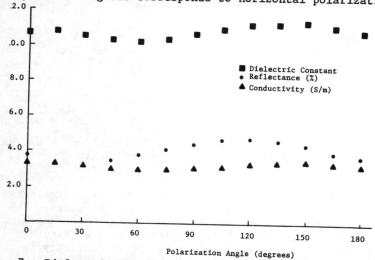

Fig. 7. Dielectric constant, reflectance, and conductivity versus
polarization angle for Reading face cleat anthracite.

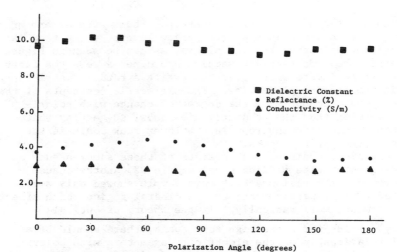

Fig. 8. Dielectric constant, reflectance, and
conductivity versus polarization angle
for Reading butt cleat anthracite.

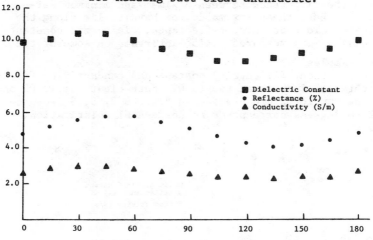

Fig. 9. Dielectric constant, reflectance, and
conductivity versus polarization angle
for Reading bedding plane anthracite.

For all of the samples the maximum and minimum values of reflectance
and dielectric constant are 90° apart as the model predicts. The
sample which shows the greatest change of dielectric constant as a
function of polarization is the bedding plane sample of Figure 9.
Likewise the sample with the smallest change of permittivity is the
butt cleat sample of Figure 8. The maximum-minimum difference in
reflectance as a function of polarization, referred to as

bireflectance, increases as coal becomes increasingly anisotropic (rank increases). Therefore, bireflectance is a good indicator of coal rank.

It has also been shown through calculations[13] that the dielectric constant variations of high rank coals (anthracites) as a function of polarization follow quite well the reflectivity's optical anisotropy trends predicted from analytical models.

In order to take advantage of the electromagnetic wave polarization effects (especially of the conductivity) detection and monitoring systems should be designed which utilize a polarization which exhibits the lowest losses as the wave travels through the medium. For the bituminous coals that were investigated, this would mean vertical polarization. Two of the most common electromagnetic elements (antennas) which are used to radiate and receive electromagnetic waves, especially for underground applications, are the linear dipole and circular loop. Figure 10 demonstrates their orientations to attain vertical and horizontal polarizations.

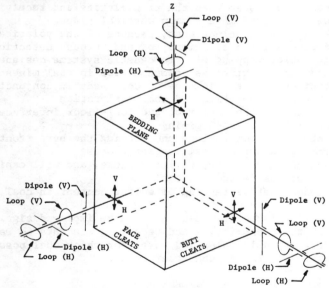

Fig. 10. Coal sample and relative orientation of electromagnetic radiators for vertical and horizontal polarizations.

CONCLUSIONS

The study has demonstrated several interesting relationships. First, the propagation direction within a coal bed seems to have very little effect on the electrical properties. On the other hand, the polarization direction has a definite bearing on the

196

electrical parameters. In general, pyrite layers cause the conductivity to be higher for all polarizations and, in particular,
to be highest when the wave is polarized in a direction parallel
to the pyrite bands (when viewed edge-on). This seems to be the
primary cause of conductivity anisotropies in lower rank (bituminous) coals.

The anthracite coals have significantly higher conductivities
(3.0 S/m versus .156 to .672 S/m) and dielectric constants (9.93
versus about 4) than the bituminous coals. In addition to the
pyrite effects present in the bituminous coals, the anthracites
have the same type of anisotropy of electrical properties that
they display for optical reflectances. Most of the results seem to
support the theoretical predictions; and in particular, the model
of optical anisotropy closely predicts the variations in the
electrical properties.

The polarization phenomenon of coal found in this investigation was one of the factors that led to previous unexplained
marked system improvements in electromagnetic transmission within
an underground coal seam when the transmitting and receiving loop
antennas were oriented in a common vertical plane.

These findings conclude that, because of the polarization
phenomenon of coal, the efficiency of underground detection,
monitoring, and/or mapping electromagnetic systems designed for
    a.   underground wireless communication in coal mines
    b.   monitoring the thickness of coal beds in conjunction
        with mine operation and coal exploration
    c.   monitoring the coal-roof and coal-floor interface for
        automation of coal mining
    d.   detection, monitoring, and mapping the burn-front of an
        underground coal gasification process
    e.   detection of voids, pyrite pockets, and well casings
        buried inside unmined coal seams
    f.   nondestructive testing of sulfur content in coal

and other coal-related applications, can be improved significantly
by orienting the transmitting and receiving elements to radiate
and receive vertically-polarized waves for which coal possesses
more permeable electrical characteristics.

ACKNOWLEDGEMENTS

The author would like to thank all of his co-workers who
participated and contributed in all phases of this investigation.
These include John L. Jeffrey, Phillip W. Shepard, Dr. Francis
T.C. Ting, and William F. Kardosh. The first phase of the work
was sponsored by the Morgantown Energy Technology Center of the
Department of Energy while the second phase was supported by the
Energy Research Center of West Virginia University.

REFERENCES

1. L. Hartshorn and W.H. Ward, J. of IEE(London) 79, 567-609 (Nov. 1936).
2. G.V. Keller and P.H. Licastro, U.S. Geol. Surv. Bull. 1052H, 257-285 (1959).
3. B.F. Howell and P.H. Licastro, The Amer. Mineralog. 46, 269-288 (March-April 1971).
4. M. Sucher and J. Fox, Handbook of Microwave Measurements Vol. 2 (Polytechnic Press of the Polytechnic Institute of Brooklyn, 1963).
5. C.A. Balanis, W.S. Rice, and N.S. Smith, Radio Sci. 2, 413-418 (April 1976).
6. C.A. Balanis, Microwave J. 14, 39-44 (March 1971).
7. D.T. Paris and K.F. Hurd, Basic Electromagnetic Theory (McGraw-Hill, N.Y., 1969), p. 323.
8. C.A. Balanis, J.L. Jeffrey, and Y.K. Yoon, IEEE Trans. Geoscience Electron. GE-16, 316-322 (October 1978).
9. C. Karr(Ed.), Analytic Methods for Coal and Coal Products (Academic Press, 1978), Chapter by F.T.C. Ting.
10. V. Hevia and J.M. Virgos, J. of Microscopy 109, 23-28 (Jan. 1977).
11. ASTM(1975a), Standard D-388, ASTM Standard Manual, Part 26, 212-216.
12. H.B. Lo, Unpublished Thesis, West Virginia Univ. (1977).
13. C.A. Balanis, P.W. Shepard, F.T.C. Ting, and W.F. Kardosh, IEEE Trans, Geoscience and Remote Sensing GE-18 (July 1980).
14. M.P. Groenewege, J. Schuyer, and D.W. van Krevelen, Fuel 34, 339-344 (July 1975).

INSTRUMENTATION FOR TRANSPORT AND SLURRIES

Nancy M. O'Fallon
Argonne National Laboratory, Argonne, IL   60439

ABSTRACT

Many advanced fossil energy processes require the continuous
transport of crushed coal into high pressure vessels. The most
common methods of doing this are sequenced pneumatic injection
from pressurized lock hoppers and pumping of a coal-liquid slurry.
Operation of these and other coal feeding systems is outlined and
standard instrumentation for monitoring coal feed is reviewed.
Applicable new and developmental instrumentation is discussed.

INTRODUCTION

Coal gasification, liquefaction, pressurized fluidized-bed combustion, and magnetohydrodynamics all require the feeding of crushed
coal into vessels at high pressures. Figure 1 shows simplified
schematics of these processes as well as oil shale retorting and indicates the major problem area for instrumentation and control.
There are, in fact, many streams containing solids within the processes, at high temperatures as well as high pressures, which need
monitoring and control.[1,2] Pressures may range from a few atmospheres to a few hundred atmospheres. Demonstration and commercial
plants will utilize typically 5,000 to 50,000 tons per day of coal.
Pilot plants are smaller by a factor of 100 or so. Monitoring of
these flows is needed for understanding and successful scale up of
the processes and, of course, monitoring instrumentation is an integral part of a control system.

The quantity of interest for control purposes is the mass flow
rate, $\rho v A$; where ρ is the density, v the flow speed, and A the pipe
cross sectional area. Since few sensing techniques respond to mass
flow directly, it is generally necessary to combine two measurements for a calculated mass flow determination. In addition, with
a multi-phase stream, the proportions of individual constituents
must be measured.

SOLIDS FEEDING SYSTEMS

One of the two most widely used methods of feeding solids into
a pressurized system employs a lock hopper system as illustrated in
Figure 2, which is a simplified schematic of the Synthane coal gasification process.

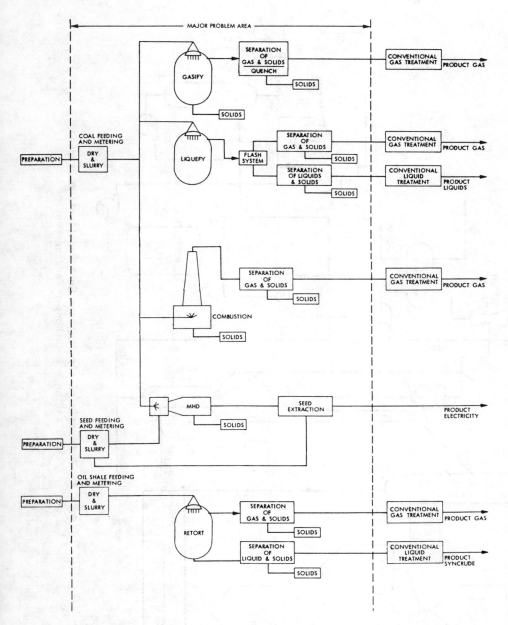

Figure 1. Schematic Overview of Advanced Fossil Energy Processes

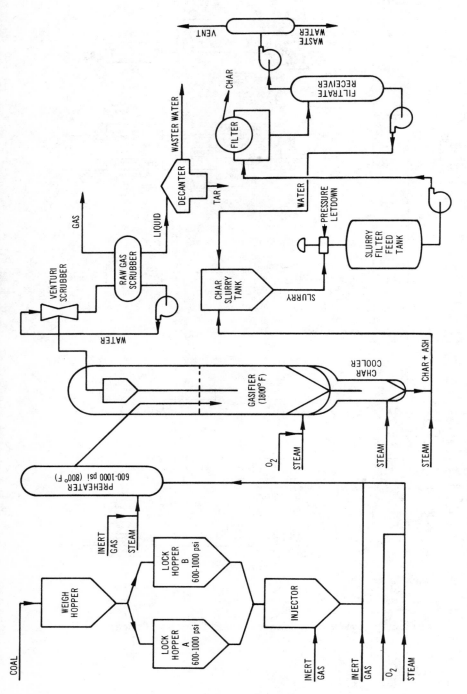

Figure 2. Schematic of the Synthane Coal Gasification Process Illustrating a Lock Hopper Feeding System

Crushed coal is fed into the weigh hopper at atmospheric pressure. Then lock hopper A, also at atmospheric pressure, is filled from the weigh hopper. A valve at the top of A is closed, B is isolated from the injector hopper, and the inert gas pressurizing B is allowed to expand into A. B is returned to atmosphere while A is brought to system pressure and the weigh hopper is filled. A then fills the injector while B is filled from the weigh hopper. The roles of B and A are reversed, and so on. Coal from the injector is picked up by the conveying gas and transported into the system, typically, with a speed of 15 to 45 m/sec at 1 or 2 percent by volume.

Problems associated with this kind of system are the valves which are required to open and close repeatedly in the presence of the abrasive solids, absence of a means of short term measurement of the flow, dilution of the product with the inert gas, and the expense of supplying and pumping the inert gas. Present means of monitoring the flow are with a weighing system at the front end or calculating the coal usage in terms of the rate of use of the other reactants. Operation and limitations of these methods will be discussed in the next section.

The second widely used method of feeding solids is to mix them with a liquid to form a slurry, which is then pumped to pressure and injected into the pressurized vessel. Figure 3 shows this schematically. Flow speeds are typically a few m/sec with solids

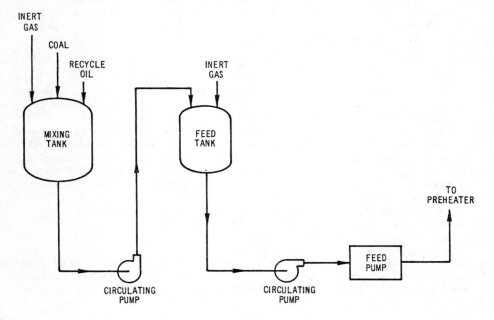

Figure 3.   Schematic of a Slurry Feed System

loadings up to 50 or more percent by weight. Serious problems with pump durability are very common in a slurry feed system.[3] Flow measurement is generally found to be inadequate,[4] especially for heated slurries, and energy is required to vaporize the carrier liquid. Present means of flow monitoring are weigh systems, head type flowmeters, and (for water-based slurries) magnetic flowmeters. For low temperature slurries, active sonic flowmeters are sometimes used.

There are a few other methods of feeding crushed solids into pressurized systems which should be mentioned although they are not in common use. Scientists at the Jet Propulsion Laboratory developed a feeder, based on plastics extruders, in which agglomerating coals are pressurized and plasticated in a screw type extruder and sprayed as fine particles into the pressurized system.[5] The device is energy expensive, but this is somewhat offset by the fact that it feeds only coal into the system. Thus a conveying gas or liquid is not present to be dealt with. Work was also done at General Electric[6] on an extrusion feeder using crushed coal mixed with a tar binder. The mixture was compacted into a "log" by the extruder and chopped as it emerged from the die at the extruder outlet. Flows are monitored by the weigh system at the front end and by the extruder screw speed and power use.

Finally, crushed coal is fed into the MHD combustor and the flash hydroliquefaction or gasification systems at Rockwell International by designing the tubing of the feed system to have a smooth interior with no sharp bends so that differential pressure moves the coal in continuous plug flow.

FLOW MONITORING INSTRUMENTATION

A very common way of monitoring the mass flow rate is a weighing system, either a weighed conveyor (or gravimetric feeder) in which the rate of loading and belt speed are controlled to deliver solids at the desired rate or a weighed hopper in which the rate of weight change is monitored. The gravimetric feeder is capable of very good accuracy, as good as ± 1/2%. It is, however, expensive and is limited to use at atmospheric pressure. In terms of monitoring flow into pressurized systems where there is considerable capacity between the point of measurement and the point of injection into the pressure vessel, it is useful only for giving an average over an hour or longer. The weighed hopper is limited in accuracy by the tare weight relative to the weight of solids in the tank. In a pressurized or hot system, it is difficult to isolate the tank from forces due to connecting equipment. These considerations plus complications due to vibrations limit the usefulness of the weighed hopper to averages over an hour or more.

Positive displacement meters or pumps provide a means of monitoring volumetric flow by counting the filling and emptying of chambers of known volume. While these are very accurate for clean liquids, the accuracy deteriorates rapidly in the pressure of abrasive solids, which change the calibration and cause leakage at the seals. A positive displacement pump is often a good first indication of the flow rate.

Head type meters, illustrated in Figure 4, introduce a restriction into the flow and measure the resultant pressure drop, which is related to the flow through Bernoulli's Equation. These

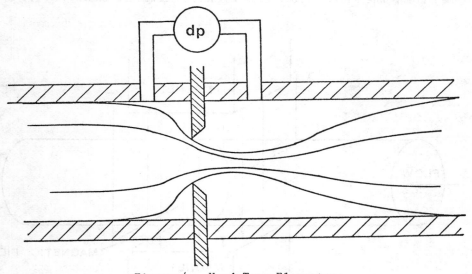

Figure 4. Head Type Flowmeter

include the orifice plate, which is pictured, and the Venturi meter, in which the flow restriction has a smoothly changing cross section. Limitations are the small rangeability of about 3:1, the requirement of a straight upstream run of several pipe diameters, and the requirement that the Reynolds number be high. The solids cause plugging problems with the pressure taps; hence the taps must be purged or sealed. An eccentric orifice, which has the opening along the bottom edge of the pipe, is recommended for measurement of slurries. Some success has been achieved in hot slurry flow measurement using a quadrant-edge orifice, which has a rounded leading edge, and permits operation at Reynolds numbers down to 1000 or so. Even so, the lifetime of the plate is only a few hours. The Venturi meter introduces a smaller pressure drop in the line, but it is expensive and is limited to Reynolds numbers above about 20,000.

The segmental wedge is another head type flowmeter which has been around for many years but only recently has been marketed aggressively. In this flowmeter, the restriction is a triangular bar across the top of the pipe. It is said to permit operation down to Reynolds numbers of 400 or less with little deviation from the square root relationship between velocity and pressure drop. Segmental wedge flowmeters have been installed in the H-Coal Pilot Plant, and will be tested as the plant begins operation using coal.

Water based slurries are not common in fossil energy systems. When they are used, the magnetic flowmeter is a good candidate for measuring the slurry flow. The magnetic flowmeter shown in Fig. 5

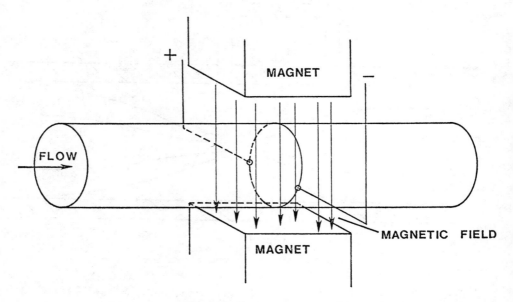

Figure 5. Magnetic Flowmeter

sets up a magnetic field across a conducting stream and measures the induced EMF, which is proportional to the flow speed according to Faraday's Law. The magnetic flowmeter has good rangeability and accuracy, and it is non-intrusive. It is relatively expensive and is not available in high pressure or high temperature models.

The coriolis flowmeter has recently become available for measurement of mixed-phase mass flows.[7] As indicated in Fig. 6, the

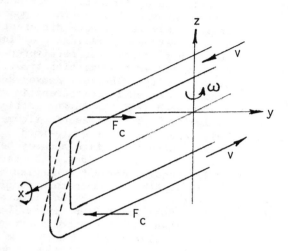

Figure 6. Coriolis Flowmeter

instrument consists of a U-shaped tube through which the process
material must pass. An angular momentum, ω, is imparted to the
instrument as shown. The flowing material in the opposite legs of
the U-shaped tube will experience coriolis forces, Fc, in opposite
directions as indicated. This moment produces a rotation of the
tube which is proportional to the mass flow. In the instrument
the angular momentum is actually oscillatory and produced by a
tuning fork. Thus the end of the tube will oscillate about the
horizontal axis and the magnitude of this oscillation can be picked
up optically. The device has undergone tests for the flow of
solids in gas at the University of Tennessee Space Institute (UTSI)
and the results were very promising. Further tests are indicated
for solid-gas and solid-liquid systems. The question of erosion
during long-term operation is yet to be addressed. Presently the
device is limited to 2 inch pipe sizes but could be manufactured
in larger sizes.

Sonic/ultrasonic flowmeters are beginning to be used for
measurement of multi-phase flow, although commercially available
instruments are limited to temperatures below 300°F or so. This
kind of instrument has several advantages, one being that it is

non-intrusive and thus not likely to cause a plug or to have the sensor damaged by the process material. There are passive instruments in which the transducer simply listens to the flow noise and interprets changes in the noise as an indication that something has changed in the system. These may vary from a simple monitoring of the total noise to a spectral analysis in which different portions of the spectrum are related to various operating conditions. There are active instruments in which a transmitter/receiver pair detects the change in upstream and downstream transmission time. This can then be related to the flow velocity. The most common active ultrasonic flowmeter is the doppler shift instrument in which the frequency shift of waves scattered from the moving particles is related to the particle velocity.[8] The ultrasonic flowmeters are fairly expensive, but their rangeability is quite good. Their accuracy is a function of the integration time and may be brought under 1% if one is willing to wait of the order of 10 seconds. Slurry applications generally require lower frequencies than clean liquid – generally below 1 MHz. A high temperature doppler slurry flowmeter is under development at Argonne National Laboratory for use in the Solvent Refined Coal Pilot Plant in Ft. Lewis, Washington. This instrument makes use of standoff wave guides in order to protect commercially available transducers from the high temperature. Pilot plant tests are underway at the present time. Further developments may include the use of high temperature transducers which were developed at Argonne for use in nuclear reactors.

A signal processing technique which is, in fact, applicable to many different types of sensors to produce a flow velocity measurement is cross correlation.[9,10] It involves taking the covariance of signals from axially separated sensors as a function of time delay on the upstream signal. The covariance will peak at a time corresponding to the transit time from the first sensor to the second, assuming that the fluctuations in the variable being sensed are propagating with the process stream. This techique was used successfully in tests on a toluene and char slurry line at the HYGAS Coal Gasification Pilot Plant using both acoustic sensing and capacitive sensing.

Capacitive instruments measure the effective dielectric constant of the process stream through capacitive plates imbedded in a non-conducting liner within the pipe.[11] In the case of a two-phase process stream for which the dielectric constants of the individual phases are known, it is possible to deduce the relative amounts of the two phases present. This is, in effect, a measurement of the density of the phase of interest. Segmenting the sensing electrode in the axial direction can give a flow velocity measurement by cross correlation of signals from the individual portions of the electrode. Thus a single set of sensing elements can yield a computed value of mass flow. A good deal of work has been done on the technique in England.[9] In this country there is a commercial unit which measures only the solid-to-liquid or solid-to-gas ratio in the process stream. The instrument tested by Argonne at HYGAS included the cross correlation for velocity measurement and was able to monitor the solid-to-liquid ratio to yield a flow velocity within a few seconds during tests at the HYGAS pilot plant. A high temperature version of this instrument is being designed.

Another technique under development for monitoring flow velocity is a Pulsed Neutron Activation (PNA) technique.[12] This involves irradiating the process stream through the pipe with a pulsed 14 MeV neutron generator to activate elements within the process stream, such as ^{16}O, and detecting the radioactive tag at a downstream station. Initial measurements have taken place at Argonne using a slurry flow loop with a 14 MeV neutron generator, and the results are promising. A system based on this technique may prove to be too costly for routine multiple installations within a plant. However, since it does not require physical penetration of the piping, it may be very useful as a movable calibration standard for installed flow instrumentation.

ACKNOWLEDGEMENTS

The Argonne National Laboratory work discussed in this paper was supported by the U. S. Department of Energy – Fossil Energy. The Jet Propulsion Laboratory made Figure 1 available.

REFERENCES

1. N. M. O'Fallon, R. A. Beyerlein, W. W. Managan, H. B. Karplus, and
    T. P. Mulcahey, "A Study of the State-of-the-Art of Instrumentation
    for Process Control and Safety in Large-Scale Coal Gasification, Liqu
    faction, and Fluidized-Bed Combustion Systems," ANL-76-4 (January 19

2. Lawrence Mattson and Will Schaefle, "Coal Handling and Feeding Chal-
    lenges and Future Prospects," Presented at the AIChE 86th National
    Meeting (April 1979).

3. W. R. Williams, J. R. Horton, W. F. Boudreau, and M. Simon-Tov, "Sur
    of Industrial Coal Conversion Equipment Capabilities: Rotating Comp
    nents," ORNL/TM-6074 (April 1978).

4. E. F. Brooks and C. W. Clendening, "Assessment of Instrumentation fo
    Monitoring Coal Flowrate and Composition," Final Report for EPA Con-
    tract No. 68-02-2613, Task 2 (August 1978).

5. William Schatz, "Development of the Plasticating Coal Pump," Present
    at the AIChE 86th National Meeting (April 1979).

6. D. E. Woodmansee, A. H. Furman, and J. K. Floess, "Development of an
    Extruder Feed System for Fixed Bed Coal Gasifiers," EPRI Report AF-9
    (January 1979).

7. W. E. Baucum, "Status of the Evaluation of a Coriolis Effect Mass Fl
    Meter for Dense Phase Coal Flows," Proceedings of the 1979 Symposium
    Instrumentation and Control for Fossil Energy Processes, ANL 79-62,
    CONF-790855, 210 (1979).

8. H. B. Karplus and A. C. Raptis, "Slurry Flow Measurements Using an
    Acoustic Doppler Flowmeter," Proceedings of the 1979 Symposium on
    Instrumentation and Control for Fossil Energy Processes, ANL 79-62,
    CONF-790855, 184 (1979).

9. M. S. Beck, J. Coulthard, P. J. Hewitt and D. Sykes, "Flow Velocity
    Mass Flow Measurement Using Natural Turbulence Signals," Proceedings
    the International Conference on Modern Developments in Flow Measureme
    Harwell (1971).

10. K. G. Porges, F. R. Lenkszus, R. W. Doering, W. W. Managan, C. L.
    Herzenberg, C. A. Nelson, and N. M. O'Fallon, "On-Line Correlation
    Flowmetering in Coal Conversion Plants," Proceedings of the 1979
    Symposium on Instrumentation and Control for Fossil Energy Processes,
    ANL 79-62, CONF-790855, 164 (1979).

11. W. W. Managan and R. W. Doering, "Capacitive Transducers for Mass Fl
    Measurement, An Overview," Proceedings of the 1979 Symposium on Inst
    mentation and Control for Fossil Energy Processes, ANL 79-62, CONF-
    790855, 226 (1979).

12. C. L. Herzenberg, "Use of Small Accelerators in Coal Analysis and Co
    Slurry Flow Measurements," IEEE Transactions on Nuclear Science, NS-
    1568 (1979).

MOSSBAUER SPECTROSCOPY FOR PYRITE ANALYSIS IN COAL

Lionel M. Levinson
General Electric Company, Corporate Research and Development
Schenectady, New York

ABSTRACT

The Mössbauer spectroscopic technique has been adapted to provide a reliable measurement of pyritic sulphur in coal. The Mössbauer method is quick, nondestructive, and adaptable to automation. It can be applied to processed as well as raw coals.

Several difficulties hindering the use of Mössbauer spectroscopy for pyritic measurements have been resolved. Sample homogenization techniques consistent with the spectroscopic requirements were developed. A sample holder compatible with a 5 to 10 g coal sample was designed. Two calibration curves were evaluated: one based on a mixture of pyrite and graphite, the other on $HC\ell$ - leached coal. A simple correction method for the effects of ash absorption was demonstrated.

Differences between the Mössbauer and ASTM wet chemical methods for analyzing pyritic sulphur in coal were evaluated. Good agreement was shown, provided residual undissolved pyritic in the ASTM method is taken into account.

Analysis of spectrometer capability indicates a routine sample could be analyzed in about 10 minutes.

1. INTRODUCTION

Most U.S. coals contain sulphur impurities in both inorganic (FeS_2) and organic forms. In view of the quite different chemical and physical properties of these sulphur-bearing contaminants, a particular coal cleaning method will usually be largely directed towards the removal of only one of these two forms of sulphur. It follows that process development and control for removal of inorganic (FeS_2) requires techniques for the accurate measurement of the (FeS_2) content in raw and processed coal. At present FeS_2 content in coal is determined using the wet chemical technique outlined in ASTM D2492-77. Briefly, the process is based upon the assumption that pyritic iron is insoluble in hydrochloric acid but soluble in nitric acid, and that the iron soluble in hydrochloric acid is also soluble in nitric acid. It then follows that the difference $Fe(HNO_3)-Fe(HCl)$ should be equivalent to the pyritic iron. A schematic depiction of the pyrite analysis procedure is given in Fig. 1.

The wet chemical technique for pyrite analysis has produced results at times believed questionable. For example, various authors[1-3] have noted that it is possible that a fraction of the pyrite particles could be completely surrounded by the coal (even after crushing) in the test sample and would therefore not be removed by nitric acid extraction. Secondly, many coals contain other iron-bearing minerals. Some of these might dissolve in hydro-

*Sponsored by the Electric Power Research Institute.

ANALYSIS OF PYRITIC SULFUR IN COAL:
ASTM D2492-77

SULFATE SULFUR DETERMINATION:

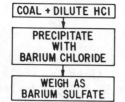

SULFATE SULFUR IS SOLUBLE
IN DILUTE HCl (2:3 H_2O)

PYRITIC AND ORGANIC FORMS
OF SULFURS ARE INSOLUBLE
IN DILUTE HCl

PYRITIC SULFUR DETERMINATION:

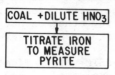

BOTH SULFATE AND PYRITIC SULFUR
ARE DISSOLVED IN HNO_3 (1:7 H_2O);

THEREFORE THE TITRATION
MUST BE PERFORMED WITH
RESPECT TO IRON

THE DIFFERENCE BETWEEN THE
NITRIC ACID AND HYDROCHLORIC
ACID SOLUBLE IRON IS PYRITIC
IRON

Fig. 1. Schematic of pyrite analysis per ASTM D2492-77.

chloric acid but be only partly soluble in nitric acid.[2] Both
above potential problems could cause significant inaccuracies in the
ASTM results.

In addition to these difficulties we note that the ASTM wet
chemical technique is (a) destructive and (b) verified in detail only
for raw coal samples. The destructive nature of the analysis procedure is a significant drawback when dealing with a material as
heterogeneous as coal. In particular, it becomes difficult to discriminate whether (sometimes small) variations in measured sulphur
content arise from real effects in processing procedures or simply
from sampling errors. In this respect a non-destructive monitor of
pyrite content is preferable insofar as a given sample may be
characterized, processed, and then remeasured.

Point (b) above is of non-negligible significance taking into
account the high interest in coal processing techniques. Specifically, it is unwise to apply the ASTM technique, which depends
sensitively on differential solubilities, to Fe-S forms resulting
upon coal processing.

A major objective of this study has been to examine the potential development of Mössbauer Spectroscopy as a routine laboratory
tool for the quantitative measurement of pyrite in coal, and in
Section 2 we will outline briefly the application of the Mössbauer
technique to coal. In this Introduction we merely note some advan-

tages of the method.

1. The technique is non-destructive.
2. It is applicable both to processed coal, and to raw coal. Any chemical changes in the nature of the in-organic sulphur as a result of processing will be evident.
3. The technique is sensitive only to Fe in coal and can differentiate between various iron-bearing im-purities in coal (e.g., between FeS_2 and siderite, magnetite, ferrous sulphate, etc.[2]).
4. The measurement is unaffected by the presence or absence or organic sulphur.
5. Carbon (coal) surrounding the FeS_2 does not affect the measurement.
6. The technique can be designed to measure relatively large amounts (tens of grams), thereby ameliorating sampling difficulties.
7. The method is quick, simple, and easily adapted to automation.

The advantages of the Mössbauer technique have been previously out-lined by the author and I. S. Jacobs[4], and independently and concurrently by other workers [5,6], who also suggested the applica-tion of Mössbauer Spectroscopy to the study of coal.

While adopting somewhat different approaches (see Section 3), the consensus of all these studies is that Mössbauer Spectroscopy is in fact a desirable alternative to the ASTM technique for measure-ment of pyrite in coal.

2.   MOSSBAUER SPECTROSCOPY OF COAL

The use of Mössbauer Spectroscopy to obtain information regard-ing the nature of solid matter is well established. The technique is now about 20 years old and has application in fields as diverse as chemistry, archeology, biology, metallurgy, mineralogy and physics. A variety of general books on the Mössbauer effect and its applications are available[7-9].

Many, but not all, nuclei exhibit a Mössbauer effect. By far the most widely studied Mössbauer transition occurs in the nucleus ^{57}Fe, a naturally occurring, stable isotope of iron with relative isotopic abundance of 2.19%. The popularity of ^{57}Fe in Mössbauer studies arises both from experimental measurement ease and from the abundance and importance of iron in many technical fields.

The use of Mössbauer spectroscopy to measure pyrite (FeS_2) in coal is based upon the application of the technique to ^{57}Fe. Sulphur does not exhibit a Mössbauer effect.

Coal is an almost ideal material for Mössbauer Spectroscopy. The reason for this derives from the fact that the organic component of coal is nearly transparent to the 14.4 keV γ-ray. Consequently, trace (≥ 0.1 wt%) amounts of Fe-bearing impurities in coals are

detectable by simply increasing the amount of coal in the γ-ray
path. A typical Mössbauer absorber has a 1–10 mg/cm^2 of natural
Fe in the absorber. For a coal with about 1 wt% Fe, we use a coal
sample with thickness 0.1 – 1 gm/cm^2.

Coal contains a variety of iron-bearing phases. A partial list
of the naturally occurring minerals is given in Table 1. A more
extensive list can be found in Ref. 6. Most of the coals studied
in this investigation were bituminous and had pyrite (or marcasite)
as their predominant iron-bearing impurity. The spectrum of
mineral pyrites is given in Fig. 2. As is evident the material
exhibits a quadrupole doublet. We obtain quadrupole splitting, ε =
0.606 ± 0.005 mm/sec, and isomer shift, δ = 0.301 ± 0.005 mm/sec
in good agreement with other measurements.

A "typical" coal will contain pyrite as the major iron-bearing
phase with smaller or greater amounts of sulphates or Fe-bearing
clays and impurities. Fig. 3 gives the spectrum of fresh* PSOC
287 (Bevier seam) coal. Note that in this sample there is essentially no other iron-bearing phase (to within about 0.1 wt%).
Fig. 4 gives the spectrum of PSOC 270 (American seam) coal. This
sample has about 0.3 wt% sulphate sulphur in addition to 1 wt%

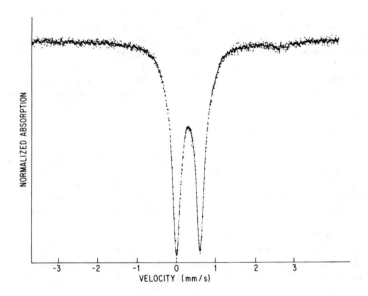

Fig. 2. Mössbauer spectrum of mineral pyrite.

*Note that if left in air in finely ground form some trivalent
 iron sulphate (Jarosite) is formed, presumably by slow oxidation
 of the pyrite.

Compound	Magnetic State	Site	ε (mm/s) quadrupole splitting	δ (mm/s) isomer shift**	H(kg)
Pyrite (FeS$_2$, cubic)	Para		0.614 ± 0.002	0.303 ± .0002	
Marcasite (FeS$_2$, orthorhombic)	Para		0.50 to 0.51	0.25 to 0.29	
Fe$_2$(SO$_4$)$_3$	Para	Fe^{3+}	0.0	0.46	
FeSO$_4$	Para		2.94	1.25	
FeSO$_4\cdot$H$_2$O	Para		2.708 ± 0.002	1.261 ± 0.002	
FeSO$_4\cdot$4H$_2$O	Para		3.17	1.23	
FeSO$_4\cdot$7H$_2$O	Para		3.20	1.31	
Jarosite ((Na,K)Fe$_3$(SO$_4$)$_2$(OH)$_6$)	Para		1.207 ± 0.004	0.368 ± 0.004	
Troilite(FeS)	Antiferro-magnetic		0.07	0.76	310
Clay minerals (various)	Para	Fe^{3+}	0.54 to 0.70	0.31 to 0.37	
Shale containing 85% illite	Para	Fe^{2+}, Fe^{3+}	2.68 ± 0.01 / 0.63 ± 0.01	1.12 ± 0.01 / 0.34 ± 0.01	
Chlorite	Para	Fe^{2+}	2.52	1.13	
Montmorillonite	Para	Fe^{2+}(Fe^{3+})	2.87 (0.06)	1.13 (0.33)	
Kaolinite	Para	Fe^{3+}(no Fe^{2+})	0.0 to 0.4	0.3 to 0.4	
Siderite(FeCO$_3$)	Para		1.81	1.23	

*Source: Reference 6
**δ with respect to Fe metal

TABLE 1. Short List of Mössbauer Parameters of Some Iron Bearing Minerals in Coal (from Ref. 6).

214

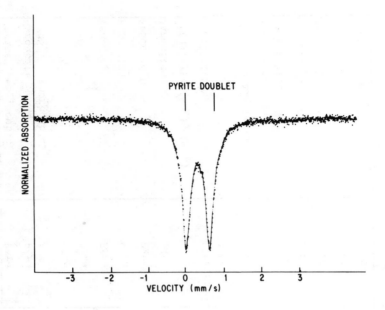

Fig. 3. Mössbauer spectrum of freshly ground PSOC 287. The doublet corresponds to pyrite.

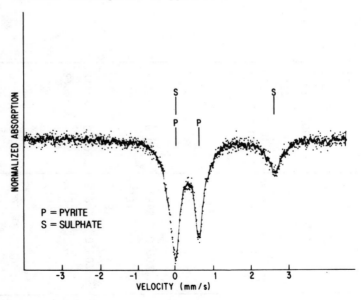

Fig. 4. Mössbauer spectrum of PSOC 270. This sample contains ferrous sulphate, $FeSO_4 \cdot H_2O$ as well as pyrite. S = sulphate, P = pyrite.

pyritic sulphur. The sulphate sulphur is in the form of the ferrous sulphate szomolnokite, $FeSO_4 \cdot H_2O$. Its Mossbauer spectrum is a doublet with a significantly greater quadrupole splitting than pyrite (Table 1). The left-most peak of the szomolnokite overlaps the left-most peak of pyrite.

In Fig. 5 we give the spectrum of PSOC 506 (Upper Sunnyside seam) coal. The coal is atypical of most of the coals examined in this report insofar as the dominant Fe-bearing phase is high in siderite/ calcite. Nevertheless the presence of some pyrite is clearly apparent. We estimate the pyrite content of this sample to be about 0.3 wt %. (We note in passing that another sample of PSOC 506 from the same batch, was reported as having no pyrite sulphur and 0.3 wt % sulphate sulphur by the ASTM procedure. Hence either all the pyrite was oxidized to sulphate in the latter sample, or else the "pyrite" in the sample was in fact soluble in HCl!)

Examination of the spectra of Figs. 3-5 leads us to conclude that the presence of pyrite in coal is usually quite detectable at a $\geq$ 0.1 wt % level. Since the intensity of the pyrite absorption peak is a monotonic function (4-9) of the amount of pyrite present in the absorber it becomes in principle possible to utilize the Mossbauer technique to measure the pyrite content of a given sample of coal.

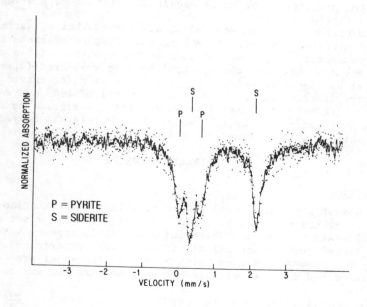

Fig. 5. Mössbauer spectrum of PSOC 506. This (atypical coal) contains mostly siderite/calcite as its dominant Fe-bearing phase. S = siderite, P = pyrite.

A number of experimental techniques can be applied to determine the pyrite content from the Mossbauer spectral intensity. Montano[5] using an expression due to Lang[10] gives a first principles estimate of the Mössbauer pyrite intensity in terms of the product FeS_2 content and the Mössbauer fraction f (also called the Debye-Waller Factor). Montano obtains an estimate of f from the temperature dependence of the Debye-Waller factor in a particular coal. While this method avoids the necessity of using a calibration standard, it requires the approximation of treating the phonon spectrum in pyrite as being describable in terms of a continuum Debye model. The validity of such an approximation is not clear.

Huffman and Huggins[6] use a method based upon the determination of the effective Mössbauer absorption thickness of pyrite as evaluated by a least square program, comparing the experimental area with a theoretical expression based upon a sum of Bessel Functions. The effective thickness of a particular coal sample is then compared with the effective thickness of pyrite found from a set of calibration experiments. This method is attractive insofar as it uses an exact expression for the Mössbauer line shape of samples of arbitrary thickness. It does, however, suffer from the difficulty of not taking into account non-resonant absorption of γ-rays by mineral matter in the coal. This can be a significant effect, especially for samples with high ash content or low Fe content. Additionally it has the practical problem of requiring a rather sophisticated computer fitting routine.

The approach we have adopted is similar to that of Huffman and Huggins insofar as we feel it is more realistic to evaluate the coal pyrite content by comparison of an observed spectrum with a calibration spectrum. However, the spectral area is evaluated (where possible) by simple numerical integration (area under curve). Each laboratory is required to evaluate its own calibration curve which would be particular to a specific experimental design. The effect of non-resonant absorption by ash, sample holder, etc. is corrected for by the method to be outlined in Section 3.

3. CALIBRATION CURVES

Introduction

The most straightforward way to establish a correlation between spectral intensity and pyritic sulphur content is via a set of standard calibration samples. While this, a priori, appears a reasonably simple procedure, substantial difficulties were experienced for a variety of reasons. We give below a list of the major problems encountered and discuss our attempts at resolving the various difficulties.

1. No commercial source of pure synthetic FeS_2 exists. Mineral pyrite is heterogeneous and impure and can contain varying impurities.
2. One would really like the pyrite particle size to be very small (< 20 μ). This is difficult to achieve. In addition very fine pyrite is susceptible to oxidation.

3. There can be some difficulty in obtaining a truly homogeneous sample. Sample heterogeneity probably arises
from density differences between carbon (graphite),
pyrite and impurity mineral particles in the calibration sample. The net result could be a non-uniform
pyrite dispersal in the graphite matrix and random and
non-uniform segregation of the mineral impurities within
the calibration samples (we have observed some signs of
this - the x-ray density of the absorber sometimes
varies as we scan various portions of a nominally
uniform sample).

4. Mineral pyrite is not formed by the same geological
process as is pyrite in coal. Thus one may question
whether or not the Mossbauer intensity for a given
quantity of pyrite in coal is the same as that of the
same amount of mineral pyrite.

Inhomogeneity of Mineral Pyrite

Mineral pyrite is typically impure and inhomogeneous. We have
verified this by crushing some crystals of "Colorado Pyrite" to -325
mesh and submitting 5 samples for chemical analysis. No special
effort was made to homogenize the pyrite by milling. The results
are given in Table 2, and from the data it is clear that our mineral
pyrite has up to about 10% impurity. Thus a calibration curve
based on this mineral would be inaccurate. Clearly it is essential
to mill the pyrite to randomize impurities and to use as large a
sample as possible per calibration measurement.

TABLE 2

Chemical Analysis Results for 5 Samples of Crushed "Colorado Pyrite"

Sample #	Wt % Fe	Wt % S
1	44.1	49.6
2	42.4	--
3	44.3	--
4	39.6	--
5	45.5	--
Nominal	46.5	53.5

Particle Size

The presumption is made that the pyrite calibration sample (and
coal sample) are uniformly dispersed. Consider however the situation
for an absorber having a pyrite content of 10 mg/cm^2. In this case,
a uniform dispersal of pyrite would result in a layer of about 20 μ
thick. If, however, the pyrite particle size were >> 20 μ we would
have the situation depicted in Fig. 6. The large particle case
depicted is undesirable since the absorber is locally thick, i.e.,

218

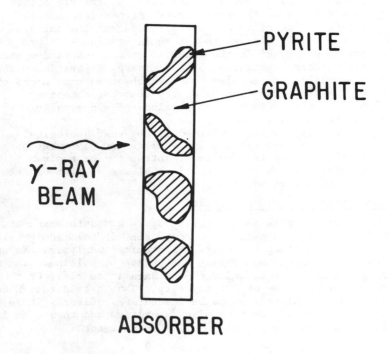

Fig. 6. Schematic depiction of Mössbauer absorber with overly
         large absorber particle size.

we have self-absorption effects which cause derivations from
linearity in the absorption intensity vs pyrite content curve.
Analyses of this so-called "granularity" problem can be found in
Refs. 11 and 12. Empirically we have tested samples and find little
effect on grinding finer than ~ 325 mesh (particle size < 50 μ).

Calibration Absorber Homogeneity

     It is difficult to ensure the homogeneity of a mixture of
pyrite and graphite. Experimentally we can obtain an indication of
sample non-uniformity by monitoring the γ-ray transmission through
various parts of the absorber. This is easily done using a rate-
meter to monitor the output of the spectrometer and the averaging
sample holder, described below.
     We believe that the homogeneity problem might make it advan-
tageous to use a HCl-leached coal sample as an alternative calibration standard. The HCl treatment will usually remove all non-

pyritic Fe in the sample, and in fact it is easy to check from the Mössbauer spectrum that pyrite is the only Fe-containing phase in the HCl-leached coal. The coal sample is then analyzed for <u>total</u> Fe by any standard analytical technique. The advantage of <u>this</u> procedure is that the coal will often have (or can be so chosen to have) finely divided pyrite. The crushed coal used in the calibration will then have fine pyrite particles intimately attached to coal particles. Thus sample segregation due to density differences (and granularity effects) would be reduced. We would also avoid the issue of using mineral pyrite as a calibration material for pyrite formed during coalification (see below).

Mineral Pyrite and Pyrite in Coal

We may question the use of mineral pyrite as a calibration standard for pyrite in coal. For example, it has been reported[13] that the X-ray diffraction pattern of coal pyrite has somewhat different line intensity ratios compared to mineral pyrite. Suggestions have been made that the pyrite in coal is either more disordered (i.e. less crystallized), or very fine particle, or both. These phenomena probably do occur in some coals. We may consider what the result of such effects would be on the Mössbauer absorption.

The Mössbauer absorption intensity is known[7-9] to be proportional to the so-called recoil-free fraction f,

$$f = \exp \{-4\pi^2 \; <x^2> \; /\lambda^2\} \tag{1}$$

In (1), λ is the wavelength of the Mössbauer γ radiation, and $<x^2>$ is the mean square amplitude of atomic vibration in the direction of the γ-ray. It follows that for "soft" materials, $<x^2>$ is large and f is small. If the pyrite in coal were "softer" than mineral pyrite, then a calibration curve based on assuming identical f factors for these materials would be in error. Since $<x^2>$ would be greater for a disordered or fine particle system, f would be small and the estimated pyrite content too low.

From (1)

$$f_{LN_2}/f_{room} = \exp \left[4\frac{\pi^2}{\lambda^2}(<x^2>_{room} - <x^2>_{LN_2}) \right] \tag{2}$$

where LN_2 denotes liquid nitrogen temperature. Evaluations of $<x^2>$ are possible in terms of various models of lattice dynamics. We will not attempt these calculations since the models are only approximate. Instead it is sufficient to note that $<x^2>_{T_1} - <x^2>_{T_2}$ will be larger for a "softer" material where say T_1 = room temperature, T_2 = liquid nitrogen temperature. Hence if we compare

$$\frac{f_{LN_2}}{f_{room}} = \frac{\text{absorption at liquid nitrogen temperature}}{\text{absorption at room temperature}} \tag{3}$$

we can establish whether or not pyrite in coal is soft.

To check this hypothesis we have set up a system to measure the Mössbauer spectra of a coal (PSOC 287) and mineral pyrite at room and liquid temperatures. In Table 3 we give data at room and LN_2 for these materials. Great care was taken to ensure that the absorber sample configuration was undisturbed between the room and LN_2 measurements. The degree of repeatability (~1%) is apparent on comparing "before" and "after" room temperature measurements. From the data it is clear that the pyrite in PSOC 287 is not soft since f_{LN_2}/f_{room} is the same (within experimental error) for both samples. Thus we can rule out the "soft pyrite in coal" hypothesis.

TABLE 3

Change in Spectral Intensity Upon Cooling Mineral Pyrite and Coal to Liquid Nitrogen Temperature

Run #	Sample	Temperature	Relative Area	$\frac{\text{area } (77^{\circ}K)}{\text{area (room)}}$
1105	PSOC 287 (Bevier seam coal) 96.6 mg/cm^2 -200 mesh	room	11.40	1.109
1106		$77^{\circ}K$	12.65	–
1107		room	11.42	1.108
1108	9.78 mg/cm^2 crushed FeS_2 (pyrite) -325 mesh +101.5 mg/cm graphite	room	11.05	1.100
1109		$77^{\circ}K$	12.16	–
1110		room	10.89	1.116

Effect of Ash

As noted in the Introduction, it is necessary to take account of non-resonant absorption of the γ-ray by ash (and even the pyrite) in some samples. The non-resonant absorption correction becomes significant for high ash or high FeS_2 contents in the absorber specimen. The effect of this absorption is graphically illustrated in Fig. 7 where we plot (open circles) "as-measured" resonant absorption of coal PSOC 287 vs amount coal in the absorber. Note the bend-over of the curve at higher coal densities. This phenomenon derives from the effect of non-resonant scattering by the ash (22 wt % for this coal). Briefly, the bend-over arises from the fact that the ash reduces the ratio of resonant (Mössbauer Effect) to non-resonant (background noise) γ-rays. This ratio can be

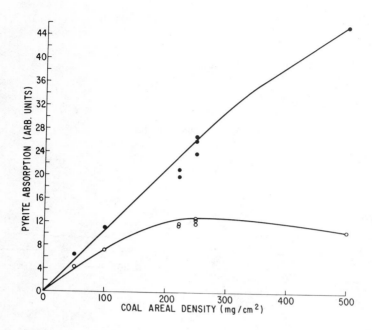

Fig. 7. Mössbauer absorption intensity vs. amount of coal in
absorber. Open circles = as measured. Solid circles =
corrected for non-resonant absorption.

presumed constant only if the number of scattering nuclei in the
beam is small.

In Fig. 8 we give a pulse height spectrum of the γ-ray beam
after passage through a typical sample. The 14.4 keV peak
(Mössbauer γ-ray) is strong but sits upon the background shown.
The ratio of resonant to non-resonant γ-ray is given by the ratio
of peak to background areas which will vary for different ash contents in the beam.

To correct for the ash scattering we have written a computer
program to estimate the background curve and evaluate the ratio R =
(peak area)/(peak + background areas). The corrected curve (= "as
measured" absorption/R) is given by the solid circles in Fig. 7.
Note that the bend-over is removed and a monotonic, almost linear
relation exists between the corrected spectral area and the coal
areal density.

Averaging Sample Holder

To reduce sampling errors it is advantageous to use as large
a sample as feasible for analysis. We have constructed an "averaging" sample holder, a schematic of which is given in Fig. 9.
Basically the coal (or calibration) absorber is contained in an

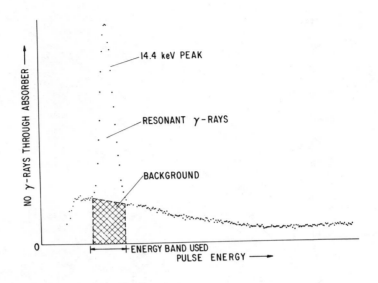

Fig. 8.   Pulse height spectrum of γ-ray beam after passage through a coal sample.

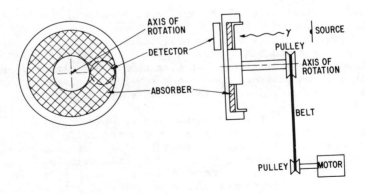

Fig. 9.   Schematic diagram of averaging sample holder.

annulus, the center of the annulus being to the side of the γ-ray beam. The annulus rotates during the course of the analysis such that the γ-ray beam intersects every part of the sample. The size of the annulus and its location in the beam is such that no part of the absorber is not probed. Additionally to minimize potential vibrational disturbance we have chosen to rotate the annulus with a low duty cycle, i.e., the annulus is stationary say 99% of the time and then rotates slightly to move to a new position.

The adequacy of the averaging sample holder technique is verified by the data of Fig. 10 where we present a plot of absorption intensity versus sample weight for HCl-leached PSOC 287 coal. Note the excellent linearity of the curve. The coal sample was milled prior to measurement. The area of the sample holder is approximately 50 cm². Thus the total weight of sample analyzed in Fig. 10 varies from 1.25 to 12.5 grams.

Sample Preparation Technique

    1.   <u>Coal</u>   In view of the results obtained we have adopted the following as preparation technique for coal absorbers for Mossbauer Spectroscopy.

        i. Mill the coal using say Al_2O_3 balls in a polyethylene

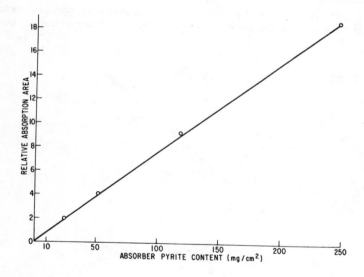

Fig. 10. Variation in absorption intensity with absorber thickness using averaging sample holder. The data are corrected for the effect of non-resonant absorption by the ash.

jar.

ii.   Screen coal to -60 mesh and remill residue if
      necessary.

iii. Dry coal at 105°C under N_2 to prevent pyrite
      oxidation.

iv.   Tumble coal in a polyethylene jar with a twisted
      metal strip to 60 min.

v.   Cone and select an appropriate amount of coal
      (~ 5 - 10 gms typically) for analysis in the
      averaging sample holder.

2. Calibration Standards

i.   Mill mineral pyrite using Al_2O_3 balls in a
      polyethylene jar.

ii.   Screen to -325 mesh (or smaller) and remill residue
      if necessary.

iii. Cone sample and select particles for chemical
      analysis to determine true Fe content.

iv.   Tumble FeS_2 with an appropriate amount of spectros-
      copically pure -100 mesh graphite.

v.   Use averaging sample holder.

Calibration Curve

We have prepared a variety of FeS_2 calibration samples using
the techniques described above. Samples were either crushed to

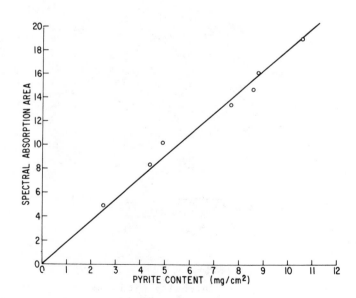

Fig. 11.   Calibration curve of Mössbauer absorption area vs. mineral
           pyrite content.

<50μ or <20μ size before analysis. No significant differences were found. The resultant calibration curve is given in Fig. 11. The points plotted are corrected for the effect of non-resonant absorption and for mineral impurities in the pyrite. A rather good linear relation is obtained, indicating the success of the sample preparation procedure outlined above.

4. COMPARISON WITH WET CHEMICAL ASTM ANALYSIS

To properly compare the Mössbauer and ASTM techniques we have chosen to carry out Mössbauer and wet chemical analyses for pyrite using the _same_ samples of well homogenized coals. The experimental procedure adopted is as follows:

1. Mill coal to -60 mesh.
2. Dry at 105°C under nitrogen.
3. Obtain a representative sample by coning.
4. Measure coal "as is" by Mössbauer.
5. Leach with HCl as per ASTM D2492-77. The filtrate is then analyzed for S and Fe ("sulphate analysis").
6. Dry the residual coal and reweigh (we have verified that handling losses are 1%).
7. Carry out Mössbauer spectroscopy on the HCl-leached coal.
8. Leach with HNO₃ per ASTM D2492. The filtrate is analyzed for Fe ("pyritic sulfur").
9. Dry residual coal and reweigh.
10. Carry out Mössbauer spectroscopy on the HNO₃-leached coal.

In the above procedure about 5-10 gms of coal are utilized. The _entire_ sample is analyzed, both chemically and in the Mössbauer setup using the "averaging" technique. Thus, sampling errors should be minimized.

The relevant results are given in Table 4. A number of features should be emphasized with regard to the data:

a. The reproducibility is good (generally 5-10%) on repeated runs of the same sample. Repeats on different samples of the same batch are also consistent provided a sufficiently large sample size is used and sufficient care is taken to homogenize the sample.

b. Samples of PSOC 287 and Joanne 8178 contain small amounts of ferric (trivalent) iron sulphate (Jarosite). We have carried out least squares constrained computer fits of the spectra and estimate that the amounts of Jarosite are about 10% of the pyrite content in these samples. We have excluded the Jarosite from the data of Table 4.

c. All the other samples have greater or lesser amounts of $FeSO_4 \cdot H_2O$ (Szomolnokite). This phase is similarly excluded from the data of Table 4.

d. All samples show same (5-15%) decrease in absorption

226

Table 4. Mossbauer and ASTM Results For a Variety of Coals.

Spectrum No.	Type of Sample		Mössbauer Pyrite Absorption(1	ASTM Analysis		Notes
				Wt% Sulphates(2	Wt% Sulphates(2	
1169	PSOC 287,	as received	7.89			
1170	Bevier Seam Coal	as received	7.49			
1173		HCl treated	6.75			
1174		HCl treated	6.57			
1175		HCl treated	6.44			
1183		HCl treated	6.73			
1188		HNO3 treated	ND	0.51	1.86	
1171	PSOC 2161,	as received	6.58			
1176	Kentucky No. 14	as received	7.01			
1182	Coal	as received	6.55			
1184		HCl treated	6.34			
1190		HNO3 treated	ND	0.67	1.82	
1191	Joanne 8178,	as received	2.76			
1193(3	Pittsburgh	HNO3 treated	0.11			
1195(4	Seam Coal	as received	2.70			
1198		HCl treated	2.38			
1200(5		HNO3 treated	0.61	0.12	0.50	0.69 when corrected by Mössbauer spectroscopy
1192	W. Virginia 891	as received	4.41			
1194	Waynesburg,	HCl treated	3.95			
1197(5	Upper Bench	HNO3 treated	0.42	0.20	0.82	1.07 when corrected by Mössbauer spectroscopy
1199	W. Virginia 1514,	as received	3.86			
1201	Waynesburg,	HCl treated	3.52	0.21	1.08	
1203	Lower Bench Coal,	HNO3 treated	ND			

Information for Table 4:
1) Excludes any trivalent iron sulphate (Jarosite), or divalent iron sulphate. Sample thickness = 100 mg/cm^2. Units measure spectral area.
2) By ASTM chemical analysis.
3) Treated directly with HNO_3.
4) New sample.
5) Not boiled.
Notes: 1. ND = no pyrite phase detected.
2. Mossbauer Data given in arbitrary units.
3. Table 5 summarizes key parts of data with all values in wt % sulphur.

intensity upon treatment with HCl. There are 2 possible reasons for this decrease.
1. Some of the pyrite in coal is soluble in HCl.
2. There is an unknown phase in coal, soluble in HCl with Mössbauer absorption parameters indistinguishable from the Mössbauer parameters of pyrite. We have no evidence for this latter hypothesis.
We would propose to avoid the above issue by measuring the pyrite absorption upon HCl-leached coal. Such a leaching is simple to perform and is necessary in any event to check for sulphate in coal (which could be, for example $CaSO_4$ or other sulphates not detectable using Mössbauer). It also has the real advantage of simplifying the analysis by removing most potential interfering phases (e.g., iron carbonates, illites, other iron-bearing clays, etc.)
e. Boiling the coals tested in HNO_3 removes all the Fe but soaking the coal overnight in HNO_3 leads to a significant (~ 25%) error. We note that the latter procedure is permissible by the ASTM pyrite analysis procedure.
In Figs. 12, 13, and 14 we give the Mössbauer spectra of Joanne 8178 as received, HCl leached and HNO_3 leached (not boiled), respectively. The position of the ferric sulphate peaks are marked S-S in Figure 11 and are just discernable by sighting along the graph at an angle. The intensities of the sulphate peaks were determined by using a constrained least-squares iterative computer fitting program which we have specially adapted for this purpose. The data of Table 4 are corrected for the sulphate peaks, i.e., the sulphate intensities are not included in the column labeled "Mössbauer Pyrite Absorption". Note also that the sulphate, as expected, is not discernable in the HCl-leached sample (Fig. 13). More importantly Fig. 14 indicates that treatment of the coal overnight in the ASTM recommended HNO_3 solution does not remove all the pyrite.
In Figure 15 we plot the Mössbauer absorption vs. pyritic sulphur for the HCl treated samples of Table 4. Where appropriate (i.e., where the HNO_3 did not completely dissolve the FeS_2) we have plotted the corrected value of the pyritic sulphur from Table 4.

228

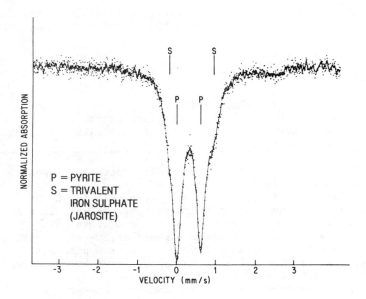

Fig. 12. Normalized Mossbauer spectrum of Joanne 8178, as received.
P = pyrite, S = trivalent iron sulphate (Jarosite).

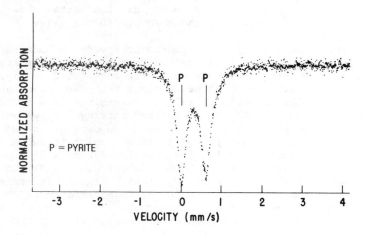

Fig. 13. Normalized Mössbauer spectrum of Joanne 8178, HCl-leached.

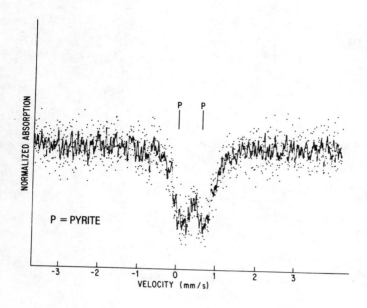

Fig. 14. Normalized Mössbauer spectrum of Joanne 8178, leached overnight in HNO$_3$ (not boiled) P = pyrite.

It is interesting to compare the calibration curve of Figure 15 obtained from Mössbauer and ASTM chemical analysis as corrected by Mössbauer, with the calibration curve previously obtained. Figure 15 predicts a Mössbauer absorption of 3.5 units per 100 mg/cm^2 of coal where the pyritic sulphur content of the coal is 1.0 wt %. Such a coal would have 1 mg/cm pyritic sulphur or equivalently 1.87 mg/cm^2 pyrite in coal = 3.4, i.e., the agreement is <u>excellent</u>, giving some confidence in both results.

We can also compare the Mössbauer and ASTM data as in Table 5. The "Mössbauer" column gives the weight % pyrite in the HCl-leached coal (i.e., sulphate extracted) as determined from the coal absorption spectrum and the calibration curve of Fig. 11. The "ASTM" column is the data using the wet chemical technique. The "ASTM corrected" column gives the ASTM data corrected where necessary for incomplete dissolution of the pyrite by the HNO$_3$. The agreement is generally excellent with the corrected ASTM data, giving some confidence in the Mössbauer technique.

5. COUNTING TIME

Sample analysis time is a matter of concern for the practical application of Mössbauer affect spectroscopy to the measurement of pyrite in coal. In this Section we shall consider the time neces-

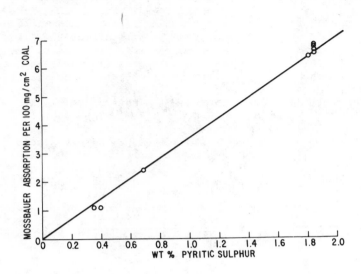

Fig. 15. Mössbauer absorption intensity vs. pyritic sulphur for
HCl-leached samples of Table 4.

TABLE 5. Comparison of Pyrite Sulphur Content by Mossbauer and
ASTM Methods

Coal Type	Wt Percent Pyritic Sulphur		
	Mossbauer	ASTM	ASTM Corrected
PSOC 287	1.96	1.86	–
PSOC 216	1.89	1.82	–
Joanne 8178	1.71	0.50	0.69
West Virginia 891	1.17	0.82	1.07
West Virginia 1514	1.05	1.08	–

sary to obtain an adequate measurement of pyrite content in a "typical" coal.

Since the Mössbauer spectrum is obtained by counting the number of γ-rays absorbed by the sample at various source velocities, it is clear that for extremely short counting times, random statistical fluctuations in the ^{57}Co decay process will affect the signal-to-noise ratio. A minimum counting time can be determined by requiring that the measurement error due to statistical fluctuations not be significant compared to other experimental errors. The calibration curve of Fig. 11 indicates experimental fluctuations in the 5% range. We would thus require that statistical errors be <5%. To evaluate the effect of count-rate-induced variations we have carried out a series of experiments on a single sample of coal in which data are accumulated for various counting times. The sample chosen was obtained from PSO7 287 by milling to -200 mesh and leaching with HCl to remove non-pyritic Fe. Ash content of this coal was about 20% and pyritic sulphur was about 2 wt %. Absorber thickness was 118 mg/cm^2.

In Table 6 we list the absorption areas and counting times for this sample. The absorption area was evaluated both using non-linear least squares iterative computer fitting and also by simple

TABLE 6. Effect of Decreased Counting Time Upon Accuracy of Mössbauer Absorption Area Determination

Spectrum No.	Counting Time (Hours)	Absorption (Computer)	Absorption (Integration)	3σ
1222J	63.0	7.07	6.86	.02
1222	20.3	7.18	6.82	.04
1222M	16.2	6.96	6.81	.04
1222D	16.3	7.13	6.84	.04
1222R	16.4	6.99	6.78	.04
1222G	4.0	7.11	6.88	.08
1222C	2.2	7.13	6.68	.1
1222B	1.0	7.29	6.96	.2
1222K	1.0	7.04	6.61	.2
1222L	1.0	7.2	6.53	.2
1222E	0.2	6.59	6.33	.3
1222F	0.42	7.09	6.96	.3
1222A	0.33	7.39	7.06	.3
1222H	0.05	6.91	6.01	.6
1222I	0.05	7.43	5.68	.6
1222N	0.05	6.23	8.51	.6
1222P	0.05	7.25	5.12	.6
1222Q	0.05	8.99	6.15	.6

numerical computation of the area under the curve. In the right-
most column we list the relative "3 sigma" value for the absorption
area. This is a measure of the statistical error (derived from the
computer fitting routine). It has no significance as far as true
experimental error is concerned except as a lower limit.

Some idea of the meaning of this error limit will be evident
from the following approximate analysis: we consider spectrum
#1222L. For this sample the average count rate was $N=2.5 \times 10^4$
counts per channel, there were 1024 channels, and the integrated
peak absorption area $A = 1.2 \times 10^5$ counts. A three sigma counting
error of the total counts would be $E = 3 \sqrt{1024N} = 1.5 \times 10^4$.
Actually, this value should be doubled since when we evaluate the
absorption we are subtracting 2 numbers, each subject to counting
statistical fluctuations. Hence, the relative statistical error
for the spectrum is approximately $2 E/A = 0.25$. The computer value
(Table 6) is 0.2.

In Figure 16 we plot the values of the absorption area versus
counting time. The statistical error is comparable with the
observed scatter. The distribution in pyrite content evident from
Fig. 16 is in fact quite small (note scale). We also note that a
counting time of 1 hour gives an experimental spread of $\pm$ 2% in the
value obtained. Since our experimental arrangement had a rather
weak γ-ray source (count rate $\approx 8 \times 10^3$ hz in the 14.4 keV line) we
could do significantly better using available electronics and a

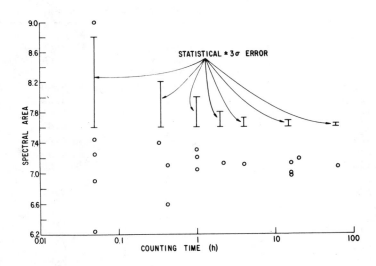

Fig. 16. Mössbauer absorption vs. counting time for a fixed sample.

stronger source. A count rate of 4×10^4 hz is quite feasible and would imply a spectrum measurement time of about 10 minutes. This value could probably be reduced even further by optimization of the spectral scan region or using a special high rate nuclear channel. However, 10 minutes per analysis would seem to be quite reasonable for routine work. In addition it is possible to configure a Mossbauer Spectrometer to measure more than one sample at a time which would further decrease the measurement time.

6. SUMMARY

This study has demonstrated that Mössbauer Spectroscopy can be employed as a viable technique to monitor the amount of pyrite sulphur in coal. The Mössbauer method is quick, nondestructive and adaptable to automation. It is straightforwardly applicable to weathered and processed coals as well as raw coals.

A number of difficulties hindering the development of a reliable measurement technique have been addressed and satisfactorily resolved. A simple correction method for coal ash absorption has been outlined; a sample holder capable of measuring 1-10 grams of coal has been constructed and tested; sample homogenizing techniques consistent with the spectroscopic requirements have been evaluated.

Two different calibration curve development methods have been evaluated. The first is based upon the measurement of mixtures of pyrite and spectroscopic graphite, the second upon the measurement of HCl-leached coal. The calibration curves are self consistent and both methods give similar results.

A comparison of pyrite analysis by Mössbauer and by ASTM methods shows agreement provided the ASTM method is corrected for residual undissolved pyrite as necessary. The Mossbauer method also indicates some dissolution of pyrite by HCl. This might be reported as sulphate by the ASTM method but the issue is not finally resolved. An estimate of measurement errors and spectrometer capability indicates that 10 minutes/sample is a viable measurement time for routine sample analysis.

On the basis of this study we would recommend that Mössbauer spectroscopy be utilized as a routine measurement of pyrite content in coal. Initially the technique could be complementary to standard wet chemical methods. As enough confidence and experience develops with use, thought should be given to its utilization as a primary method for measurement of pyritic sulphur in coal.

ACKNOWLEDGEMENTS

The author thanks I. S. Jacobs and J. J. Renton for helpful discussions during the course of this work. J. Cocoma provided competent technical assistance. J. J. Renton and L. A. Riva supplied coal samples used in this study.

234

REFERENCES

1. A. H. Edwards, J. M. Jones and W. Newcombe, Fuel 43, 55 (1964).
2. M. S. Burns, Fuel 49, 126 (1970).
3. R. T. Greer, In "Scanning Electron Microscopy/1977", O. Johari Ed., Vol. 1, pp. 79-93. I. T. T. Research Institute, Chicago.
4. L. M. Levinson and I. S. Jacobs, Fuel 57, 592 (1978).
5. P. A. Montano, Fuel 56, 397 (1977).
6. G. P. Huffman and F. E. Huggins, Fuel 57, 592 (1978).
7. G. K. Wertheim, "Mossbauer Effect: Principles and Applications", Academic Press, (1964).
8. L. May, "An Introduction to Mossbauer Spectroscopy", Plenum Press, (1971).
9. "Applications of Mossbauer Spectroscopy", R. L. Cohen Ed., Vol. 1, Academic Press (1976).
10. G. Lang, Nucl. Instr. Methods 24, 425 (1963).
11. J. D. Bowman, E. Kankeleit, E. N. Kaufmann and B. Persson, Nucl. Instr. Methods 50, 13 (1967).
12. A. J. Stone, Nucl. Instr. Methods 107, 285 (1973).
13. J. Renton, private communication.

SPECTROSCOPIC STUDIES OF CO CHEMISORPTION
ON TRANSITION METALS AND STUDIES OF THE
KINETICS OF CH_4 SYNTHESIS*

John T. Yates, Jr., D. W. Goodman, R. D. Kelley, and T. E. Madey
Surface Science Division, National Bureau of Standards
Washington, D. C. 20234

ABSTRACT

The reaction of CO and H_2 to produce CH_4 has been studied at
high pressures over Ni(100) and Ni(111) surfaces using a micro-
catalytic reactor system and Auger Spectroscopy. It has been found
that the rate per surface Ni atom is identical on the two crystal
planes and is very similar to the specific rate on high area supported Ni catalysts. In addition, the activation energy is independent of Ni catalyst structure. Auger spectroscopy has been used
to characterize the surface carbon produced by CO decomposition on
Ni, and it is concluded that a carbidic form of carbon is an intermediate in the catalytic synthesis of CH_4 over Ni.

*Work supported by DOE - Division of Basic Energy Sciences

0094-243X/81/70235-1$1.50   1981 American Institute of Physics

MECHANISMS OF CATALYZED GASIFICATION OF CARBON

D. W. McKee

Corporate Research and Development Center
General Electric Company, Schenectady, New York 12301

ABSTRACT

Various mechanisms have been invoked in the past to explain
the effects of catalysts on the gasification reactions of carbon.
However in many cases the observed catalytic behavior can be
interpreted in terms of oxidation-reduction cycles occurring on the
carbon surface. This paper describes how the techniques of thermal
analysis, together with the predictions of thermodynamics, have
proved useful in identifying the sequential steps involved in the
catalytic processes.

INTRODUCTION

It has been known for many years that small amounts of a wide
variety of inorganic impurities can effectively catalyze the
reactions of carbon with gases such as oxygen, carbon dioxide,
steam and hydrogen[1,2]. The catalyzed gasification of carbon has an
extensive literature which however is often confusing and
contradictory. In recent years there has been a revival of interest
in this area in connection with the gasification of coal and, with
the advent of new experimental techniques and the detailed study
of individual catalysts, some degree of understanding has been
brought to this complex but important field[3]. This paper presents
a brief summary of results obtained in the author's laboratory on
the mechanistic aspects of the catalytic gasification reactions.
Graphite, in the form of natural single crystals for hot stage
microscopy studies or high purity powder for kinetic measurements,
was chosen for these investigations as it is a relatively well-
defined reproducible crystalline material. It must be recognized
however that the behavior of graphite may not be identical with
that of other carbonaceous materials, such as coal-derived chars
or other porous carbons which show considerable inherent
variability and complexity in physical structure.

THE GASIFICATION OF REACTIONS OF CARBON

The reactions of relevance and some of the species that have
consistently proved to be active catalysts are listed in Table I.
The oxidation reactions (A) and the hydrogenation reaction (D) are
exothermic, whereas the Boudouard reaction (B) and the steam
gasification reaction (C) are endothermic. The kinetics of the
reactions however differ considerably. For example, at a typical
temperature of 800°C, in the absence of catalysts, the rate of the
$C-O_2$ reaction is about five orders of magnitude faster than the
$C-H_2O$ and $C-CO_2$ reactions and about eight orders of magnitude
faster than the $C-H_2$ reaction. The uncatalyzed steam reaction (C)
is generally slightly faster than the $C-CO_2$ reaction (B).

As indicated, many different types of inorganic materials act as catalysts for these reactions. Reactions (A) are promoted readily by oxides of the transition metals, by metallic silver and the platinum group metals. They are also catalyzed by certain salts (particularly the oxides and carbonates) of the Group IA alkali metals and the Group IIA alkaline earths, which also catalyze reactions (B) and (C). The hydrogenation reaction (D), which is the least thermodynamically favorable of the series, is mainly catalyzed by metals. The details of catalyst behavior for the individual reactions will be discussed in the following sections.

THE CATALYZED $C-O_2$ REACTION

A unique feature of the catalyzed oxidation of graphite, which is also often observed with the other gasification reactions, is the localized pitting and channeling of the graphite substrate commonly associated with the movement of catalyst particles on the graphite basal plane surface during the reaction. This effect has been much studied on graphite single crystals, using controlled atmosphere hot stage optical microscopy[4,5] and electron microscopy[6,7]. The driving force for the particle motion is still somewhat mysterious and has been often discussed. Figure 1 shows

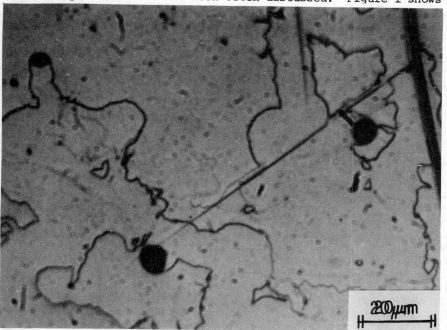

20 μm

the channels produced by mobile Co/CoO particles on the surface of

Fig. 1. Channeling produced on natural graphite flake by Co/CoO catalyst particles during oxidation at 700°C in O_2[8]

TABLE I
Carbon Gasification Reactions

		ΔH°_{298K} kJ/mole	ΔG°_{1200K} kJ/mole	Catalysts
A	$C + O_2 = CO_2$	−393.7	−396.3	Oxides of V, Mn, Fe, Co,
	$C + 1/2\ O_2 = CO$	−110.6	−217.9	Ni, Cu, Ag, Pt metals, Na_2CO_3, BaO_2
B	$C + CO_2 = 2CO$	+172.5	−39.4	Fe, Ni, Co, Pt metals Na_2CO_3, $BaCO_3$
C	$C + H_2O = CO+H_2$	+131.4	−36.3	Fe, Ni, Co, Pt metals K_2CO_3, $BaCO_3$
D	$C + 2H_2 = CH_4$	−74.9	+41.4	Fe, Co, Ni, Pt metals

crystalline graphite during oxidation at 700°C[8]. In general, in the case of catalyzed oxidation, the smallest particles exhibit the most vigorous motion with agglomerates above about 10 μ in size being immobile and inactive. The moving particles often resemble liquid droplets and show surface tension effects and coalescence, even at temperatures substantially below the bulk melting point of the catalyst phase. Channeling, which generally becomes noticeable at temperatures exceeding one half the melting point of the catalyst (i.e. above the Tammann temperature where lattice atoms become mobile) often takes place preferentially in the <11$\bar{2}$0> directions on the graphite basal plane and usually begins when the moving particles encounter a step on the surface. Very detailed studies of these effects have been made by controlled atmosphere electron microscopy[7,9].

In the past there have been two general mechanisms proposed to account for the catalytic effects of metals, oxides and salts in the C-O_2 reaction. The electron-transfer theory, proposed by Long and Sykes in 1950[10], postulates that electron transfer to or from the catalyst entity to the carbon substrate results in a redistribution of π-electrons, a weakening of the C-C bonds at edge sites on the graphite layer planes and an increase in C-O bond strength during the catalyzed oxidation reaction. Such electron transfer processes can certainly play a role in determining the electronic properties of graphite intercalation compounds and there have been recent attempts[11] to account for the catalytic behavior of alkali metal carbonates on the basis of an electron transfer or electrochemical mechanism[12]. The alternative oxygen-transfer mechanism regards active catalysts as oxygen carriers which undergo oxidation-reduction cycles on the carbon surface during the reaction. As originally proposed by Kröger et al. in 1931[13], an intermediate compound, such as a metal oxide, may promote the transfer of oxygen from the gas phase to the carbon substrate. As shown in the discussion of individual catalysts which follows, the concept of specific oxidation-reduction cycles can explain many of the features of the behavior of known catalysts and the approach may have predictive value.

Transition Metal Oxide Catalysts

Standard free energies of formation of a number of metal oxides and also of CO at pressures of 1, 10^{-3} and 10^{-5} atms. (10^5, 10^2, 1 Pa) are shown in Figure 2 as functions of temperature. Reduction of metal oxide to metal by reaction with carbon,

$$MO + C = M + CO$$

is possible in the case of Cu, Pb, Ni, Co and Fe oxides at temperatures above a threshold of 700°C or below. With low ambient partial pressures of CO, the reduction of ZnO is also possible. All these metal oxides are active catalysts for carbon oxidation[8,14], whereas more stable oxides such as Al_2O_3 or MgO, which are not reduced by carbon below 1000°C, are inactive. In other cases, for example with V_2O_5, Co_3O_4, Pb_3O_4 and As_2O_5, reduction to a lower oxide by reaction with carbon is thermodynamically feasible. These oxides are also very effective as carbon oxidation catalysts. In many cases it is possible to identify the intermediate oxide species formed during the catalyzed oxidation reaction. Thus, if a

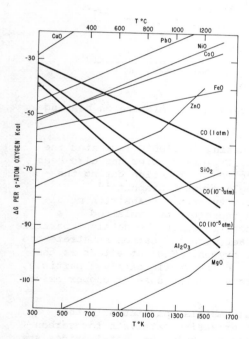

Fig. 2. Standard free energies of formation of metal oxides and CO as a function of temperature and P_{CO}.

mixture of V_2O_5 and graphite powder is heated in an inert atmosphere, a rapid reaction occurs at 675°C, the melting point of V_2O_5, as shown by the thermogram in Figure 3. X-ray analysis

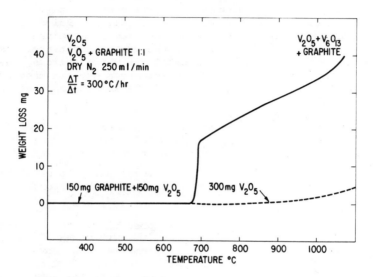

Fig. 3. Weight changes of a V_2O_5-graphite mixture on heating in N_2[8]

of the residue obtained after heating to 1100°C indicated the presence of the lower oxide V_6O_{13}. This species has also been identified by selected area electron diffraction during the V_2O_5-catalyzed oxidation of single crystal graphite[15].

The observed catalytic activity of the transition metal oxides in the $C-O_2$ reaction can therefore be explained by a cyclic oxidation-reduction process in which the catalyst particle is sequentially reduced by contact with the carbon substrate and then re-oxidized by the ambient oxygen. The net effect is the gasification of the carbon beneath the mobile catalyst particle, as shown schematically in Figure 4 for the case of copper oxide.

Alkali Metal Salts

Salts of the Group IA alkali metals, particularly the carbonates and oxides, are very effective catalysts for carbon oxidation. In these cases it seems likely that metal oxides are the active species during the catalyzed reaction. Thermodynamic considerations[16] indicate that in the presence of solid carbon, the Na_2CO_3/Na_2O equilibrium,

$$Na_2CO_3 + C + O_2 = Na_2O + 2CO_2$$

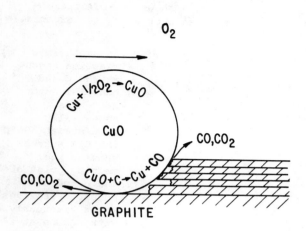

Fig. 4. Schematic representation of
the catalytic behavior of a mobile CuO
particle during the oxidation of graphite.

is shifted towards
the oxide so that
above 600°C free
Na_2O can form over a
wide range of P_{O_2} and
P_{CO_2}. Formation of
oxide by this
reaction can be
demonstrated by
heating the alkali
carbonates in oxygen
in the presence and
absence of added
carbon. Table II
shows the results of
analysis of the
residues remaining
after 900°C fusion of
the Group IA
carbonates in O_2 both
alone and with added
graphite. Small
amounts of oxides were formed by thermal decomposition of the
carbonates at this temperature, but substantial concentrations
were produced in the presence of the graphite. Similar
experiments with alkali metal sulfates and chlorides showed that
small amounts of free oxides were formed on heating these salts
with graphite in an O_2 atmosphere at 900°C.

Table II

Formation of oxides from alkali carbonates

Carbonate M_2CO_3	M_2O concn. in original M_2CO_3	M_2O concn. after heating M_2CO_3 in O_2 for 4 h at 900°C	M_2O concn. after heating M_2CO_3 + graphite (1:1) in O_2 for 4 h at 900°C
Li	< 0.2 wt.%	2.1 wt.%	4.6 wt.%
Na	< 0.2	0.2	1.1
K	< 0.2	0.6	7.6
Rb	< 0.2	1.6	6.6
Cs	< 0.2	3.2	7.7

242

The Group IA monoxides react with O_2 at elevated temperatures with the formation of peroxides. Figure 5 shows TGA-DTA thermograms for a sample of Na_2O heated in O_2 at increasing temperature.

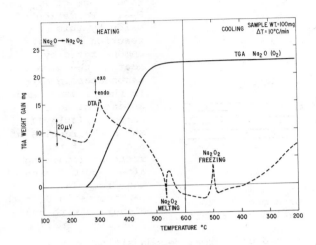

Fig. 5. TGA-DTA heating and cooling curves for Na_2O heated in flowing oxygen[16]

An exothermic reaction began at 250°C and conversion to Na_2O_2 was essentially complete at 500°C the DTA curves showing the melting point of the peroxide phase. Peroxides react vigorously with carbon at this temperature so that a likely sequence of reactions for the alkali metal oxide-catalyzed gasification of carbon is

$$M_2O + n/2O_2 = M_2O_{1+n}$$

$$M_2O_{1+n} + nC = M_2O + nCO$$

$$\overline{C + 1/2\ O_2 = CO}$$

THE CATALYZED C-CO_2 REACTION

Alkaline Earth Carbonates

The carbonates of the Group IIA alkaline earth metals, especially those of Ba and Sr, are active catalysts for the C-CO_2 reaction at temperatures above about 700°C[17]. Figure 6 shows Arrhenius plots of gasification rates vs. 1/T in 1 atm. (10^5Pa) CO_2 for pure graphite and graphite doped with 5 percent of the Group IIA carbonates. At this concentration the Ba and Sr salts increased the reaction rate by about three orders of magnitude and reduced the apparent activity energy from 383 to 245 kJ/mole. The effects of Ca and Mg carbonates were less dramatic.

When equal amounts of $BaCO_3$ and graphite powder were heated

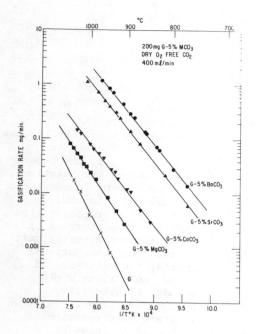

together in an inert atmosphere, a rapid weight loss occurred above 900°C, as shown in the thermogram in Figure 7. The weight loss resulted from the solid state reaction,

$$BaCO_3 + C = BaO + 2CO$$

which proceeded to completion above 1000°C. Whereas $SrCO_3$ showed similar behavior, $CaCO_3$ was reduced by carbon only to a small extent and $MgCO_3$ decomposed completely to MgO below 600°C without reacting with carbon. As the Ba and Sr oxides once formed are readily converted back to the carbonates by reaction with ambient CO_2, it appears likely that the catalyzed gasification reaction involves the sequential steps,

Fig. 6. Effects of additions of 5 wt.% Mg, Ca, Sr and Ba carbonates on the gasification rates of graphite in 1 atm. CO_2. Rates vs. $1/T°K$.[17]

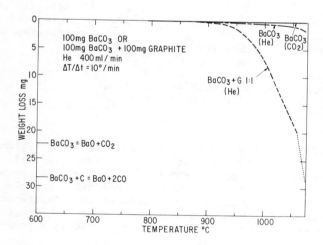

Fig. 7. Thermogravi-metric heating curves for $BaCO_3$ in He and CO_2 and a 1:1 $BaCO_3$-graphite mixture in He. Weight losses vs. temperature[17].

$$MCO_3 + C = MO + 2CO$$

$$MO + CO_2 = MCO_3$$

$$C + CO_2 = 2CO$$

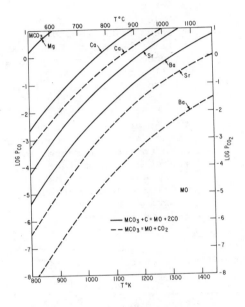

Figure 8 shows the equilibrium stability regions of oxide and carbonate corresponding to the occurrence of these two reactions, as functions of temperature, P_{CO} and P_{CO_2} (in atms.). These curves are calculated from the free energies of the reactions assuming unit activity for the various phases; the existence of mutual solubility would broaden the stability boundaries. The solid lines indicate that reaction of carbonate to oxide by reaction with carbon is possible over a wide range of P_{CO} and temperature for Ba and Sr. For example, with $P_{CO} = 10^{-3}$ atm. (10^2Pa), $BaCO_3$ can be reduced to BaO at all temperatures above

Fig. 8. Equilibrium stability regions of MO and MCO_3 corresponding to the reactions, $MCO_3+C=MO+2CO$ (solid curves) and $MCO_3=MO+CO_2$ (dashed curves) as functions of P_{CO}, P_{CO_2} (atms.) and temperature for M=Mg, Ca, Sr, Ba[17].

700°C. The dashed curves show the thermal stability regions of the pure salts. For example, with $P_{CO_2} = 1$ atm. (10^5 Pa), $BaCO_3$ will exist at all temperatures shown. Hence a cyclic process based on these two reactions is thermodynamically possible in the presence of CO_2 and low partial pressures of CO. On the other hand, $MgCO_3$ will not exist at all in the temperature range of interest, so it is not surprising that this salt is a poor catalyst for the $C-CO_2$ reaction.

Alkali Metal Salts

The Group IA alkali metal carbonates are among the most active

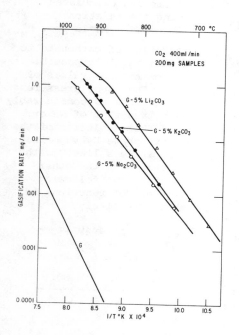

catalysts known for the $C-CO_2$ reaction. Studies using high purity graphite[16] have consistently shown that Li_2CO_3 is the most active species, with K and Na carbonates being somewhat less effective, as shown by the Arrhenius plots in Figure 9. The effect of the anion on the catalytic activity of a series of potassium salts is illustrated in Figure 10. The carbonate, sulfate and nitrate which tend to form oxide by thermal decomposition or by reaction with carbon, are the most effective additives, whereas stable species such as halides and silicates are less active as catalysts.

Fig. 9. Effects of additions of 5 wt.% Li, Na and K carbonates on the gasification rates of graphite in 1 atm. CO_2. Rates vs. $1/T°K$.

In these cases several distinct reaction paths are possible to account for the observed catalytic effects. For example, Figure 11 shows the equilibrium stability regions of oxide and carbonate based on the occurrence of the reactions,

$$M_2CO_3 + C = M_2O + 2CO$$
$$M_2O + CO_2 = M_2CO_3$$

as functions of temperature, P_{CO} and P_{CO_2}. As with the alkaline earth salts (Figure 8), this sequence of reactions is possible for low values of P_{CO} and P_{CO_2} = 1 atm. (10^5Pa) over a wide range of temperature for all the alkali metal salts. The first reaction is favorable in the case of Li and this, combined with the low melting point of the Li_2CO_3 phase, may explain the exceptionally high catalytic activity of the Li species. However there are other possible reaction paths. Vaporization of free alkali metal during the catalyzed $C-CO_2$ reaction is often observed[16,18] and, as shown by the equilibrium stability diagram

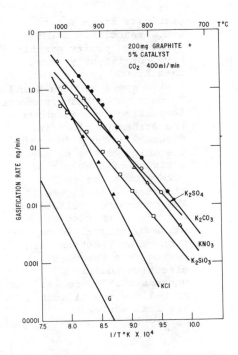

Figure 12, calculated using the relation,
$$\Delta G = \Delta G^\circ + RT \ln P_{CO}^3 \cdot P_M^2$$
reduction of carbonate to alkali metal is possible at low ambient values of P_{CO}. The vapor pressure of Li metal is considerably less than that of Na or K so that depletion of the active phase by vaporization is less likely with Li than with the other metals. The following sequence of reactions is therefore likely,

$$M_2CO_3 + 2C = 2M + 3CO$$
$$2M + CO_2 = M_2O + CO$$
$$\underline{M_2O + CO_2 = M_2CO_3}$$
$$C + CO_2 = 2CO$$

Fig. 10. Catalytic effects of potassium salts (5 wt.%) on the gasification of graphite in 1 atm. CO_2. Rates vs. $1/T^\circ K$.

Group VIII Metals

Fe, Co and Ni and also the metals of the Pt group are active catalysts for the $C-CO_2$ reaction and have been the subject of detailed investigations[19,20]. From magnetic susceptibility measurements, Walker and coworkers[2] demonstrated that in the case of Fe, the metal is the most active catalytic entity. The metal surface promotes the dissociative chemisorption of CO_2 to give chemisorbed oxygen atoms[21] which may then diffuse to the metal-carbon interface and there react with the carbon to yield gaseous CO,

$$Fe + CO_2 = Fe-O_{ads} + CO$$
$$Fe-O_{ads} + C = Fe + CO$$
$$\underline{\phantom{Fe-O_{ads} + C = Fe + CO}}$$
$$C + CO_2 = 2CO$$

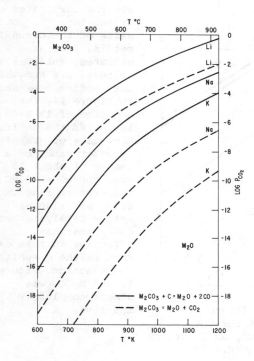

Localized gasification of the carbon takes place in the immediate vicinity of the metal particles. Walker et al.[2] have also shown that as the metallic phase is progressively oxidized to wüstite $Fe_{0.95}O$ and then to magnetite Fe_3O_4 the catalytic activity declines sharply. However a deactivated oxidized iron catalyst can be rejuvenated by reduction by CO, H_2 or even by carbon at elevated temperatures. Iron minerals that occur naturally in coal may also act as catalysts in this reaction. Figure 13 illustrates the catalytic effect of added siderite ($FeCO_3$) and pyrite (FeS_2) on the gasification of graphite in 1 atm. (10^5 Pa) CO_2 as a function

Fig. 11. Equilibrium stability regions of M_2O and M_2CO_3 corresponding to the reactions, $M_2CO_3+C=M_2O+2CO$ (solid curves) and $M_2CO_3=M_2O+CO_2$ (dashed curves) as functions of P_{CO}, P_{CO_2} (atms) and temperature for M=Li, Na, K.

of temperature. These minerals which are decomposed at gasification temperatures, can react with the carbonaceous phase to give free metallic Fe, which is probably the active entity during the catalyzed reaction.

THE CATALYZED C-H_2O REACTION

The effects of catalysts on the reaction of carbon with water vapor or steam are of particular importance at the present time because of the current interest in the steam gasification of coal. Among the most active, and most frequently studied, additives are the metals of the Fe and Pt groups and the salts (especially the carbonates) of the alkali and alkaline earth metals. Other additives, for example the oxides of Cu, Mn, V, Pb, Ti, Cr, B, Mo and Sb have been found to be moderately active[22] but the behavior of these materials has not been studied in detail.

Alkaline Earth Salts

Among the most active catalysts for the C-H_2O reaction

248

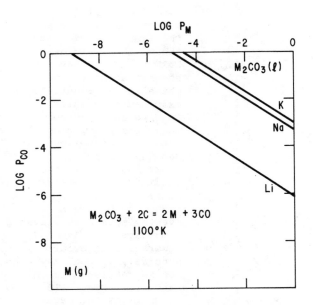

LOG P_M

$M_2CO_3(\ell)$

K

Na

Li

$M_2CO_3 + 2C = 2M + 3CO$

1100°K

M(g)

Fig. 12. Equilibrium stability regions of M_2CO_3 and M (M=Li, Na, K) for the reaction, $M_2CO_3+2C = 2M + 3CO$, as functions of P_{CO} and P_M (atms.) at 1100°K.

are the carbonates of Sr and Ba[23,24]. Other salts of the Group IIA metals, such as nitrates, sulfates and halides, are somewhat less active. As shown in Figure 14, the addition of 1% $BaCO_3$ increased the gasification rate of graphite by water vapor by a factor of about 25 over the 800-1000°C temperature range. $SrCO_3$ was also an active catalyst, but $CaCO_3$ was less effective in this case than in the graphite-CO_2 reaction (Figure 6) and $MgCO_3$ was almost completely inactive. A comparison of the gasification rates for pure graphite and for graphite doped with 0.5% $BaCO_3$ is shown in Figure 15. As indicated, the uncatalyzed graphite-H_2O reaction was about five times more rapid than the uncatalyzed graphite-CO_2 reaction in the temperature range investigated. However, on addition of the $BaCO_3$, both catalyzed reactions were found to occur at the same rate and with the same activation energy. These results suggest that the same rate determining step was operating in both cases and, by analogy with the C-CO_2 reaction previously discussed, it appears likely that the catalytic effect of the alkaline earth carbonates involves a sequence of reactions with the intermediate formation of oxide or hydroxide. Thus,

$$MCO_3 + C = MO + 2CO$$

$$MO + H_2O = M(OH)_2$$

$$M(OH)_2 + CO = MCO_3 + H_2$$

$$C + H_2O = CO + H_2$$

These reactions are thermodynamically favorable in the case of Sr and Ba, with the solid state reaction between the salt and carbon controlling the overall kinetics. $CaCO_3$ is less stable in the presence of low partial pressures of CO_2 and consequently this salt is a better catalyst in the C-CO_2 reaction than in the C-H_2O

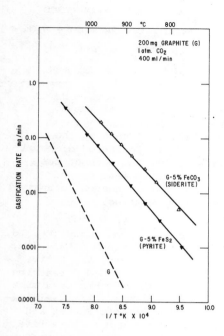

Fig. 13. Catalytic effects of
the minerals siderite ($FeCO_3$)
and pyrite (FeS_2) on the
gasification of graphite in 1 atm.
CO_2. Rates vs. $1/T°K$.

reaction. $MgCO_3$ will have very
limited stability in the 800-
1000°C range and hence this salt
is a poor catalyst for both
reactions.

Alkali Metal Salts

Salts of the Group IA alkali
metals, especially the carbonates
and hydroxides, are also very
active catalysts for the $C-H_2O$
reaction. Detailed studies with
graphite have indicated that the
order of effectiveness is Li >
K > Na between 700 and 1000°C[25],
as shown in Figure 16. A number
of intermediate species may
participate in the catalyzed
reaction in these cases and it is
likely that several different
reaction paths are followed
simultaneously. The carbonates
tend to be hydrolyzed in the
presence of water vapor with the
formation of hydroxides. Figure
17 shows stability regions for
the carbonates and hydroxides as
functions of P_{CO_2} and P_{H_2O} (in

atms.) at 1100°K (827°C). Thus in 1 atm. CO_2 (10^5 Pa), the stable
species present will be carbonate, whereas with $P_{CO_2} < 10^{-6}$ atm.
(0.1 Pa) and P_{H_2O} = 23 mm. (3 x 10^3 Pa) i.e. the conditions used in
these experiments, the dominant species will be hydroxide. These
predictions are confirmed by the experimental rate data shown in
Figure 18 for graphite doped with, initially, 5 percent Li_2CO_3 or
LiOH. Both catalysts were more active in the graphite-H_2O reaction
than in the graphite-CO_2 reaction, but in the CO_2 environment LiOH
behaved like Li_2CO_3, whereas in water vapor Li_2CO_3 behaved like
LiOH. In both cases a cyclic process is possible with the
intermediate formation of metal oxide or free alkali metal. The
equilibrium stability diagram shown in Figure 19 indicates that
LiOH can be reduced to free metal by reaction with carbon over a
wide range of P_{CO} and P_{H_2O} at 1100°K(827)°C. A similar reaction
can occur with the carbonate, as shown in Figure 12. Hence the
catalytic reaction sequence may involve the separate processes,

250

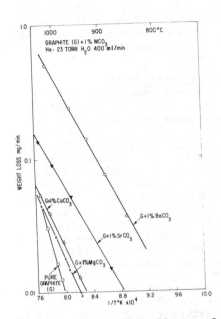

$$M_2CO_3 + 2C = 2M + 3CO$$

or $$2MOH + C = 2M + CO + H_2O$$

$$2M + 2H_2O = 2MOH + H_2$$

$$2MOH + CO = M_2CO_3 + H_2$$

$$C + H_2O = CO + H_2$$

There are however other possibilities, such as the intermediate formation of metal oxide, as in the case of the alkaline earth metals, and further work is needed to unravel these complexities.

Group VIII Metals

Both the iron group (Fe, Co, Ni) and the platinum metals (Ru, Rh, Pd, Os, Ir, Pt) are active catalysts for the $C-H_2O$ reaction and their behavior has been much studied. For the steam gasification of graphite[22], Fe, Ni and Co have been found to be active

Fig. 14. Catalytic effects of wt.% Mg, Ca, Sr and Ba carbonates on the gasification of graphite by water vapor. Rates vs. $1/T^\circ K$[23].

catalysts between 600 and 1000°C but only when the metal is kept in the reduced state by means of hydrogen added to the gas phase. The results of thermogravimetric experiments at 930°C with graphite impregnated with 1% Fe are shown in Figure 20 for three different gaseous atmospheres. Initially, with flowing nitrogen saturated with water vapor (23 mm., 3×10^3 Pa), no gasification of the graphite was detectable in a period of 30 mins. When wet hydrogen was substituted for the nitrogen a rapid gasification of the graphite took place but the rate was markedly reduced when the water vapor was removed from the gas stream. Thus the Fe-catalyzed graphite-H_2O reaction was considerably more rapid than the graphite-H_2 reaction at this temperature. Similar effects were observed with Ni and Co which were rather less active catalysts than Fe under the same conditions. In all three cases the metal appeared to be the catalytically active entity, whereas the oxides were inactive. Thermodynamic calculations indicated that Fe, Ni and Co would exist in the metallic state in wet H_2 at 1000°C, whereas in wet N_2 these additives were present as oxides.

It is known[26] that metals such as iron chemisorb H_2O dissociatively even at low temperatures with the formation of chemisorbed O and H atoms,

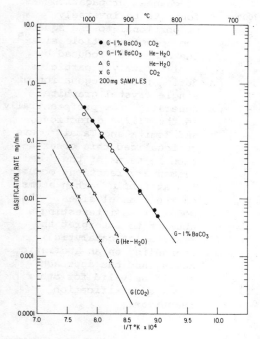

$$3Fe + H_2O = Fe-O_{ads} + 2(Fe-H_{ads})$$

This dissociative chemisorption may be the rate determining step in the overall carbon gasification reaction so that the higher the affinity of the metal for oxygen, the more rapid will be this oxygen transfer step and the greater the catalytic activity of the metal. Thus the catalytic activity should decrease in the order Fe > Co > Ni, as found experimentally. Similarly the Pt metals may also act as sites for the dissociation of the H_2O molecule to form labile chemisorbed species which subsequent react with the carbon substrate.

THE CATALYZED C-H_2 REACTION

Below 1000°C the kinetics of the reaction of the reaction of carbon with

Fig. 15. Comparison of the rates of the uncatalyzed graphite-CO_2 and graphite-H_2O reactions with the rates in the presence of 0.5 wt.% $BaCO_3$ catalyst. Rates vs. $1/T°K$[17].

hydrogen are very slow in the absence of catalysts. The most effective catalysts are the metals of Group VIII, the Pt metals being generally more active than those of the Fe group. In mechanistic terms the catalyzed hydrogenation of carbon is probably the simplest of the catalyzed gasification reactions. Metal particles appear to promote the reaction by acting as sites for the dissociation of molecular H_2. The resulting chemisorbed H atoms then migrate along the metal surface to the carbon substrate by a "spillover" process[27] and subsequently react with labile carbon atoms at preferred sites on the surface. In general the activity of the metals as catalysts for the C-H_2 reaction parallels the sequence of activity observed in gas phase hydrogenation or hydrogenolysis reactions, although minor differences in behavior may be related to variations in the degree of dispersion of the particles on the carbon substrate. Using the technique of controlled atmosphere electron microscopy, Baker et al.[28] have shown that with Pt the larger the catalyst particle the faster the rate of reaction, as measured by the rate of propagation of channels on the graphite surface. This is in contrast to the Pt-catalyzed oxidation of graphite where the rate

252

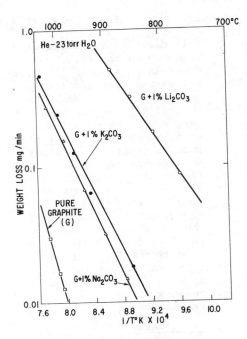

Fig. 16. Catalytic effects of 1 wt.%
Li, Na and K carbonates on the gasification of graphite by water vapor. Rates
vs. $1/T^{\circ}K$[25].

of channel propagation was
found to be inversely
proportional to the square
root of the particle size[29].
The channels produced by
mobile catalyst particles
during the hydrogenation of
single crystal graphite
generally grow preferentially
in the <1120> directions,
and Tomita and Tamai[30] have
rationalized this observation in terms of the
elementary reactions occurring at labile carbon atoms
on the channel sides. This
model is an interesting
attempt to interpret the
features of catalytic
channeling on an atomistic
basis, and the approach
should be valid for other
types of gasification
reactions.

CONCLUSIONS

Although many details of the complex effects of catalysts on
the rates of carbon gasification reactions still remain obscure,
the application of the techniques of thermal analysis and a
consideration of the thermodynamics of the possible elementary
processes have recently shed light on the behavior of individual
catalysts. In particular, specific oxidation-reduction cycles
have been invoked to interpret the catalytic effects of transition
metals, oxides and alkali metal salts in the various types of
carbon gasification reactions.

ACKNOWLEDGEMENT

The author is indebted to the Editors of Carbon (Pergamon
Press) and Fuel (IPC Press) for permission to reproduce data and
figures.

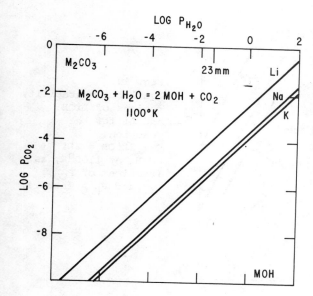

Fig. 17. Equilibrium stability regions of M_2CO_3 and MOH (M=Li, Na, K) for the reaction, $M_2CO_3+H_2O= 2MOH+CO_2$, as functions of P_{CO_2} and P_{H_2O} (atms.) at 1100°K.

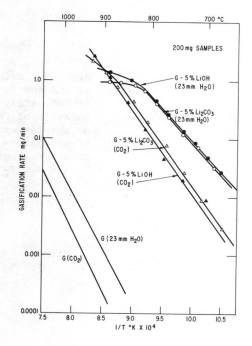

Fig. 18. Catalytic effects of LiOH and Li_2CO_3 additives on the gasification rates of graphite (G) in CO_2 (1 atm.) and water vapor (23 mm). Rates vs. $1/T°K$.

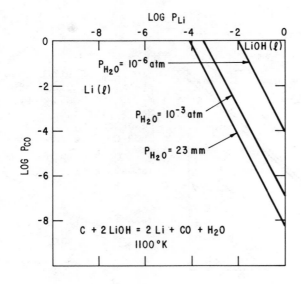

Fig. 19. Equilibrium stability regions of LiOH and Li for the reaction, C + 2LiOH = 2Li + CO+H$_2$ at 1100°K, as functions of P_{CO}, P_{Li} and P_{H_2O} (atms.).

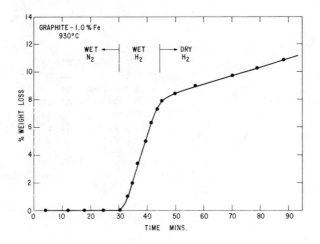

Fig. 20. Effect of atmosphere on the catalytic behavior of Fe. Weight losses of graphite-% Fe in wet N$_2$, wet H$_2$ and dry H$_2$ at 930°C[22].

REFERENCES

1. P. L., Walker, Jr., F. Rusinko, Jr., L. G. Austin, Adv. in Catalysis, 11, 133 (1959).
2. P. L. Walker, Jr., M. Shelef, R. A. Anderson, Chemistry and Physics of Carbon, P. L. Walker, Jr., ed. (Marcel Dekker, New York, 1968.) Vol. 4, p. 287.
3. D. W. McKee, Chemistry and Physics of Carbon, P. L. Walker Jr., and P. A. Thrower eds., (Marcel Dekker, New York, 1980. Vol. 16, p. 1.
4. J. M. Thomas, Chemistry and Physics of Carbon, P. L. Walker Jr., ed., (Marcel Dekker, New York, 1965.) Vol. 1, p. 122.
5. D. W. McKee, Carbon 8, 131 (1970).
6. P. S. Harris, F. S. Feates, B. G. Rueben, Carbon 12, 189 (1974).
7. R. T. K. Baker, Catal. Rev.-Sci. Eng. 19 (2), 161 (1970).
8. D. W. McKee, Carbon 8, 623 (1970).
9. R. T. K. Baker, J. A. France, L. Rouse, R. J. Waite, J. Catalysis 41, 22 (1976).
10. F. J. Long, K. W. Sykes, J. Chim. Phys., 47, 361 (1950).
11. F. H. Franke, M. Meraikib, Carbon 8, 423 (1970).
12. B. P. Jalan, Y. K. Rao, Carbon 16, 175 (1978).
13. C. Kroger, B. Neumann, E. Fingas, Z. Anorg. Chem., 197, 321 (1931).
14. R. T. K. Baker, P. S. Harris, Carbon 11, 25 (1973).
15. R. T. K. Baker, R. B. Thomas, M. Wells, Carbon 13, 141 (1975).
16. D. W. McKee, D. Chatterji, Carbon 13, 318 (1975).
17. D. W. McKee, Fuel 59, 308 (1980).
18. D. A. Fox, A. H. White, Ind. Eng. Chem., 23, 259 (1931).
19. H. Marsh, B. Rand, Carbon 9, 63 (1971).
20. Y. Tamai, H. Watanabe, A. Tomita, Carbon 15, 103 (1977).
21. G. Blyholder, L. D. Neff, J. Phys. Chem., 66, 1464 (1962).
22. D. W. McKee, Carbon 12, 453 (1974).
23. D. W. McKee, Carbon 17, 419 (1979).
24. K. Otto, L. Bartosiewicz, M. Shelef, Carbon 17, 351 (1979).
25. D. W. McKee, D. Chatterji, Carbon 16, 53 (1978).
26. D. J. Dwyer, G. W. Simmons, Surface Sci., 64, 617 (1977).
27. M. Boudart, A. W. Aldag, M. A. Vánnice, J. Catalysis 18, 46 (1970).
28. R. T. K. Baker, R. D. Sherwood, J. A. Dumesic, J. Catalysis, to be published (1980).
29. R. T. K. Baker, R. D. Sherwood, J. A. Dumesic, J. Catalysis, 62, 221 (1980).
30. A. Tomita, Y. Tamai, J. Phys. Chem., 78, 2254 (1974).

CHEMICAL AND PHYSICAL ASPECTS OF REFINING COAL LIQUIDS

By

Y. T. Shah [1], G. J. Stiegel [2], S. Krishnamurthy [1,3]

[1] Department of Chemical and Petroleum Engineering
University of Pittsburgh
Pittsburgh, PA 15261

[2] Pittsburgh Energy Technology Center
U.S. Department of Energy
Pittsburgh, PA 15236

[3] Presently with:
Mobil Research and Development Corporation
Paulsboro, N.J. 08066

ABSTRACT

Increasing costs and declining reserves of petroleum are forcing oil importing countries to develop alternate energy sources. The direct liquefaction of coal is currently being investigated as a viable means of producing substitute liquid fuels. The coal liquids derived from such processes are typically high in nitrogen, oxygen and sulfur besides having a high aromatic and metals content. It is therefore envisaged that modifications to existing petroleum refining technology will be necessary in order to economically upgrade coal liquids. In this review, compositional data for various coal liquids are presented and compared with those for petroleum fuels. Studies reported on the stability of coal liquids are discussed. The feasibility of processing blends of coal liquids with petroleum feedstocks in existing refineries is evaluated. The chemistry of hydroprocessing is discussed through kinetic and mechanistic studies using compounds which are commonly detected in coal liquids. The pros and cons of using conventional petroleum refining catalysts for upgrading coal liquids are discussed.

INTRODUCTION

Liquid fuels derived from the direct liquefaction of coal are currently expected to enter the commercial market by the end of the 1980's as a substitute feedstock for petroleum refineries and other petroleum dependent industries. Coal-derived liquids, however, have been shown to be quite different in composition than petroleum fuels. In general, coal liquids possess higher concentrations of polynuclear aromatics and oxygen- and nitrogen-containing compounds, their relative amounts being dependent upon the initial coal liquefaction technique (ie., catalytic, noncatalytic, pyrolytic, etc.). Due to

these differences, the chemistry for upgrading these fuels may be
quite different from that of petroleum and could therefore have
considerable impact on the final processing scheme and process
economics.

In this paper, the fundamental physical and chemical aspects of
upgrading coal-derived liquids are explored. The compositions of
several different coal liquids are presented. A comparison of the
compositions of the various coal liquids establishes the basis for
the remaining discussion. The compatibility of coal-derived liquids
with petroleum feedstocks is discussed due to the possibility of
blending these two feedstocks in the initial stages of refining. A
discussion of the stability of coal liquids is presented. Model
compounds studies on heteroatom removal which have relevance to the
upgrading of coal liquids are reviewed. The kinetics and reaction
mechanisms involved therein as well as the role of catalysts are
discussed. Finally, a review of the upgrading of coal liquids is made
along with recommendations for future research.

PHYSICS OF COAL LIQUIDS UPGRADING

Characterization:

Due to the complex chemical structure and elemental composition
of coal itself, liquids derived from it by direct liquefaction
processes possess a wide spectrum of hydrocarbons. The type of
liquefaction process employed plays a major role in the composition of
the final product. For a given coal, catalytic liquefaction
processes generally yield products with lower molecular weight compounds and fewer heterocyclics than noncatalytic processes. This is
clearly due to the ability of the catalyst to promote hydrogen
transfer reactions that result in increased hydrogen comsumption.
Even for a given liquefaction process and coal feedstock, changes in
processing conditions such as reaction pressure and slurry space
velocity, can result in substantial changes in the final product
quality and composition. In this regard, an optimum set of
processing conditions must be established to minimize total production cost. The use of coals of different ranks (ie., lignites,
subbituminous and bituminous coals) in a given liquefaction process
will also result in different products due to differences in the basic
structure and composition of various coals which affect their react-
vity during coal liquefaction. Since coals are heterogeneous in nature,
coals from even the same mine and/or seam can yield products of different
composition and quality under identical reaction conditions. With coal
liquids of different composition being produced continuously on a
commercial scale, it would be impracticable, if not impossible, to
design a coal liquids refinery for a particular feedstock. Instead,
the processes should be designed around the major constituents in all
coal liquids. The discussion which follows attempts to identify such
components from several coal liquids.

Mass spectroscopic data of several coal-derived liquids (i.e.,

SRC-I and recycle solvents, SYNTHOIL, COED light and heavy oil, CO-Steam product) are presented in Tables 1 and 2. The data of Schiller[1], shown in Table 1, illustrates differences in composition of coal-derived fuels derived from different liquefaction processes and from different feed coals for a given process. The analyses of the products from the SRC-I process using Amax, Pittsburgh #8 and Illinois #6 coals show some differences in the concentration of aromatics such as tetralins, naphthalenes and acenaphthalenes/biphenyls. Differences are also present in the concentrations of heterocyclic components such as quinoline, acridine, phenols and dibenzofuran.

The data also show that over 97% of the SRC-I recycle solvent is distillate material whereas the SYNTHOIL and CO-Steam products contain less than 60% distillate. As shown in Table 1, the products from these two processes contain higher molecular weight components in the distillate, and therefore the molecular weights of the whole products are probably much higher than those of the SRC-I recycle solvents. The higher molecular weight products are generally more viscous; a comparison of SRC-I recycle solvent and SYNTHOIL bears this out.

The data presented in Table 2 were taken from the work of Krishnamurthy et al.[2]. These coal-derived liquids were all produced from Western Kentucky coals; however, the two COED oils were hydro-treated after the liquefaction step. Data for raw COED oil has been presented by Cusamano et al.[3,4]. As shown, the hydrotreated heavy COED oil contains fewer naphthenic components and more high molecular weight polyaromatics than the light oil. The SRC-II recycle solvent also possesses substantial quantities of polyaromatic components.

The prominent compounds in coal-derived liquids can be ascertained by comparing the data presented in Tables 1 and 2. The more common aromatics are benzenes, tetralins, naphthalenes, anthracenes, phenanthrenes, fluorenes, acenaphthylenes, and pyrenes. Oxygen-containing heterocyclics which appear to be common among coal liquids are phenols, dibenzofurans, benzonaphthofurans and indanols. Compounds containing nitrogen are only presented in Table 1 and the more prevalent of these are quinolines, acridines, carbazoles and indoles. Sulfur-containing compounds are not presented in either Tables 1 or 2; however, the available literature indicates that the more common components are thiophenes, benzothiophenes, dibenzo-thiophenes and benzonaphthothiophenes.[5] Chemical structures for the above compounds are presented in Figures 1 through 4.

The data in Tables 1 and 2 present only the readily detectable components in coal liquids; however many more do exist in smaller concentrations. To illustrate this, a study performed by Aczel et al.[6,7] on SYNTHOIL, the results of which are summarized in Table 3, showed the coal-derived liquid to contain up to 150 individual isomers, more than 300 compounds types, and about 2500 separate carbon number homologs. These include one to eight ring condensed furans, difurans,

trifurans, thiophenes, dithiophenes, thiophenofurans, hydroxy-
aromatics, dihydroaromatics, hydroxyfurans, pyrroles, pyridines,
hydroxy nitrogen compounds, and assorted heterocompounds with up to
five heteroatoms per molecule. In another study, Kershaw[8] isolated,
using fluorescence spectroscopy, twelve polyaromatic ring systems
containing up to seven aromatic rings.

Compatibility of Coal Liquids and Petroleum Fuels:

As coal liquefaction progresses towards commercialization,
coal liquids will initially supplement decreases in petroleum
imports. The quantities produced, however, may not be sufficient
to justify dedicated refineries, and besides, these initial supplies
will most likely be produced via several liquefaction processes.
This, together with the use of different coals will result in
products with different compositions and physical properties that
will therefore have to be upgraded to different extents.

In blending petroleum fuels derived from crudes of different
origin, compatibility of the blending stocks is an important parameter. This will be an even more important consideration in the
blending of coal liquids and petroleum stocks because of the
differences in their composition and physical properties. Materials
which are incompatible yield residues upon mixing and/or storage.
The methods and criteria employed to determine compability of
fuels varies considerably among investigators; however, frequently
employed techniques involve comparing the amount of residue produced from a given blend with that from a reference blend   or
determining the total amount of residue deposited upon blending.

In general, fuels containing similar components are more likely
to be compatible than those containing dissimilar components.
Table 4 presents a comparison of three coal liquids and three
petroleum fuels. The petroleum fuels can be seen to possess
higher concentrations of hydrogen and sulfur and lower concentrations of nitrogen and oxygen. Table 4 also shows the petroleum
fuels to contain more saturated compounds and fewer polynuclear
aromatics than coal liquids. Figure 5 presents a plot of the
aromaticity of various fuels versus their carbon to hydrogen
atomic ratio. It is clear from this figure that coal liquids can
contain up to twice the C/H atomic ratio of petroleum fuels due to
their higher aromaticity.

Investigations of the compatibility of coal-derived liquids
with petroleum fuels are scarce. The most recent and currently
applicable investigations have been conducted by Bendoraitis et al. [9],
Cabal et al. [10] and Stein et al. [11,12]. In these studies, coal
liquids produced by the SYNTHOIL, H-COAL, and SRC-I processes were
evaluated for compatibility with petroleum-derived No. 2 distillate
fuel, No. 6 fuel oil, heavy coker gas oil, and Fluid Catalytic Cracker
(FCC) clarified slurry oil. Each petroleum and coal-derived liquid was
characterized by Gradient Elution Chromatography (GEC) in order to

determine its composition[9].

Table 5 presents a summary of some of the results obtained in their investigations. As shown, the full range products of SYNTHOIL and H-COAL liquids are incompatible with the petroleum fuels, whereas the -783K fractions are completely compatible. A comparison of the GEC analyses for the coal liquids with those of the petroleum fuels reveals a similarity in the composition of the -783K fraction of the coal liquids and the petroleum fuels. In particular, the C/H ratio and aromatic content of this fraction of the coal liquids more closely approximate the petroleum fuels than the whole product. The full range coal liquids contained more asphaltenic material than the petroleum fuels and distillation was effective in removing these components from the raw coal liquid and rendering them compatible with petroleum fuels.

The above investigators also found compatibility to be a function of the concentration of coal liquid in the blend[9,10]. Using SYNTHOIL and heavy coker gas oil, they showed that at high concentrations of SYNTHOIL, the coker gas oil and SYNTHOIL were compatible; however, at low concentrations the two liquids were incompatible. Results from their study which are shown in Figure 6 indicate that small amounts of petroleum fuels in coal liquids may not have adverse affects during processing, but the processing of small amounts of coal liquids in petroleum may not be possible.

The compatibility of petroleum fuels and SRC-I derived from Illinois #6 Burning Star, Monterey, and Wyodak coals were also investigated by the above authors. The SRC-I products were found to be incompatible with all petroleum fuels except for the FCC clarified slurry oil where partial solubility was observed. This was attributed to the fact that this petroleum fuel contains more aromatic components with more of these components concentrated in the higher GEC fractions than the other fuels.

Raw and hydrotreated H-COAL (+478K), SRC-I recycle solvent, and SRC-I/recycle solvent mixture were also investigated in these studies for compatibility with the same petroleum fuels. In all cases, the blends containing raw coal liquids were incompatible. However, as shown in Table 5, hydrotreating the coal liquids prior to blending significantly reduced the formation of residue. Of the hydroprocessing severities employed, blends containing moderately hydrotreated coal liquids resulted in the lowest yield of residue. These results suggest that an optimum set of hydroprocessing conditions exist which minimize the incompatibility of coal liquids with petroleum fuels. Figure 5 can also be used to provide a rough estimate of the amount of hydrogen required to convert an incompatible fuel into a compatible one.

In another study, coal liquids (480-810K fraction) derived from Illinois #6 coal via the Exxon Donor Solvent process were found to be incompatible with several commercial petroleum heavy fuel oils[13]. However, hydrotreating of the coal liquids was reported to result in compatibility with petroleum fuels over the range of concentrations from 1 to 50 wt%.

Stability of Coal Liquids:

The stability of coal liquids upon storage and/or heating is an important factor which can seriously affect the further pro-cessing and final product specifications. Several investigations[14-21] have focused on these problems to some degree using coal liquids produced via the SYNTHOIL, H-COAL, COED and SRC-I processes or coal carbonization tars. The raw coal liquids in the above studies invariably were found to be unstable upon storage. Significant increases in viscosity and higher molecular weight components have been observed along with the formation of gum and other deposits.

The aging characteristics of SYNTHOIL have been investigated by Finseth et al.[14] and Karn et al.[15]. These authors evaluated the effects of storage time, temperature and atmosphere, agitation, and light on product stability. Both studies found the viscosity of the coal liquid to increase substantially with time when stored under an oxygen atmosphere, whereas it increased only slightly under nitrogen. The results of Finseth et al.[14] are shown in Table 6 for storage under both oxygen and nitrogen atmospheres. The data show the oil content of the original liquid to decrease upon storage while the asphaltene content increases. It is also interesting to note that the benzene insoluble fraction and the oxygen content of the coal liquids increase substantially upon storage under oxygen. This indicates that oxygen is definitely involved in the aging process. The following two reaction mech-anisms were proposed by Finseth et al.[14] to account for the observations:

1. Coupling of small neutral molecules by oxygen

$$A + B \xrightarrow{O_2} A - O - B + H_2O \qquad (1)$$

2. Coupling of radicals with radicals or neutrals

$$A\cdot + B\cdot \longrightarrow AB \qquad (2)$$
$$A\cdot + B \longrightarrow AB\cdot \qquad (3)$$
$$A\cdot + B \longrightarrow AB' + C\cdot \qquad (4)$$

Mechanisms 1 accounts for the observed increase in the oxygen con-centration whereas mechanism 2 accounts for the observed increase in viscosity of the coal liquid stored under nitrogen. ESR examination of the samples suggested that radical-radical recombination reaction is not a preferred route. The validity of the other reactions could not be confirmed. They concluded that the aging mechanism involved the condensation of small heteroatom-rich molecules to form higher molecular weight insolubles and asphaltenes in the presence of oxygen.

Karn et al.[15] also showed that lower storage temperatures and
lower initial viscosities of stored liquids reduces the rate of
increase of viscosity. The effect of agitation on samples stored
under nitrogen was negligible, whereas under an oxygen atmosphere,
agitation caused significant increases in viscosity. No effect of
laboratory light on the liquid viscosity was observed.

Similar changes in viscosity and toluene insoluble content of
coal liquids stored under air and inert atmospheres were observed
by Krishna and Ehsan [17] for a carbonization tar. In their study, a
substantial decrease in the concentration of hydroxyl groups and an
increase in the number of carbonyl groups was observed. A slight
decrease in the concentration of carboxyl groups was also observed.
The conversion of phenolic "OH" to a quinonoid type structure due
to oxidation was offered as a possible explanation.

The reaction kinetics of coal tar autoxidation were investigated
by Lin et al. [16]. The ease and extent of these reactions were
found to be dependent on the composition of the liquid, the reaction
temperature, and the oxygen partial pressure. The reaction of neutral
oils was found to involve long-chain radicals and was proportional
to the concentration of hydrocarbon reactant. The oxidation of tar
acids and bases was observed to be first order in both reactant
concentration and oxygen partial pressure. The reaction of the
whole liquid was found to be diffusion controlled.

Additional studies [19,20] have revealed that hydrotreatment of
the coal liquid prior to storage results in stable products. This
is probably due to the saturation and removal of trace amounts of
unstable species which react to form high molecular weight components.

CHEMISTRY OF COAL LIQUIDS UPGRADING

Hydroprocessing:

Coal liquids derived by the donor solvent processes possess
high nitrogen, sulfur and oxygen contents. This is in contrast to
petroleum fuels wherein sulfur is invariably the heteroatom of
concern. Nitrogen compounds are detrimental to the activity of
cracking and hydrocracking catalysts. They also have deleterious
effects on the properties of fuels, besides leading to undesirable
combustion products, such as NO_x, and instability during storage.

Sulfur compounds, although not present to the extent of nitrogen
and oxygen compounds, are undesirable in that they can also poison
catalysts and result in the formation of noxious fumes of SO_x upon
combustion. The effect of oxygen compounds is not clearly understood
at the present time. The removal of oxygen occurs competitively
under conditions of nitrogen and sulfur removal and the ultimate
product of hydrodeoxygenation, namely water, is a known catalyst poison[22]

The removal of nitrogen, sulfur and oxygen heteroatoms is
generally practiced in the presence of hydrogen at 1000-2000 psi,
temperatures of 300-450° C, and a hydroprocessing catalyst containing
combinations of Co,Ni,Mo,W on Al_2O_3 or Al_2O_3-SiO_2 supports. The
inportant reactions that occur concurrently under the above conditions
are:

Nitrogen containing compounds + H_2 $\longrightarrow$ Hydrocarbons + NH_3

sulfur containing compounds + H_2 $\longrightarrow$ Hydrocarbons + H_2S

oxygen containing compounds + H_2 $\longrightarrow$ Hydrocarbons + H_2O

In addition to the above reactions a certain degree of saturation of aromatics inevitably occurs. The complexity of coal liquids
precludes a detailed study of the kinetics and mechanisms of the
above reactions as well as the role of catalysts in such reactions.
An understanding of such processes can therefore be gained through
studies made on individual compounds that occur predominantly
in coal liquids. Numerous studies made on representative heterocyclic
compounds have been reported in the literature [23]. A few pertinent
studies are described below.

Hydrodenitrogenation:

The nitrogen containing heterocyclics that occur in coal
liquids can be classified into basic and non-basic compounds.
Examples of basic compounds are pyridine, quinoline, and acridine
and non-basic compounds are pyrrole, indole, and carbazole. The
commonly observed mechanism for nitrogen removal is the hydrogenation of the heterocyclic ring, which under most circumstances
has been observed to be reversible, followed by hydrogenolysis and
subsequent release of nitrogen as NH_3. Although the removal of
nitrogen from five membered heterocyclics is generally easier than
their six membered counterparts, hydrogenation of the heterocyclic
ring was found to be a necessary prerequisite to nitrogen removal
from indole as reported by Stern [24]. The mechanism proposed by
Stern is shown below.

McIlvried [25] postulated the following mechanism for the hydro-denitrogen of pyridine over sulfided Co-Ni-Mo/Al_2O_3 catalysts.

$$\text{(pyridine)} \rightleftharpoons \text{(piperidine)} \longrightarrow C_5H_{11}NH_2 \longrightarrow C_5H_{12} + NH_3$$

The beneficial effect of presulfiding catalysts upon their activity towards each step in the reaction network for pyridine has been discussed by Goudriaan et al. [26].

With increasing number of aromatic rings in a heterocyclic compound the reaction network assumes further complexities. This is portrayed by Shih et al. [27] and Katzer and Sivasubramanian [28] for the reaction networks of quinoline and acridine in Figures 7 and 8, respectively. It may be observed in Figure 8 that the hydrogenated heterocyclic intermediate which was found to be in equilibrium with the primary reactant for the other model compounds is absent. This could be due to its high reactivity. Two other features of these networks merit consideration. The carbon-nitrogen bond scission in an aromatic ring does not occur without ring hydrogenation and secondly, the rate constants in each network clearly indicate the nonselective nature of commerical hydroprocessing catalysts for nitrogen removal. Thus, currently available catalysts for nitrogen removal cause indiscriminate hydrogenation of aromatics resulting in high hydrogen consumption during the upgrading step. This is in contrast to sulfur removal wherein catalysts show high selectivity towards direct sulfur removal as discussed below.

Hydrodesulfurization:

The removal of sulfur from thiophene has been the subject of several investigations [29-33]. Figure 9 shows the reaction network proposed by Owens and Amberg [32,33] for the hydrodesulfurization of thiophene over chromia and CoMo/Al_2O_3 catalysts. Butadiene was not observed in the products when CoMo/Al_2O_3 was used. The absence of hydrogenated thiophenes indicates the -C-S- bond cleavage to be the first step in the network. In contrast to the above results, the formation of hydrogenated intermediates has been reported by several workers [34,35]. This suggests that the desulfurization of thiophene proceeds via two routes, the main route being the one shown in Figure 9.

The reaction network postulated by Houalla et al. [36] for the hydrodesulfurization of dibenzothiophene over presulfided CoMo/Al_2O_3 is shown in Figure 10. The direct removal of sulfur without any ring hydrogenation was found to be favored on this catalyst. Such a behavior was found to be exhibited to a lesser extent on NiMo and NiW catalysts supported on Al_2O_3.

Hydrodeoxygenation:

The oxygen compounds present in coal liquids are predominantly composed of phenols and furans. The removal of oxygen from phenol has been reported to proceed through direct dehydroxylation [37] as well as through the formation of cyclohexanol [38]. The hydrodeoxygenation of dibenzofuran has been recently examined by Krishnamurthy et al.[39]. These authors found, as shown in Figure 11, that the removal of oxygen proceeded via direct oxygen extrusion as in the case of dibenzothiophene as well as through the formation of hydrogenated intermediates. The NiMo/Al$_2$O$_3$ used in their investigation favored the removal of oxygen through the formation of hydrogenated intermediates.

A comparison of heteroatom removal from dibenzothiophene, dibenzofuran and carbazole [23] upon NiMo/Al$_2$O$_3$ indicates that whereas the removal of sulfur can be effected through minimal ring hydrogenation, the removal of nitrogen is inevitably accompanied by extensive hydrogenation. The hydrogenation accompanying oxygen removal is intermediate between that accompanying sulfur and nitrogen removal. This is probably due to the bond strengths of the carbon atoms bonded to the respective heteroatom.

Other Upgrading Processes:

Among the other upgrading processes in the refining industry, the main problem of concern is not as much the chemistry involved, as it is the maintenance of the activity of the catalysts employed. The modes of catalyst deactivation during coal liquids upgrading is the subject of a subsequent section. The effect of catalyst deactivation upon cracking, reforming, and hydrotreating catalysts will be discussed below.

It is imperative that hydrotreating of coal liquids precedes catalytic cracking or reforming in an upgrading scheme. This is due to the inability of the catalysts employed in the latter processes to withstand irreversible metal poisons as well as heteroatom-containing polyaromatics. The deactivation in the latter processes is primarily due to the coke formation on the catalyst. Polycyclic aromatic hydrocarbons resist cracking and therefore contribute to coke formation on the catalyst surface. This problem is more prominent in coal liquids due to its higher aromaticity. Although hydrotreating prior to catalytic cracking may minimize coke formation, it does not obviate the need to develop improved catalysts that resist deactivation and also possess the ability to crack refractory components in the feedstocks.

Catalytic reforming has shown great potential in the upgrading of petroleum feedstocks to high octane gasoline. The extension of such a process to coal liquids has obvious advantages. The catalysts employed in the process are typically composed of one or more

metals of Group VIII B of the periodic table supported on Al_2O_3.
The susceptibility of such catalysts to metals poisoning needs
little elaboration here. Hydrotreating prior to reforming is therefore a necessity and may in fact require a lower net hydrogen
consumption in the combined hydrotreating reforming step when
compared with the hydrotreating step alone. This is due to the
generation of molecular hydrogen in the reformer from dehydrogenation reactions. Deactivation of reforming catalysts due to
coke formation can be minimized in the reformer by maintaining a
suitable H/C atomic ratio in the coal liquid feedstock.

Catalyst Deactivation:

Deactivation of catalysts used in the upgrading of coal liquids
can be classified into three types: a) metals deposition, b) coke
formation and c) formation of refractory reaction intermediates.
Metals deposition on catalysts cover the active surface and cause
plugging of the pores. This is an irreversible process in contrast
to coke deposition and therefore adversely affects catalysts life.
The types of metals that occur in coal liquids are quite different
from those that occur in petroleum fuels. Table 7 provides a
comparison of the important metals that occur in both coal liquids
and petroleum fuels. Nickel and vanadium are present to a considerable extent in petroleum fuels whereas titanium is the most
important metal to be contended with in coal liquids. Although the
metals in petroleum fuels occur as organometallic compounds, the
nature of the metallic species in coal liquids has not been ascertained. It is envisaged that the metals content of coal liquids
have to be reduced to levels acceptable for catalytic cracking and
reforming. This is therefore best accomplished concurrently with
heteroatom removal and hydrogen enrichment in the hydrotreater.
Experimental studies so far have shown that catalyst deactivation
in the hydrotreater is fairly rapid[40]. Deactivation due to
coke deposition is also expected to be more pronounced in the case
of coal liquids when compared with petroleum fuels due to the
higher aromatic content of the former. It is therefore apparent
that the catalyst employed for hydrotreating petroleum fuels will
not function efficiently for coal liquids. In fact, commercial
hydroprocessing catalysts that have been used in the petroleum
industry have been found to rapidly deactivate when processing coal
liquids. The development of improved hydroprocessing catalysts is
therefore imperative for the commercialization of upgrading processes for coal liquids.

Surface Chemistry:

In order to design improved hydrotreating catalysts it is necessary to understand the surface chemistry of the pertinent reactions. Very little is understood at the present about the surface
chemistry involved during heteroatom removal on commercial hydroprocessing catalysts. Desikan and Amberg[29] have examined the nature of active
sites for the hydrodesulfurization of thiophene on $CoMo/Al_2O_3$

catalyst using pyridine as a poison. They found the catalyst to possess active sites with two levels of acidity a) strongly acidic sites which were primarily responsible for hydrogenation of olefins formed through desulfurization of thiophene and b) weakly acidic sites wherein most of the desulfurization was postulated to occur. Desulfurization was also expected to occur to a lesser extent on the strongly acidic sites. Poisons such as H_2S, pyridine and NH_3 were found to adsorb preferentially on the strongly acidic sites thus inhibiting the hydrogenation of olefins. The above investigators hypothesized that the adsorption of thiophene occurs through chemical interactions of the sulfur heteroatom with catalyst sites. This mechanism is commonly referred to as one-point adsorption since the remainder of the thiophenic ring is assumed to stand upright from the sulfur bonded to the surface. Similar schemes have been proposed by Lipsch and Schuit [22] and Kolboe [41]. Kwart and coworkers [42,43] have found the above scheme to be inadequate to explain the steric hindrance caused by methyl substitution of dibenzothiophenes on a $CoMo/Al_2O_3$ catalyst. Specifically, as shown in Figure 12, they found 4,6 dimethyl dibenzothiophene to be about 28 times less reactive than dibenzothiophene whereas the former compound is about two times less reactive than 4-methyl dibenzothiophene under the same conditions. This is in contrast to calculations based on steric hindrance which predict differences of several orders of magnitude between the reactivities of these compounds. Similar findings have been reported for substituted benzothiophenes [44] and substituted thiophenes [30]. Kwart et al. [43] have recently proposed the following model for thiophene which could be extended to benzothiophenes and dibenzothiophenes. This model, shown in Figure 13, assumes the thiophenic molecule to be chemisorbed in such a manner that the 1 - 2 bond is coordinated at an anion vacancy with the sulfur from thiophene interacting with a sulfur atom on the surface resulting in the formation of a dihydrogenated intermediate. Subsequent product formation is postulated to occur through several alternate pathways which are discussed in detail elsewhere [43].

Another model on the lines of the one point adsorption model has been proposed by Cowley [45]. He hypothesized that the thiophenic molecule was adsorbed flatly on the catalyst surface in a π complex. The removal of sulfur was presumed to be completed in this manner with the π complex being converted to a σ complex wherein the sulfur orbitals normal to the plane of the molecule bond themselves to the Mo^{3+} center at the anion vacancy. The different models proposed above for the adsorption of thiophene, which are equally acceptable, are indicative of the difficulty involved in the evaluation of such mechanisms.

The nature of active sites on other hydroprocessing catalysts such as $NiMo/Al_2O_3$ and NiW/Al_2O_3 has not been examined at the present

time. There is a paucity of information even about the nature of
adsorption of nitrogen-containing molecules on $CoMo/Al_2O_3$ catalysts.
In a recent study on the hydrodenitrogenation of quinoline on
$NiMo/Al_2O_3$ catalyst, Shih et al.[27] observed that the primary
hydrogenation reactions occurred on the metal sites, whereas hydrogenolysis reactions involved interactions with the support. In
view of the increasing difficulty of hydrodenitrogenation when
compared with hydrodesulfurization, studies on the surface chemistry of such reactions are well warranted.

CONCLUSIONS AND RECOMMENDATIONS

The composition of coal liquids differs considerably from
petroleum fuels in that the former possess higher concentrations of
heteroatoms, metals and polynuclear aromatic components. Although
much of the technology developed for refining petroleum is applicable
to upgrading coal liquids, the actual chemistry involved differs
somewhat because of the compositional as well as physical differences
of the two materials.

The compatibility of coal liquids and petroleum fuels must be
investigated further using coal liquids produced via the SRC-II,
EDS, and H-COAL processes. Subsequent processing must also be
evaluated to determine the effects of blending petroleum and coal
liquids upon downstream processing. The stability of coal liquids
needs to be studied further to ascertain the chemical nature of
these materials under typical processing conditions. An evaluation
of the stability of compatible blends of coal liquids and petroleum
fuels is also necessary.

Of all the areas involved in upgrading coal liquids, catalyst
development is probably the most important since it offers the
greatest improvement in product quality as well as process operability
and economics. From the above discussion, it is apparent that the
hydrotreating catalysts that are optimal for sulfur removal may
not be as efficient for nitrogen removal. What is needed are
catalysts which can minimize coking and selectively coordinate and
catalyze the -C-N-bond scission reaction without prior hydrogenation. It is speculated that sulfur removal occurs on the catalyst
without ring hydrogenation because of the sulfided state of the metallic
components of the catalyst. If this is true, then transition metal
nitrides may be promising candidates for catalyzing denitrogenation
reactions. In addition, mixed transition metal oxides, oxysulfides, as
well as borides and carbides of Mg, Co, Mn, Zn, and other transition
metals may partly fulfill these requirements[4,46,47]. The development
of bifunctional HDS/HDN catalysts should be pursued. The effects of
presulfiding and the presence of NH_3, O_2, H_2O, CO, and CO_2 on the
acidity and activity of these catalysts should be evaluated.

Very few investigations of the interaction of heteroatoms have appeared in the literature [23]. An examination of these interactions in greater detail would provide an understanding of the deactivation effects of reactants, intermediates, and products on the catalyst. Of particular importance are the effects of H_2S, NH_3, H_2O, and O_2 on heteroatom removal.

New catalysts which selectively hydrocrack polynuclear aromatics without excessive hydrogenation would substantially improve the hydro-cracker operation with coal liquids. Catalysts having a controlled ratio of cracking to hydrogenation activity should be pursued. The develop-ment of these catalysts should encompass novel inorganic materials and novel preparative procedures. As for catalytic cracking, catalysts which selectively crack the large polycylic compounds and are resistant to coking and other poisons would be desirable. Present reforming catalysts used in the petroleum industry are adequate for coal liquids.

As for refining coal liquids, conventional petroleum refining tech-nology is sufficient at the present time. To date, however, most of the processing has been conducted in individual units and the inter-actions of other processes on optimum operating conditions and product distribution have not been considered. Only when coal liquids are processed in integrated pilot scale refineries will such information become available. Once this is obtained, the overall ecomomics of refining coal liquids can be evaluated.

270

REFERENCES

1. Schiller, J.E., Hydrocarbon Processing, 57 (1), 147 (1977).
2. Krishnamurthy, S., Y.T. Shah, and G.J. Stiegel, Fuel (in press).
3. Cusamano, J. A., R.A. Dalla Betta, and R.B. Levey, "Catalysis in Coal Conversion", Academic Press, New York (1978).
4. Cusamano, J.A., R.A. Dalla Betta, and R.B. Levey, ERDA Report No. FE-2017-1, Catalytic Associates, Inc. (1976).
5. Akhtar, S., A. G. Sharkey, J.L. Shultz, and P.M. Yavorsky, ACS Div. of Fuel Chem. Preprints, 19(1), 207 (1974).
6. Aczel, T., R.B. Williams, R.J. Pancirov, and J.H. Karchmer, U.S. DOE Report No. METC-8007-1 (Pt. 1) (1976).
7. Aczel, T. and H.E. Lumpkin, "Refining of Synthetic Crudes", ACS Adv. in Chem. Series No. 179 (1979).
8. Kershaw, J.R., Fuel, 57, 299 (1978).
9. Bendoratis, J. G., A.V. Cabal, R.B. Callen, T.R. Stein, and S.E. Voltz EPRI Report No. EPRI 361-1 (1976).
10. Cabal, A.V., S.E. Voltz, and T.R. Stein, Ind. Eng. Chem. Prod. Res. Dev., 16 (1), 58 (1977).
11. Stein, T.R., J.G. Bendoratis, A.V. Cabal, R.B. Callen, M.J. Dabkowski, R. H. Heck, H.R. Ireland, and C.A. Simpson, EPRI Report No. EPRI AF-444 (1977).
12. Stein, T.R., A.V. Cabal, R.B. Callen, M.J. Dabkowski, R.H. Heck, C.A. Simpson, and S.S. Shih, EPRI Report No. EPRI AF-873 (1978).
13. Fant, B.T., U.S. DOE Report No. FE-2353-20 (Vol.1) (1978).
14. Finseth, D.,M. Hough, J.A. Queiser, and H.L. Retcofsky, ACS Div. of Pet. Chem. Preprints, 24 (4), 979 (1979).
15. Karn, F.S., F.R. Brown, and A.G. Sharkey, ACS Div. of Fuel Chem. Preprints, 19 (5), 2 (1974).
16. Lin, Y.Y., L.L. Anderson, and H.H. Wiser, ACS Div. of Fuel Chem. Preprints, 19 (5), 2 (1974).
17. Krishna, M.G. and M. Ehsan, Erdol and Kohle, 29 (6), 261 (1979).
18. Brinkman, D.W., M.L. Whisman, and J.N. Bowden, U.S. DOE Report No. BETC/RI-78/23 (1979).
19. Reynolds, T.W., NASA Report No. NASA TMX-3551 (1971).
20. O'Rear, D.J., R.F. Sullivan, and B.E. Strangeland, ACS Div. of Fuels, Chem. Preprints, 25 (1), 78 (1980).
21. Given, E.N., M.A. Collura, W. Alexander, E.J. Greshovich, C.D. Engleman, J.B. Wetherington, C.W. Clump, and E.K. Levy, U.S. DOE Report No. FE-2003-27 (1977).
22. Lispch, J.M. J.G. and G.C.A. Schuit, J. Catal., 15, 179 (1969).
23. Shah, Y.T., "Reaction Engineering in Coal Liquefaction, "Addison-Wesley, Inc. (1980).
24. Stern, E.W., J. Catal., 57, 39 0 (1979).
25. McIlvried, H. G., ACS Div. of Pet. Chem. (Feb., 1970).
26. Gondriaan, F., H. German, and J.C. Vlugter, J. Inst. Pet. 59, 40 (1973).
27. Shih, S.S., J.R. Katzer, H. Kwart, and A.B. Stiles, ACS Div. of Pet. Chem. Preprints, 22 (3), 919 (1977).
28. Katzer, J.R. and R. Sivasubramanian, Catalysis Rev. Sic. and Engr. 20 (2), 155 (1979).

29. Desikan, P. and C. H. Amberg, Can. J. Chem., 41, 1966 (1963)
30. Desikan, P. and C. H. Amberg, Can. J. Chem., 42, 843 (1964).
31. Kolboe, S. and C. H. Amberg, Can. J. Chem., 44, 2623 (1966).
32. Owens, P. J. and C. H. Amberg, Can. J. Chem. 40, 941 (1962).
33. Owens, P. J. and C. H. Amberg, Can. J. Chem., 40, 947 (1962).
34. Lee, H. C. and J. B. Butt, J. Catal., 49, 320 (1977).
35. Moldavskii, B. L. and S. E. Livshits, Zh. Obshch. Khim., 4, 298 (1934).
36. Houalla, M. N. K. Nag, A. V. Sapre, D. H. Broderick, and B. C. Gates, AIChE J., 24 (6), 10105 (1978).
37. Moldavskii, B. L. and S. E. Livshits, Zh. Obshch, Khim., 3, 603, (1933).
38. Polozov, V. F., Khim. Tverd. Toplivia, 6, 78 (1932).
39. Krishnamurthy, S., S. Panvelker, and Y. T. Shah, submitted to AIChE J. (1980).
40. deRosset, A. J., G. Tan, J. G. Gatsis, J. P. Shoffner, and R. F. Swensen, U.S. DOE Report No. FE-2010-09 (1977).
41. Kolboe, S., Can. J. Chem., 47, 352 (1969).
42. Houalla, M., D. Broderick, V. H. J. deBeer, B. C. Crates, and H. Kwart, ACS Div. of Pet. Chem., 22 (3), 941 (1977).
43. Kwart, H., G.C.A. Schuit, and B. C. Gates, J. Catal., 61, 128 (1980).
44. Given, E. N. and P. B. Venuto, ACS Div. Pet. Chem., 15 (4), A183 (1970).
45. Cowley, S. W., Ph.D. Thesis, Southern Illinois University (1975).
46. Toth, L. E., "Transition Metal Carbides and Nitrides," Academic Press, New York (1971).
47. Tauster, S. J., J. Catal., 26, 487 (1972).

Table 1.   Composition of Coal Liquefaction Products[1]

SRC-I Recycle Solvents
Wilsonville, Alabama

	Amax Coal	Pittsburgh #8 Coal
Benzenes	1.66	1.05
Tetralin	4.17	8.12
Tetrahydroacenaphthene/ Dihydronaphthalene	–	–
Naphthalene	27.5	27.7
Acenaphthene/Biphenyl	7.33	6.31
Flourene/Acenaphthylene	3.16	3.25
Phenanthrene/Anthracene	5.35	4.77
Dihydropyrene	0.48	–
Pyrene/Flouranthene	1.44	1.13
Chrysene/Triphenylene	–	–
Binaphthyl	–	–
Benzopyrene	–	–
Dibenzanthracene	–	–
Tetrahydroquinoline	–	0.23
Indole	0.29	0.38
Quinoline	9.49	9.89
Phenylpyridine/Tetrahydroacridine	1.14	1.39
Carbazole	0.41	0.18
Acridine	0.93	0.89
Naphthenobenzoquinoline	–	–
Azapyrene/Benzocarbazole	–	–
Benzacridine	–	–
Phenol	27.5	27.8
Indanol	1.59	–
Dibenzofuran	6.56	9.23
Hydroxyanthracene	0.26	0.20
Benzonaphthofuran	0.48	–
Benzo(def)phenanthrene	–	–
Benz(ghi)perylene	–	–
Coronene	–	–
	99.74	97.52

a   55-65% of product to distillable and contains aromatics, 0 com-
      pounds and N compounds in the approximate ratio of 3:2:1
b   7.14% paraffins   3.84% monocycloparaffins   1.39%decyloparaffins

Illinois #6 Coal	Co-Steam Process North Dakota Lignite	Synthoil W. Virginia Bituminous Coal[a]	SRC-I Tacoma, WA W. Kentucky 9-14 Coal[b]
1.60			
5.75	0.44	1.70	
–	0.42	1.64	0.22
21.8	2.95	3.74	14.27
9.75	9.46	9.01	12.57
4.67	3.08	3.71	7.60
5.92	3.11	1.58	16.55
1.59	6.55	6.85	7.00
1.87	3.06	1.82	4.49
–	0.98	1.26	0.42
–	–	0.95	
–	0.93	1.36	
	0.06	0.40	
0.26	0.14	0.67	
0.13	0.02	0.41	
8.43	0.34	1.07	
2.74	1.06	1.77	1.36
0.59	0.62	0.68	3.00
1.38	0.52	0.52	2.56
–	0.04	0.15	4.17
–	0.65	0.42	0.09
–	0.02	0.10	0.69
21.5	4.30	9.03	0.13
–	1.69	2.30	0.88
8.53	7.29	4.50	9.78
0.59	0.55	0.29	
0.64	4.18	2.85	1.74
–	0.37	–	
–	0.10	0.71	
–	–	0.11	
97.74	52.93	59.63	87.5

Table 2. Composition of Coal Liquefaction Products[2]

	COED light oil	COED heavy oil	SRC-II Recycle Solvent
FIA Recovery	95%	61%	NP
1-ring naphthenes	68.78	10.46	
2-ring naphthenes	–	9.87	
3-ring naphthenes	5.71	9.16	
4-ring naphthenes		2.22	
Benzenes	15.58	6.35	4.31
Indenes	1.33	–	3.60
Tetralins/Indans	8.36	12.52	8.52
Naphthalenes	3.78	0.52	19.34
Fluorenes/Acenaphthylenes	–	0.12	8.32
Acenaphthenes/Biphenyl	–	–	14.63
Phenanthrene/Anthracene	–	0.20	9.32
Tetrahydroacenaphthenes	–	1.76	–
Octahydroacenaphthenes	–	–	–
Octahydrophenanthrene		5.50	–
Pyrenes		–	4.51
Hexahydropyrenes		0.55	–
Decahydropyrenes		1.70	–
Methylene Phenanthrene/ Phenyl Naphthalene			2.61
Chrysenes			.70
Phenols			13.03
Indenols			0.30
Indanols/Acetophenone			3.81
Dihydrophenol			1.09
Pehnylphenol			2.10
Dibenzofuran			3.41
Biphenol			0.40

NP: FIA (Fluorescent Indicator Absorption) analysis of this
material was not possible. Data presented was determined by
mass spectroscopy of the entire sample.

Table 3. Summary of Heteroaromatic Compound Types Identified in Coal Liquids[7].

Furans, Difurans, etc.

$C_nH_{2n-2}O$ through $C_nH_{2n-46}O$

$C_nH_{2n-6}O_2$ through $C_nH_{2n-38}O_2$

$C_nH_{2n-8}O_3$ through $C_nH_{2n-28}O_3$

Thiophenes,

$C_nH_{2n-6}S$ through $C_nH_{2n-36}S$

Dithiophenes, etc.

$C_nH_{2n-6}S_2$ through $C_nH_{2n-32}S_2$

Thiophenofurans

$C_nH_{2n-10}SO$ through $C_nH_{2n-34}SO$

$C_nH_{2n-8}SO_2$ through $C_nH_{2n-44}SO_2$

$C_nH_{2n-22}SO_3$ through $C_nH_{2n-38}SO_3$

Hydroxy compounds,

$C_nH_{2n-0}O$ through $C_2H_{2n-42}O$

Dihydroxy compounds,

$C_nH_{2n-6}O_2$ through $C_nH_{2n-40}O_2$

Hydroxyfurans, etc.

$C_nH_{2n-8}O_2$ through $C_nH_{2n-32}O_2$

Nitrogen compounds (one through eight cond. ring pyrroles, pyridines)

$C_nH_{2n-3}N$ through $C_nH_{2n-47}N$

Hydroxy-Nitrogen compounds

$C_nH_{2n-5}NO$ through $C_nH_{2n-41}NO$

$C_2H_{2n-9}NO_2$ through $C_nH_{2n-37}NO_2$

Miscellaneous Sulfur, Oxygen, and Nitrogen compounds

$C_nH_{2n-x}S_3$, $C_nH_{2n-x}S_2O$,

$C_nH_{2n-x}NO_2S$, $C_nH_{2n-x}O_4$,

$C_nH_{2n-x}SO_4$, $C_nH_{2n-x}SN$,

$C_nH_{2n-x}NO_3$, $C_nH_{2n-x}NO_4$, etc.

Table 4. Comparison of Coal Liquids and Petroleum Fuels[9]

		Coal Liquids		
Gradient Elution Chromatographic Analyses				
Cut No.	Description	SRC-I (Wt.%)	H-COAL (Wt. %)	Synthoil (Wt. %)
L	Saturates	0.04	9.31	6.60
2	MNA + DNA Oil	0.18	22.81	17.70
3	PNA Oil	0.90	17.15	12.10
4	PNA Soft Resin	10.83	13.15	17.00
5	Hard Resin	2.40	1.67	3.30
6	Polar Resin	4.47	1.92	3.70
7	Eluted Asphaltenes	23.94	11.34	24.50
8	Polar Asphaltenes	14.64	5.20	7.00
9	Polar Asphaltenes	6.14	2.14	1.30
10	Polar Asphaltenes	5.94	2.17	1.90
11	Polar Asphaltenes	13.90	5.80	2.80
12	Polar Asphaltenes	6.67	1.34	0.20
13	Non-eluted	9.95	6.00	1.90
	Ultimate Analyses (Wt. %)			
	Carbon	87.93	89.00	87.62
	Hydrogen	5.72	7.94	7.97
	Oxygen	3.50	2.12	2.08
	Nitrogen	1.71	0.77	0.97
	Sulfur	0.57	0.42	0.43

FCC Clarified Slurry Oil (Wt. %)	No. 6 Fuel Oil (Wt. %)	$1000°$ F+ Light Arabian Residuum (Wt. %)
10.37	30.50	11.0
6.81	20.30	18.6
32.04	13.20	19.7
29.89	16.10	28.0
2.95	2.30	4.5
4.35	4.20	6.1
8.78	7.90	8.9
1.98	3.80	2.3
0.54	0.30	0.2
0.14	0.08	0.04
0.06	0.06	0.06
0.04	0.02	0.02
2.05	1.20	0.6
90.52	86.4	83.88
7.64	11.2	9.97
0.76	0.30	0.48
0.44	0.41	0.40
1.16	1.96	4.19

Table 5. Compatibility of Coal Liquids and Petroleum Fuels

	Residue after Centrifuging (vol. %)			
	No. 2 Distillate Fuel	No. 6 Fuel Oil	Heavy Coker Gas Oil	Fluid Catalytic Cracker Clarified Slurry Oil
SYTHOIL (Pittsburgh Seam, Ireland Mine) [a]				
Full Range Product	18	50	15	0.4
−783 K fraction	0.05	0.05	0.05	---
H-COAL (Illinois No. 6, Burning Star Mine) [a]				
Full Range Product	26	50	30	20
−783 K fraction	0.05	0.05	0.05	---
H-COAL (+ 478 K)				
Raw	---	100	---	---
Mild Hydrotreatment	---	6	---	---
Moderate Hydrotreatment	---	0.05	---	---
SRC-I (Illinois No. 6, Burning Star Mine) [a]	Insoluble	Insoluble	Insoluble	20
(Illinois No. 6, Monterey Mine) [a]	---	---	Insoluble	---
SRC-I Recycle Solvent				
Raw	trace	---	---	---
Hydroprocessed	nil	---	0.05	---
Recycle Solvent/SRC-I (Wyodak) [b]				
Raw	12	14	---	---
Hydrotreated (203 sm^3/m^3)	5	2	---	---
Hydrotreated (364 sm^3/m^3)	3	0.3	---	---
Hydrotreated (494 sm^3/m^3)	0.3	1.0	---	---
Hydrotreated (548 sm^3/m^3)	13	11	---	---
Recycle Solvent/SRC-I (Illinois No. 6 Burning Star Mine) [b]				
Raw	50	---	50	---
Hydrotreated (311 sm^3/m^3)	---	---	15	---
Hydrotreated (317 sm^3/m^3)	15	---	4	---
Hydrotreated (503 sm^3/m^3)	---	---	9	---

[a] 1/1 blend

[b] 2/1 blend

Table 6. Analyses of an Aged Coal Liquid[14]

	Oil	Asphaltenes	Benzene Insolubles	O_2 Concentration wt. %
Original Sample	70	23	7	3.11
N_2 Storage	67	26	7	3.19
O_2 Storage	55	25	20	3.69

Table 7. Trace Metals Analyses of Coal-Derived Liquids and Petroleum[9]

Element ppm	SRC recycle solvent	SRC	H-coal	Synthoil	El Palito No. 6
Si	0.0	30.0	2.0	1348.0	3.0
Mg	0.2	4.0	1.0	33.0	4.0
Pb	0.3	1.0	0.0	5.0	2.0
Fe	4.4	140.0	20.0	375.0	6.4
Al	1.5	31.0	11.0	886.0	5.0
Ni	0.3	2.1	1.0	1.0	59.0
Ti	11.0	130.0	80.0	150.0	78.0
Ca	0.4	49.0	8.0	27.0	14.0
Na	1.6	100.0	0.8	79.0	19.0
K	0.4	8.0	0.4	116.0	na
V	0.0	6.8	2.6	1.8	275.0

280

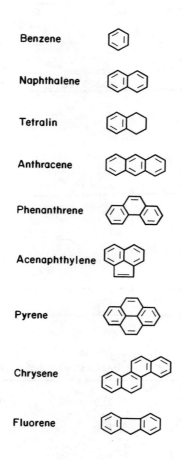

Benzene

Naphthalene

Tetralin

Anthracene

Phenanthrene

Acenaphthylene

Pyrene

Chrysene

Fluorene

Figure 1. Aromatic Compounds in Coal Liquids.

Phenol

Dibenzofuran

Benzonaphthofuran

Indanol

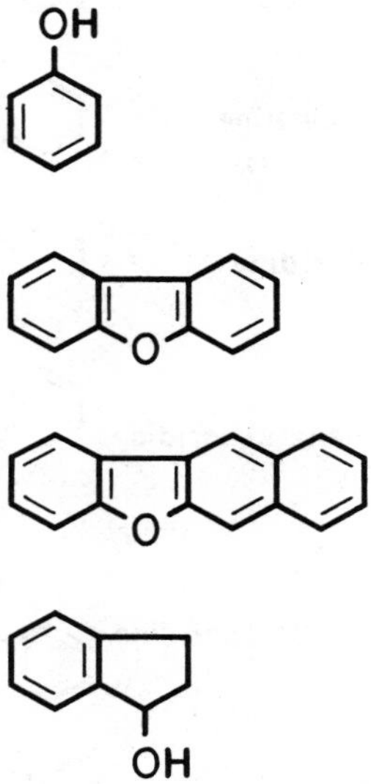

Figure 2. Oxygen-Containing Compounds in Coal Liquids.

282

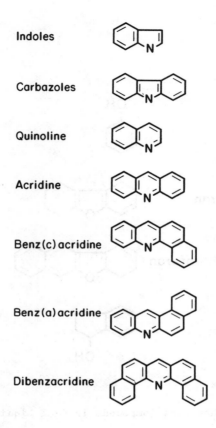

Indoles

Carbazoles

Quinoline

Acridine

Benz(c)acridine

Benz(a)acridine

Dibenzacridine

Figure 3. Nitrogen-Containing Compounds in Coal Liquids.

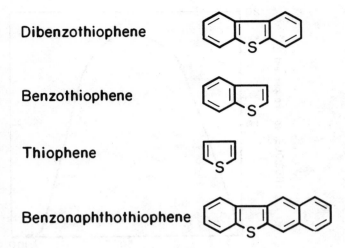

Figure 4. Sulfur-Containing Compounds in Coal Liquids.

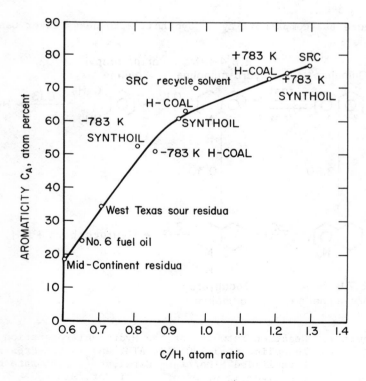

Figure 5. Aromaticity Versus C/H Atom Ratio for Various
Hydrocarbon Feedstocks[9].

284

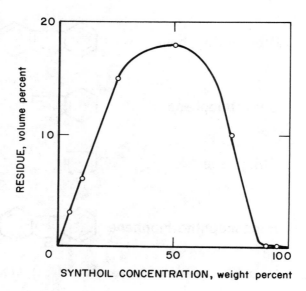

Figure 6. Compatability of Synthoil with Heavy Coker Gas Oil[9].

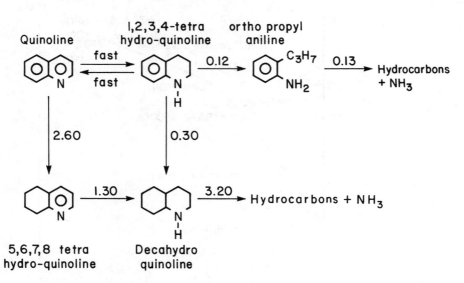

Figure 7. Reaction Network for the Hydrodenitrogenation of
Quinoline at 342°C, 136 ATM, and in the Presence of a
Presulfided NiMo/Al₂O₃ Catalyst[27]. The rate constants
are given in G. of oil/G. of Cat./Min.

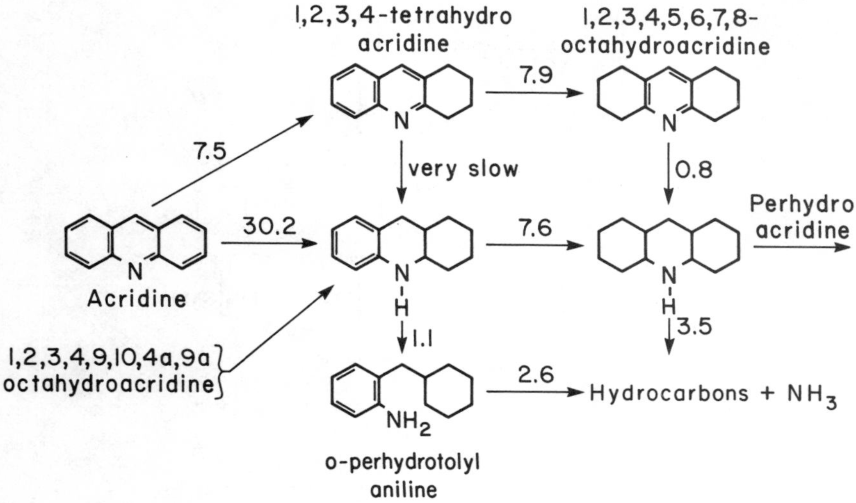

Figure 8. Reaction Network for the Hydrodenitrogenation of Acridine at 342°C, 136 ATM, and in the Presence of a Presulfided NiMo/Al$_2$O$_3$ Catalyst[28]. The Pseudo First Order Rate Constants are Given in G. of Oil/G. of Cat./ Min.

286

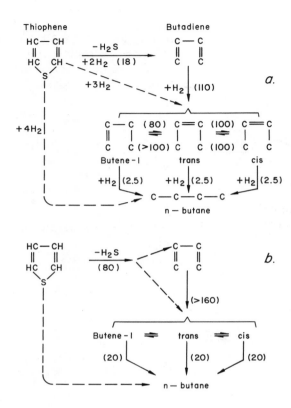

Figure 9.   Reaction Schemes and Estimates of Rates and Steps for the Hydrodesulfurization of Thiophene[32,33]. The Units for the Rate Constants are Micromoles Per Gram of Catalyst per Seconds

   A.   Chromia at 415°C

   B.   Cobalt Molybdate at 400°C

287

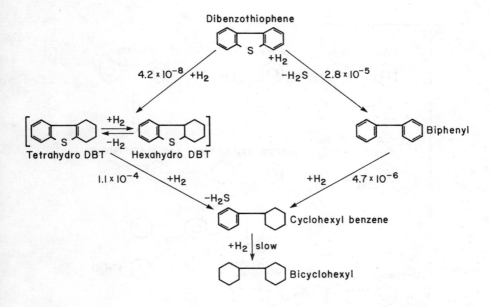

Figure 10. Reaction Network for the Hydrodesulfurization of Dibenzothiophene at 300°C, 103, ATM and in the Presence of a Presulfided CoMo/Al$_2$O$_3$ Catalyst[36]. The Pseudo First Order Rate Constants are Given in Cubic Meters/Kg. Cat./Sec.

288

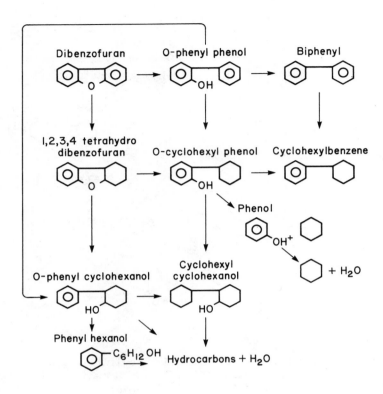

Figure 11.   Network for the Hydrodeoxygenation of Dibenzofuran[39] at 100 atm, 350°C and in the presence of a presulfided NiMo/λAl$_2$O$_3$ catalyst.

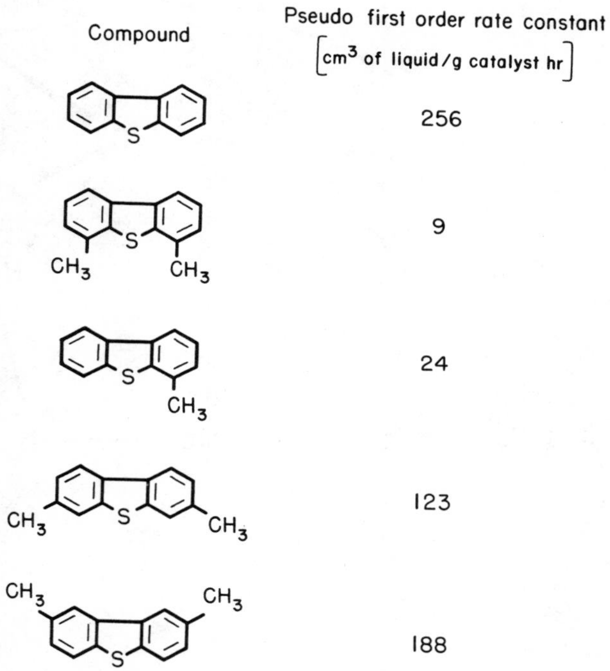

Compound	Pseudo first order rate constant $\left[cm^3 \text{ of liquid/g catalyst hr} \right]$
	256
	9
	24
	123
	188

Figure 12. Catalytic Hydrodesulfurization: Reactivities of
Dibenzo- and Methyl Dibenzothiophene[42,43]. Experimen-
tal Conditions: 1500 PSIG, 300°C, 2-5-37.5 X 10^3 G.
Cat. HR/CM, Sulfided CoMo/Al$_2$O$_3$.

290

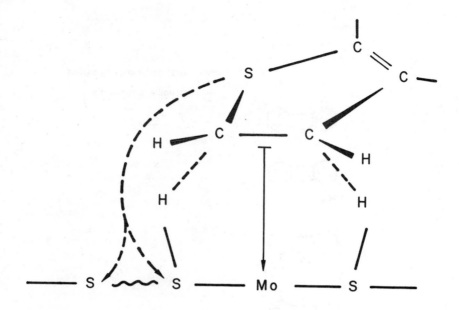

Figure 13. Suggested Structure of Chemisorbed Thiophene in the
Three-Point Mechanism. The Dashed Line Indicates a
Weaker Interaction Between Centers than the Full Line.
The Wavy Line Indicates a Longer or More Remote Bond
Between Centers.

MINERAL MATTER EFFECTS IN COAL CONVERSION*

B. Granoff
Sandia National Laboratories, Albuquerque, NM 87185

P. A. Montano
Dept. of Physics, West Virginia University
Morgantown, WV 26506

ABSTRACT

Coal is a heterogeneous, hydrogen-deficient, organic rock. In order to convert to an environmentally acceptable liquid fuel, it is necessary to: (1) add hydrogen; (2) hydrocrack to lower the molecular weight; (3) remove sulfur, nitrogen and oxygen; and (4) separate unconverted coal and mineral residues. Catalysts that are selective for increased oil production without concomitant gas formation are highly desirable. Certain naturally occurring minerals in coal, such as pyrite and clay, have been shown to enhance the liquid yield and product quality. Several high-volatile bituminous coals (KY No. 11, IL No. 6, WV, PA, etc.), with similar petrographic composition, were liquefied at 425°C for 30 min. with creosote oil as solvent. As the mineral content of the feed coal increased from 5 to 24 percent, the conversion to benzene solubles increased from 22 to 74 percent. Similar trends were observed when pyrite (pulverized to minus five microns) was added to an Illinois No. 6 coal, which was then liquefied at 425°C using SRC-II heavy distillate as the solvent. These and other mineral matter effects will be discussed, and the concept of disposable catalysts for coal liquefaction will be introduced. A brief description of the pyrite-to-pyrrhotite transformation will be given.

INTRODUCTION

Our dwindling supplies of domestic oil, coupled with a growing demand for transportation fuels has resulted in a significant increase in oil imports.[1] In Figure 1, we can see that imports increased from 24 percent of total oil consumed in 1970 to 46 percent in 1979. This reliance on foreign oil, with its associated economic, political and social problems, provides the basis for our "energy crisis." It is imperative that we reduce our dependence on imported liquid fuels.

* This work supported by the United States Department of Energy.

292

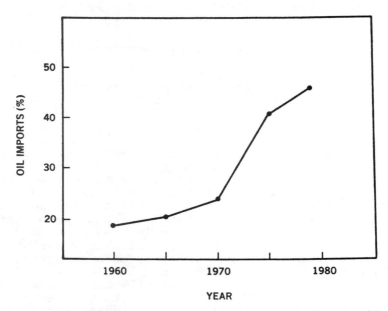

Fig. 1. Oil imports as a percentage of total oil
        consumed in the United States.

We do not have a shortage of fossil fuels; but we
do have a shortage of desirable fuels. If we look at
our proved and currently recoverable fossil fuel resources (see Table 1), it is apparent that oil and gas
(i.e., the desirable fuels) account for less than 10
percent of the total, whereas coal represents 85 percent
of our fossil fuel resource.[2] There is a large imbal-
ance between fuel consumption and fuel availability.
Approximately 76 percent of our total energy consumed is
supplied by oil and gas, but only 21 percent is supplied
by coal.[3] It is obvious, therefore, that in order to
alleviate our energy problems, we must utilize more coal
and reduce our consumption of oil and gas.

The problem with coal is that it is a hydrogen-
deficient, heterogeneous solid. To convert coal to a
desirable liquid fuel, it is necessary to add hydrogen,
remove mineral matter, remove sulfur, nitrogen and oxygen and reduce the molecular weight. Direct coal liquefaction is one valuable option for converting a wide
variety of coals into liquid products. Let us now look
into the direct liquefaction process, and introduce a
simple mechanism by which coal is converted to oil.

Table I   Fossil Fuel Resources[a]

Fossil Fuel	Proved and Currently Recoverable (10^9 BOE)
Crude Oil	29
Natural Gas	39
Oil Shale } + Bitumens }	75
Coal	802
Total	945

[a] BOE = Barrel of Oil Equivalent

COAL LIQUEFACTION

A generalized scheme for the direct hydroliquefaction of coal is shown in Figure 2. Coal, process-derived solvent and hydrogen are mixed and fed to a preheater (not shown), where liquefaction per se begins. The exit stream from the preheater enters a reactor where further conversion of the initially solubilized coal takes place. Separation of unconverted coal and mineral matter from the desired liquid products is achieved by means of distillation, filtration or solvent deashing.[4] A portion of the liquid product is recycled to provide process solvent for preparation of the coal feed slurry.

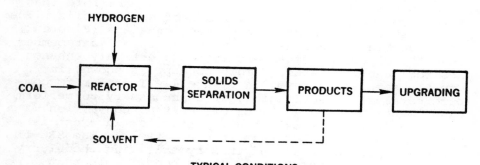

TYPICAL CONDITIONS

TEMPERATURE: 400 - 450°C

PRESSURE: 1500 - 3000 PSI

TIME: 15 - 60 MIN

Fig. 2. Generalized process scheme for the direct hydroliquefaction of coal.

In the SRC-II process (Solvent Refined Coal, distillate product), part of the unconverted coal and mineral matter residue are recycled back to the preheater along with the solvent.[5] It is believed that this slurry recycle enhances the yield of distillate products by: (1) increasing the effective residence time of the unconverted organic matter; and (2) utilizing the catalytic effects of the inherent mineral matter. In the EDS (Exxon Donor Solvent) process, the solvent is catalytically hydrogenated prior to being mixed with the feed coal.[6] A recent, beneficial, modification of the EDS process has been to recycle a portion of the vacuum bottoms.[7] This was shown to result in higher conversion of coal and in improved flexibility of the product slate. In the H-Coal (Hydrocarbon Research, Inc.) process, the feed slurry contacts a catalyst (e.g., cobalt molybdate on alumina) in an ebullated bed reactor.[8] It is believed that this catalytic hydroliquefaction results in improved liquid properties.

A simple mechanism for coal liquefaction is shown in Figure 3. The initial solubilization consumes solvent and the resulting preasphaltenes, as operationally defined by Soxhlet extraction, have been shown to be adducts of coal and solvent.[9,10] The preasphaltenes are converted to asphaltenes and oils in a series of secondary reaction steps. The preasphaltenes are soluble in pyridine or tetrahydrofuran (THF) but are insoluble in benzene or toluene. The asphaltenes are soluble in benzene or toluene, but are insoluble in pentane or hexane. The oils are soluble in pentane or hexane.

At this point, we want to determine the role that a catalyst plays in the conversion of coal to oils. The key question is: Which of the reaction steps (Figure 3) are amenable to catalysis? It is important to remember that the object of catalysis is to selectively enhance desired reaction paths.[2] This concept is shown, in greatly simplified form, in Figure 4. In order to compare catalyst selectivities, it is necessary to perform experiments under conditions of constant conversion.

One class of catalysts of particular interest is the inherent mineral matter in coal itself. The SRC-II process takes advantage of this by utilizing the slurry recycle mode of operation.[5] It is conceivable that similar mineral matter effects occur in the EDS bottoms recycle process.[7] The importance of mineral effects must be assessed for optimum conversion of coal to liquid products, and for the development of relatively inexpensive disposable catalysts. The role of the mineral matter must be understood if we are to develop direct liquefaction processes that are not coal-specific.

1. INITIAL DISSOLUTION

   COAL + SOLVENT → PREASPHALTENES

2. SUBSEQUENT REACTIONS

   PREASPHALTENES → ASPHALTENES → OILS

COMPONENT	MOLECULAR WEIGHT
PREASPHALENES	1500
ASPHALTENES	400 - 700
OILS	200 - 300

Fig. 3. The mechanism of coal liquefaction.

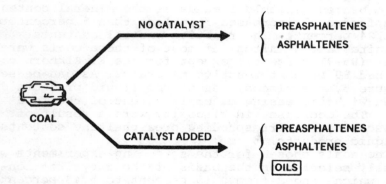

Fig. 4. The objective of catalysis is to selectively
enhance desired reaction paths.

MINERAL MATTER EFFECTS

The major minerals in coal are listed in Table 2.
Approximately one-half of the mineral matter in most
bituminous coals consists of clays.[11,12] The mineral
content may be determined by acid extraction,[13] or low
temperature ashing.[11,14]

The objective of our work has been to correlate
mineral effects with process chemistry and product composition. Of particular interest are the effects of
minerals on conversion, oil yield, heteroatom content,
viscosity, distillate boiling range and hydrogen consumption. Several experimental approaches have been

Table II   Major Minerals in Coal

Montmorillonite	)	
Illite	}	Clays
Kaolinite	)	
Pyrite	FeS_2	
Calcite	$CaCO_3$	
Quartz	SiO_2	
Gypsum	$CaSO_4 \cdot 2H_2O$	

used: (1) screening studies with coals of similar petrographic composition, but widely varying mineral content;
(2) removal of minerals from highly reactive coals; and
(3) addition of minerals to non-reactive coals.

An example of a mineral matter effect is shown in
Figure 5. Here, a number of high-volatile bituminous
coals (KY, WV, IL, and PA) were liquefied in creosote
oil solvent for 30 minutes at 430°C and 1700 psig.[15]
There was a significant increase in the conversion of
coal to benzene soluble liquids as the mineral content
of the feed coal increased from less than 5 percent to
almost 24 percent. The reactive maceral contents
(vitrinites and exinites) of most of these coals were
similar (85-91 percent), except for the KY Elkhorn coal
which had 80 percent reactive macerals. As can be seen
in Figure 5, the viscosities of the liquid products decreased with increasing mineral content of the feed
coals. The decreases in viscosity were a result of
decreases in the corresponding preasphaltene contents of
the liquid products.[16]

The coals chosen for the screening experiments were
carefully selected on the basis of their pyrite* contents, which ranged from 0.15 percent to 5.43 percent.[15]
There has been a great deal of interest in the catalytic
role of pyrite in coal liquefaction.[17-19] Pyrite is
known to occur in a number of forms (single crystal,
framboid, massive) and particle size distributions in
coal.[20] For the screening series, the correlation between conversion to liquid products (benzene solubles)
and pyrite content of the feed coal is shown in Figure
6. It is clear that there is a significant increase in
conversion with increasing pyrite content.[15]

Similar effects were observed in a study of a North
Assam coal that had been separated into several specific

---

* Pyrite is the cubic dimorph of iron disulfide. The
orthorhombic variety is known as marcasite.

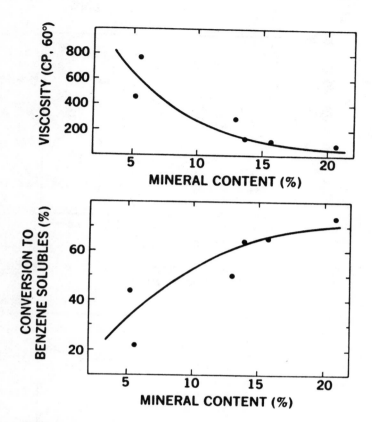

Fig. 5. Minerals in coal affect conversion and product quality.

gravity fractions.[21] The coal samples were liquefied with hydrogen in the absence of a solvent. The correlation between conversion (to benzene solubles) and mineral matter content is shown in Figure 7. These trends are similar to those found in our screening study (Figure 5).[15] In the Assam coal, the predominant iron-containing mineral was pyrite, and the relationship between conversion and pyrite content was similar to that shown in Figure 7 for total mineral content.[21]

The effect of pyrite has been demonstrated in a recent study conducted at the Pittsburgh Energy Technology Center (PETC).[22] A high-sulfur, western Kentucky coal was cleaned to remove the pyrite. The cleaned coal and the run-of-mine coal were liquefied in a continuous reactor. The viscosity of the liquid obtained from the cleaned coal was considerably higher than that obtained from the run-of-mine coal. Portions of the pyrite concentrate (gravity sink fraction) were then added to

298

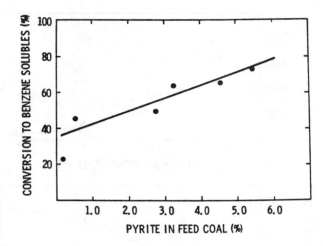

Fig. 6. Correlation between conversion and pyrite
content of the feed coal.

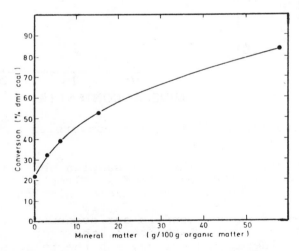

Fig. 7. Effect of mineral content on the conversion of
North Assam coal to benzene soluble liquids.
(Mukherjee and Chowdhury)

the cleaned coal, and the mixture was liquefied. As can
be seen from the data in Figure 8, the viscosity of the
liquid product decreased as the iron (i.e., pyrite) content of the feed increased. The viscosity of the liquid
from the "reconstituted coal" was virtually the same as
that from the original run-of-mine coal. This decrease
in product viscosity with increasing pyrite content of

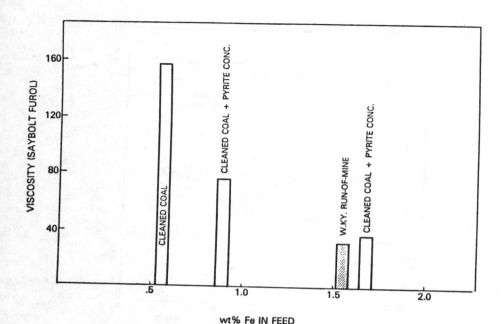

Fig. 8. Correlation between liquid product viscosity
and total iron content of feed coal (PETC).

the feed coal was similar to that observed in the
screening experiments (Figure 5).

In a recent study at The Pennsylvania State University, pyrites from Ohio, Iowa, and Pennsylvania coals,
as well as mineral pyrite from Colorado, were added to
a West Virginia coal that had a very low pyrite content.[23] The resulting mixtures were liquefied in tetralin solvent at 425°C for 30 minutes at 1400 psig. The
conversion to liquid products (ethyl acetate solubles)
as a function of the concentration of added pyrite is
shown in Figure 9. In all cases, there was an increase
in conversion with increased pyrite content of the feed.
It is interesting to note that the pyrite from the Ohio
coal was the most active catalyst, whereas the pyrite
from the Iowa coal was least active. It is not known
whether this was due to minor impurities, morphological
differences, surface area effects, etc. The key result,
however, was that pyrite enhanced conversion and that
the activities of different pyrites appeared to vary
considerably.

From the data presented above, it is clear that
pyrite has a definite effect on coal conversion and
product quality (e.g., viscosity). Can pyrite be utilized as a disposable catalyst for coal liquefaction? In
order to address this question, we have been conducting

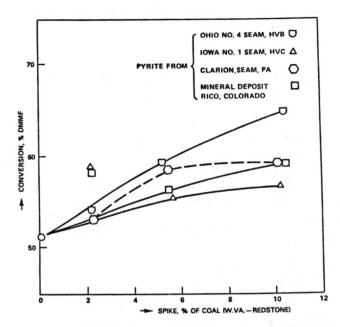

Fig. 9. The effect of various pyrites on the conversion
of West Virginia coal to liquid products.
(Penn State)

experiments in which a coal-derived pyrite concentrate
was used as a liquefaction catalyst. The pyrite samples
were made available to us by U.S. Steel Corporation
(Robena laboratory) and were pulverized to minus 5 μm
using a fluid energy mill.* Mineralogical analysis, by
low temperature ashing (LTA), revealed a total mineral
content of 98 percent and a pyrite (plus approximately
10 percent marcasite) content of 70 percent.[24] Properties of the Robena pyrite are given in Table 3.

Batch autoclave experiments were carried out with
Illinois No. 6 coal (River King Mine) in SRC-II heavy
distillate solvent at 425°C for 30 minutes at 1800 psig.
Properties of the coal and solvent are given in Tables
4 and 5, respectively. The effect of added pyrite on
the conversion of coal to benzene solubles is shown in
Figure 10. There was an increase in conversion from
73 percent, without any added pyrite, to 88 percent when
8 percent pyrite (based on coal) was added to the feed.
The corresponding distillate (850°F-) and SRC (850°F+;
THF soluble) yields are shown in Figure 11. There was

---

* This comminution was performed for us by the Jet
Pulverizer Company, Palmyra, NJ.

301

Table III   Properties of Minus 5 μm Robena Pyrite

Particle Size	1.8 μm (average)
Surface Area	4.7 m^2/g
Helium Density	4.07 g/cm^3
Mineral Content (LTA)	98.0 %
Pyrite Content (x-ray)	70.0 %

an increase in total distillate and a concomitant
decrease in SRC as the pyrite content of the feed increased. The data strongly suggest that pyrite has a
catalytic effect in the conversion of heavy, non-distillate materials to distillate-range liquids. The corresponding reduction in high molecular weight components
(e.g., preasphaltenes) could account for the reduced
liquid viscosities that were observed in the work at
PETC.[22]
In an attempt to determine whether pyrite is a
selective catalyst for the production of liquids, we
carried out experiments at constant conversion. The
objective was to compare a thermal run at 425°C (no added pyrite) with a catalytic run (5 percent Robena pyrite
added) at 400°C. In each case, the coal was Illinois
No. 6 (River King), the solvent was SRC-II heavy

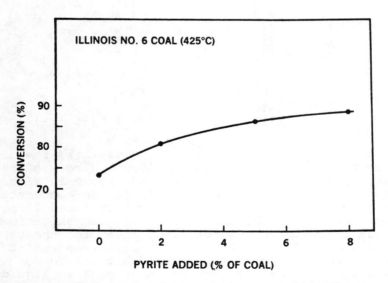

Fig. 10. Addition of Robena pyrite to Illinois No. 6
coal improves conversion to benzene soluble
liquids.

Table IV   Proximate and Ultimate Analysis of
Illinois No. 6 (River King) Coal

Proximate Analysis (wt. %)

Moisture	2.29
Ash	13.32
Volatile	38.17
Fixed Carbon	46.22

Ultimate Analysis (wt. %, dry)

Carbon	67.23
Hydrogen	4.87
Nitrogen	1.39
Chlorine	0.03
Sulfur	4.03
Oxygen (diff)	8.82
Ash	13.63

Sulfur Forms (wt. %, dry)

Pyrite	1.35
Sulfate	0.15
Organic (diff)	2.53

Table V   Properties of SRC-II Heavy Distillate

Carbon (%)	89.8	
Hydrogen (%)	7.6	
Nitrogen (%)	1.4	
Sulfur (%)	0.4	
Oxygen (%)	1.8	
Ash (%)	0.05	
Pentane Insolubles (%)	10.0	
Distillate Yield (%)	86.5	(850°F-)

distillate, the solvent-to-coal ratio was 2:1, the
pressure was 1800 psig and the residence time was 30
minutes. The data from these runs are shown in Figure
12. The conversions to benzene solubles were essentially the same: 73 percent for the thermal run and 75 percent for the catalyzed run. The C_1-C_4 gas make, however, was approximately three times lower for the catalyzed run at 400°C than for the non-catalyzed run at
425°C. It would appear, therefore, that pyrite can
selectively catalyze the conversion of coal to liquid
products without concomitant gas formation.

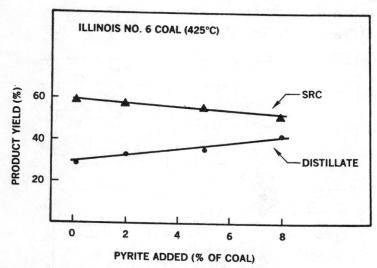

Fig. 11. Addition of Robena Pyrite to Illinois No. 6
coal improves distillate yield.

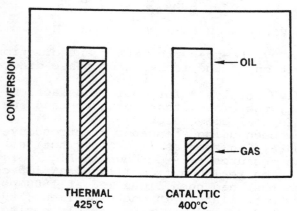

Fig. 12. Pyrite enhances the selective conversion of coal
to oil (benzene solubles).

THE PYRITE-TO-PYRRHOTITE TRANSFORMATION

Under typical liquefaction conditions (425°C, 30
min., 2000 psig), pyrite (FeS_2) is transformed into a
non-stoichiometric iron sulfide known as pyrrhotite
($Fe_{1-x}S$).[25-27] This is illustrated in Figure 13, along
with a portion of the iron-sulfur phase diagram. It is
clear that a number of different pyrrhotites can be
formed, depending on the reaction temperature and sulfur
fugacity. The suggestion has been made that pyrrhotite

304

$$(1-x)\ FeS_2 + (1-2x)\ H_2 = Fe_{1-x}S + (1-2x)\ H_2S$$

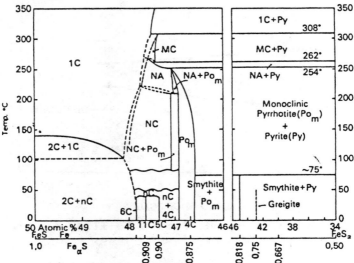

PORTION OF Fe-S PHASE DIAGRAM

Fig. 13. The pyrite-to-pyrrhotite transformation and a portion of the iron-sulfur phase diagram.

is the "active catalyst" for coal liquefaction.[28,29] It is important, therefore, that we understand the role of pyrrhotite in coal conversion.

We have been using Mössbauer spectroscopy to study the iron-containing species (i.e., pyrrhotites) in coal liquefaction residues.[25,30] Shown in Figure 14 are the data obtained on residues from Illinois No. 6 coal (River King) that had been liquefied at 1800 psig in SRC-II heavy distillate. In one experiment, the coal was heated to 325°C and was then rapidly cooled to room temperature. The Mössbauer spectrum of the residue shows that the major iron-containing species is pyrite (doublet), but pyrrhotite has already started to form. In a second experiment, coal was heated to 405°C and was held there for 30 minutes. The resulting residue shows the presence of hexagonal pyrrhotite without any remaining pyrite. The same result was obtained at 425°C. The differences in the Mossbauer spectra (405°C vs. 425°C) can be interpreted in terms of differences in the stoichiometries of the pyrrhotites in the residues.

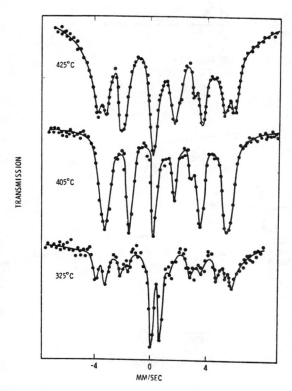

Fig. 14. Mössbauer spectra of liquefaction residues from
Illinois No. 6 coal that was liquefied in
SRC-II heavy distillate at different tempera-
tures.

The residues from Kentucky 9/14 coal (Colonial Mine)
and West Virginia coal (Blacksville No. 2 Mine) have
also been studied by Mössbauer spectroscopy.[30] The data
are shown in Figure 15, where we plot conversion to
benzene and THF solubles as a function of the iron content of the pyrrhotite in the liquefaction residues.
All runs and product workups were carried out under the
same conditions. It is seen that, as the conversion to
liquids increased, the iron in the pyrrhotite decreased.
It appears as if the most reactive coal (Illinois No. 6)
is associated with the most iron-deficient pyrrhotite.
When Robena pyrite (five weight percent based on coal)
was added to the West Virginia coal, the conversion to
benzene and THF solubles increased, and the resulting
liquefaction residue contained a pyrrhotite with a
lower atomic percent iron than did the residue from the
original run without added pyrite. Our experimental

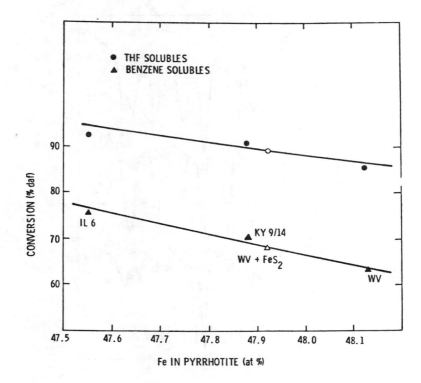

Fig. 15. Conversion of coals to THF and benzene solubles
as a function of the atomic percent iron in
the liquefaction residue pyrrhotite.

evidence suggests a correlation between the atomic percent iron in the pyrrhotites and the conversion of coal
to benzene and THF solubles.[30] We are continuing this
work in order to more clearly define the role of pyrite
in coal liquefaction, and to identify the most catalytically active iron sulfide species.

SUMMARY

We have shown that inherent mineral matter enhances the conversion of coal to liquids, and improves
the quality of the product (e.g., viscosity). Pyrite,
in particular, has been shown to be a selective catalyst
for the production of liquids without concomitant gas
formation. Pyrite is effective for the conversion of
non-distillate SRC to distillate (850°F-) liquids. During coal liquefaction, pyrite is rapidly converted to
pyrrhotite. It is not clear at this time whether

pyrrhotite is the "active catalyst", but we have found
a correlation between coal reactivity and pyrrhotite
stoichiometry. We are currently investigating the role
of sulfur in coal liquefaction and are attempting to
define the mechanism(s) of the pyrite-pyrrhotite
transformation.

REFERENCES

1. R. Stobaugh and D. Yergin, Harvard Business Review,
   Jan.-Feb. 1980, pp. 57-73.
2. H. R. Linden, "The Robert A. Welch Foundation Con-
   ferences on Chemical Research. XXII. Chemistry of
   Future Energy Resources", Houston, TX, Nov. 6-8,
   1978, p. 153.
3. H. Brooks and J. M. Hollander, Ann Rev. Energy 4, 1
   (1979).
4. R. H. Wolk, "Overview of Liquefaction Process Technology", IGT Symposium on Advances in Coal Utilization Technology, Louisville, KY, May 14-18, 1979.
5. B. K. Schmid and D. M. Jackson, Coal Processing
   Technology, 5, 146 (1979).
6. W. R. Epperly and J. W. Taunton, ibid., 5, 78 (1979).
7. L. L. Ansell, K. L. Trachte and J. W. Taunton,
   "Bottoms Recycle Studies in the EDS Process Development", Fifth Annual EPRI Contractor's Conference,
   Palo Alto, CA, May 7-8, 1980.
8. C. D. Hoertz and J. C. Swan, "The H-Coal Process",
   in Coal Conversion Technology, ed. by A. H. Pelofsky,
   ACS Symposium Series No. 110 (1979).
9. M. G. Thomas and R. K. Traeger, Preprints Div. Fuel
   Chem., Amer. Chem. Soc., 24 (3), 224 (1979).
10. M. G. Thomas and T. C. Bickel, to be presented at
    the American Chemical Society Meeting, San Francisco,
    CA, August, 1980.
11. C. P. Rao and H. J. Gluskoter, "Occurrence and Dis-
    tribution of Minerals in Illinois Coals", Illinois
    State Geological Survey, Circular 476 (1973).
12. R. S. Mitchell and H. J. Gluskoter, Fuel 55, 90
    (1976).
13. M. Bishop and D. L. Ward, ibid., 37, 191 (1958).
14. R. N. Miller, R. F. Yarzab and P. H. Given, ibid.,
    58, 4 (1979).
15. B. Granoff and M. G. Thomas, Preprints Div. Fuel
    Chem., Amer. Chem. Soc., 22 (6), 183 (1977).
16. M. G. Thomas and B. Granoff, Fuel 57, 122 (1978).
17. E. C. Moroni, "DOE Overview - Advanced Direct Coal
    Hydroliquefaction", Fifth Annual EPRI Contractor's
    Conf., Palo Alto, CA, May 7-8, 1980.
18. J. A. Guin, A. R. Tarrer, J. W. Prather, D. R.
    Johnson and J. M. Lee, Ind. Eng. Chem., Proc. Des.
    Dev. 17, 118 (1978).

308

19. T. D. Padrick, F. V. Stohl and M. G. Thomas, "The Decomposition of Pyrite: A Kinetic Study Under Coal Liquefaction Conditions", Sandia National Laboratories (to be published).
20. R. T. Greer, "Coal Microstructure and Pyrite Distribution", in Coal Desulfurization, ed. by T. D. Wheelock, ACS Symposium Series, No. 64 (1977).
21. D. K. Mukherjee and P. B. Chowdhury, Fuel 55, 4 (1976).
22. E. C. Moroni and R. H. Fischer, "Disposable Catalysts for Coal Liquefaction", 179th National Meeting of the American Chemical Society, Houston, TX, March 24-28, 1980.
23. P. H. Given, "Catalysis of Liquefaction by Iron Sulfides from Coals", U.S. DOE Project Review Meeting on Disposable Catalysts in Coal Liquefaction, Albuquerque, NM, June 5-6, 1979.
24. F. V. Stohl, Sandia National Laboratories, Private Communication (1979).
25. P. A. Montano, "Stoichiometry of Iron Sulfides in Liquefaction Residues and Its Correlation with Conversion", U.S. DOE Project Review Meeting on Disposable Catalysts in Coal Liquefaction", Albuquerque, NM, June 5-6, 1979.
26. J. T. Richardson, Fuel 51, 150 (1972).
27. Y. A. Liu and C. J. Lin, IEEE Trans. on Magnetics, MAG-12 (5), 538 (1976).
28. D. K. Mukherjee, J. K. Sama, P. B. Chowdhury and A. Lahiri, "Hydrogenation of Coal with Iron Catalysts", Proc. Symp. on Chemicals and Oil from Coal, Central Fuel Research Institute, Dhanbad, India, December 6-8, 1969.
29. P. H. Given, "The Dependence of Coal Liquefaction Behavior on Coal Characteristics", Short Course on Coal Characteristics and Coal Conversion Processes, The Pennsylvania State University, June 7-11, 1976
30. P. A. Montano and B. Granoff, Fuel 59, 214 (1980).

FUNDAMENTAL RESEARCH IN COAL COMBUSTION:
WHAT USE IS IT?

Robert H. Essenhigh
E. G. Bailey Professor of Energy Conversion
Department of Mechanical Engineering
The Ohio State University
Columbus, Ohio 43210

SUMMARY

The value of fundamental research in coal combustions is still
regarded as questionable by many concerned with utilization of
research results or with providing research support. It is recognized that the record of the utility of fundamental research has not
been outstanding to date, but it is argued here that this is very
largely a function of the complexity of the problem which still has
not been fully recognized, and the need for those computational
facilities that have only become sufficiently available in the last
decade or so. Three examples of effective modeling are discussed:
the boiler flame, the one-dimensional char flame, and the fixed-bed
gasifier that have identified important mechanisms with evident
implications for utility of the results. A particular feature of
the work discussed is the use of parsimoneous modeling which is
shown to be particularly effective in obtaining useful results and
in specifying priorities for future research. The importance of
coupling experimental verification with the parsimoneous modeling
is also emphasized. A particular problem identified, as an example
of the evaluation of research priorities, is the need to understand
the way in which the internal surface of coals and chars develop
during reaction.

1. INTRODUCTION

The combustion of coal is a process that has been a matter of
technical and technological interest for many decades. To be sure,
the level of interest has waxed and waned with the fortunes of the
coal industry. However, those fortunes today, as I have no need to
emphasize, are believed to be on the rise. In consequence, there is
at least nominal interest, once again, in the scientific base of
coal combustion technology, which is to say, the coal and carbon
combustion mechanisms, pyrolysis mechanisms, mechanisms of heat
transfer, combustion aerodynamics, mechanisms of ignition and
extinction, principles of flame holding, and the like.
    The assumption is that such information will be of value in
boiler, furnace, and burner design. At the same time, however,
there are also those who are totally sceptical of any such assumption;
they will point out the extraordinary developments and successes of
boiler design that were achieved with little knowledge of or
reference to fundamentals. Those developments, however, were mostly
achieved over a time span of roughly three quarters of a century,

with relatively small incremental improvements occurring year by year, and being tested at full scale; but this is a procedure that becomes progressively more expensive and time consuming as the scale increases. There is, therefore, major incentive to try once again to link the combustion fundamentals with boiler performance, which is the primary justification for increased research on the fundamentals. Additional pressures for this approach come, first, from increased use of hitherto unused coals, mostly western lignites, where there may be only a marginal back-log of prior empirical experience; and then also from prospective availability of chars in large tonnage from gasification and liquefaction processes. This last presents a particularly difficult problem as some chars will probably require addition of supplementary fuel if they are to be used in combustion (jointly the problems of ignition, flame holding, and burn-out). Questions already being asked are: how much fuel, of what type, and how or where should it be injected?

In developing a fundamental research program, I would like to emphasize the significance of practical applications for three reasons. First, the investigation of basic coal combustion mechanisms is still applied research, albeit at a fundamental level, since utilization is the source of the incentive for such investigations. Support for the fundamental investigations is, therefore, likely to be somewhat in proportion to the degree to which the results can be seen to have value in utilization. Second, the test of "application", even for very fundamental research, has historically usually provided an excellent basis for identifying the most significant problems and, thus, for ordering research priorities. It is, therefore, likely to be efficient, or parsimoneous, in allocating scarce research resources to a very complex problem. Third, the ultimate users of the research must be the boiler, furnace, and burner companies who all, traditionally, have been reserved in their support of such research. There are recognized wide differences of opinion within the companies' personnel, but on balance, there tends to be good correlations between age, the conservative views, and influence in the company. The most extreme view is that all "fundamental" research is a waste of time. However, if the research results cannot be used, it was, indeed, a waste of time.

There has been a tendency in the research community to regard the more conservative views towards research as due to ignorance and neanderthalic attitudes. The conservative beliefs, however, are generally based on two, three, or four decades of experiences, and it is a little high handed to write these off as of no account. In my view, I do not think they give altogether sufficient recognition to the awful complexity of the problem -- a nuclear reactor in basic engineering design is simple by comparison; but beyond that, it would be realistic to recognize that fundamental research has not yet contributed much to boiler design and operation. (There is also the possibility that, in fact, much of the fundamental work has already been done, is used, but is proprietary, and the companies have no wish to see their (nominal) competitive advantage diminished.) This would seem to place the onus on the research community to

demonstrate utility of appropriate research rather than on the users
to prove inapplicability. The point of view is important as it
determines programs, priorities, and objectives. At the same time,
it must be recognized by the users that demonstration of utility
will always remain an impossibility unless the research community
is provided with the necessary resources to engage in the appropriate
research.

In this present paper, it is neither possible to review the
present standing of all fundamental research, not is it possible to
explore too deeply exactly how the link between fundamentals and
application is to be accomplished; essentially there are still too
many blanks. Moreover, research developments across a field are
always uneven and not necessarily yet connected. I have, therefore,
selected a limited number of problems, for discussion and as a means
for indicating how research priorities can possibly be set within
the context of recognizing the constraint of ultimate utilization.
We start with a broad-brush mapping of the research field.

2. SCOPE OF THE RESEARCH PROBLEM

Coal can be burned (or gasified) in one of three ways: as lump
coal in a fixed bed on a grate; as crushed coal in a fluid bed; or
finely ground in an entrained stream.

In all cases, the phenomena involved included the following: as
the result of combined conductive, convective, and radiative heating,
the coal temperature rises to the point where the coal starts to
pyrolyze and/or react with oxygen (and with CO_2 and water vapor at
sufficiently elevated temperatures). If conditions are appropriate,
the pyrolysis products may crack either inside or outside the
particles, depositing carbon on the surface of other particles. The
balance of the pyrolysis and gasification products are then able to
burn-out if, first, there is sufficient overall oxidizer present (at
a high enough temperature) -- the problem of stoichiometry; and
second, the combustibles and oxidizer are appropriately present and
mixed -- the problem of combustion aerodynamics. The rate of
oxidation is governed jointly by the rates of combustibles production
and reaction -- the problems of pyrolysis and homogeneous reaction
kinetics. Calculations indicate that smaller particles may possibly
oxidize during pyrolysis, but the majority (by weight) probably
pyrolyze so fast that the surface is swept clear of oxygen and the
surface oxidation reaction is quenched until the volatiles efflux
dies down. When heterogeneous reaction is able to commence, it does
so by the oxygen-transfer gas (O_2, CO_2, H_2O) diffusing from the
interparticulate space to the external particle surface where a
fraction may react. The balance diffuses into the porous char
particle where it will react, with the penetration depth of the
reaction varying from nothing to full penetration to the particle
center, depending on the degree of particle porosity, gas concentration, temperature, and change of porosity during reaction -- the
problem of heterogeneous kinetics with diffusion. The rate of
pyrolysis and structure of the resultant char depends on the parent

coal, its pretreatment (if any), rate of heating, and ambient gas type during heating -- the problem of <u>fuel materials</u>. Maintenance of the flame and/or stability inside the reactor depends on a complex thermal balance, if stability is mainly thermally determined -- the problem of <u>reactor analysis and flame holding</u>; some investigators also believe that normal flame speed in a non-recirculating flame is significant -- the problem of <u>flame speed in a one-dimensional flame</u>. Since combustion is for a purpose, heat must be transferred from the flame, and the effective utilization of that heat yields the thermal efficiency of the boiler or furnace -- the problem of <u>furnace analysis</u>.

The above brief description identifies a number of specific topics that are linked by complex interactions in a complete flame. The research problem is then definable as having two parts: first, to develop the necessary information on the individual topics (pyrolysis, char reactions, etc.) that must be appropriately isolated for experimental and theoretical investigation; and second, to determine the effects of interactions, taking, for example, reaction kinetics and heat transfer together: thus, basic kinetic data are determined under "isothermal" conditions, if possible, with temperature constrained. With temperature no longer constrained, the influence of heat transfer is added to the system, not always with expectably predictable results.

The development of necessary data on individual topics, and data on topic interactions poses a separate problem of planning. Figure 1 shows one possible solution to the problem. This is designed to illustrate: identification of the four basic sciences that combustion technology and combustion engineering rest on; to illustrate the technological interactions of the basic sciences that create combustion theory, flame propagation theory, reactor theory, and furnace analysis; and to illustrate the relationship of the science and technology of combustion to the engineering applications. It also shows that there is no such thing as "applied" and "fundamental" research levels; there is, instead, a hierarchy of levels of applied research of increasing degree of fundamentality. It also illustrates the difficulty of relating bench scale reactivity tests that do not include heat transfer and aerodynamic effects to full-scale boiler operation. It is not necessarily impossible; Waggoner and Weingartner (1), for example, have reported that TGA "burning profiles" of different coals mostly fall into groups that show good correlations with boiler performance, but the causation behind the correlation is uninown, and not all TGA profiles can be interpreted. It is also possible that the correlations apply uniquely to the particular boiler and burner designs examined, and a different pattern of correlations would be found with different boiler designs.

In Fig. 1, the four basic sciences identified are reaction kinetics (governing homogeneous, heterogeneous, and pyrolysis reactions); materials properties (implying reactor containment materials -- refractories -- as well as fuels); heat transfer (conduction, convection, and radiation); and combustion aerodynamics. It is usually

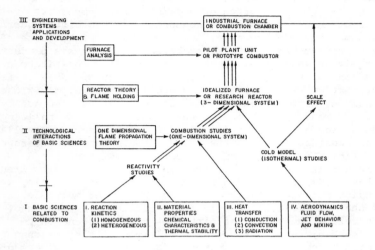

Figure 1. Schematic representation of relationship of combustion
science fundamentals to combustion engineering applications, illustrating also the concept of a hierarchy of
research levels.

assumed, but with less than total justification, that all necessary
knowledge at the basic sciences level is available.

We may accept that the heat transfer laws are fully formulated,
at least in principle; there can be difficulty in applying them,
particularly in the pulverized coal flame, because of the difficulty
of predicting or determining absorbing and scattering parameters.

- In the combustion aerodynamics, treatment of turbulence remains
a problem; likewise, the problem of predicting backmix and mixing
patterns is still largely open ended.

- In the fuel properties, which effectively overlaps with
response to thermal stress (pyrolysis), there is still no working
model of coal structure or coal constitution, beyond statistical
generalizations. This makes reliable extrapolation to new coals

impossible. In consequence, it is still necessary to do full physical, chemical, and combustion tests on each individual coal to be sure of performance.

     - In reaction kinetics, there are three aspects. Gas phase kinetics equations can be assumed to be known in principle. In practice, there is often uncertainty what components are locally present; necessary kinetic constants are too often uncertain or unknown; but, in any case, the use of a complete suite of equations to represent, for example, reactions of pyrolysis products would be largely unmanageable. Uncertainty over the constituents and concentrations of the pyrolysis products is a consequence of too little information yet available from studies of pyrolysis kinetics. It requires further work on the lines of that carried out by Howard (2) and Solomon (3), but again, the prospects of generalizing to all coals from a few are still clouded by the lack of a valid model of coal constitution. In the third aspect, that of heterogeneous kinetics -- which governs the bulk of coal combustion -- there are some significant problems that make first-principles prediction of behavior difficult and often impossible. For example, we cannot predict burnout rates and times of single or multiple sets (clouds) of char particles except in the limited circumstances of diffusion-controlled combustion of particles greater than 100 μm. Even more so is this the case for coal particles. Again, we cannot predict the ignition temperature of either single or multiple sets of char or coal particles.

     What this is saying is that we are still a long way from a "first-principles" prediction of furnace and boiler behavior because there are still some major gaps in our knowledge; and the question next at issue is how best to fill those gaps. This is where the matter of point of view towards research planning becomes of importance. The ideal is certainly the deductive approach by which predictive (first-principles) mathematical models are constructed; but this presupposes sufficient confidence in the knowledge of the basic principles and concepts, and in their interactions, that such activities are worthwhile. Viewing the gaps that exist at present, however, it is my view that deductive modeling is mostly premature (the fluid bed may be an exception) and that the more fruitful approach is by induction; this can also be regarded as a form of partial modeling. This point of view will be illustrated in what follows. First, however, we compare system scales and magnitudes.

3. COMPARISON PERFORMANCE OF COAL FIRED DEVICES

     Furnaces and boilers operate in the range of heat rates of approximately 10^5 Btu/hr (0.03 MWt) for domestic boilers up to 10^{10} Btu (3000 MWt) for utility boilers. Combustion intensities or firing densities and reaction times depend more on the firing device (fixed, fluid, or entrained bed) than on its scale. This is illustrated by the listings in Table 1 (4) which compares a number of operational design parameters. This Table shows the trade-offs possible between combustion intensity (highest for fluid bed),

Table I. Source-Ref. 4

Comparative Combustion Intensities and Related Quantities

(Note on units. SI conversions from FPS values have been rounded off.)

Parameter	Grate	Fluid Bed	Pulverized Coal
Heat rate			
Btu/hr	$(10^5)10^6 -10^8$	(Projected) Up to 10^8	$(10^6 -)10^8 -10^{10}$
MW(t)	(0.03) 0.3-30	Up to 30	(0.3) 30-3000
Volumetric combustion intensity, I_v, Btu/ft^{3}hr	25,000-75,000 (based on bed and freeboard volume)	(i)Up to 200,000 (based on bed volume) (ii)Up to 50,000 (based on bed and freeboard volume)	15,000-25,000
kW/m^3	250-750	(i)Up to 2000 (ii)Up to 500	150-250
Area combustion intensity, I_A			
Btu/ft^{2}hr	$10^5 -6 \times 10^5$	Up to 10^6	(Up to 2.5×10^6)
kW/m^2	300-1800	Up to 3	(Up to 7.5)

Parameter	Grate	Fluid Bed	Pulverized Coal
Effective reactor height, $H_e = I_l/I_v$ ft	5-25	(i) Up to 5 (ii) Up to 20	Up to 150
m	1.5-7.5	(i) Up to 1.5 (ii) Up to 6	45
Coal firing density, $J_{f,V}$ lb/ft³ hr	2-6	(~15)	1-2
kg/m³ hr	30-100	(~250)	15-30
Area firing density, $J_{f,A}$ lb/ft² hr	8-50	Up to 100	(Up to 200)
kg/m² hr	40-250	Up to 500	(Up to 1000)
POC velocity (hot) ft/sec	Up to 10	Up to 12 (15)	Up to 60 (70)
m/sec	Up to 3	Up to 3.5 (4.5)	Up to 20
POC (air) velocity (cold) ft/sec	Up to 1.5	Up to 3 (4)	Up to 10
m/sec	Up to 0.5	Up to 1	Up to 3
Combustion time, sec	Up to 5000	100-500	~1

Parameter	Grate	Fluid Bed	Pulverized Coal
Particle heating rate, °C/sec	<1	10^3–10^4	10^4
Boilers only Heat transfer coefficients, h			
Btu/ft^2 hr °F	2	50–100	10–100 / 50–500 (variable over surface)
W/m^2 K	10	250–500	50–500 (variable over surface)
Heat transfer fluxes to heat exchange surfaces			
Btu/ft^2 hr	>3500	35,000	(15–150) x 10^3
kW/m^2	>10	100	50–500 (variable over surface)
Heat exchange surface area per unit cross-sectional area of combustion chamber, Φ	~200	~30	25–250

318

combustion time (fastest for p.c. or entrained bed firing), and
heat transfer flux to heat exchange surface (lowest for grate
firing). In all these factors, however, the greatest range in
values is shown by the combustion times, which are determined mainly
by the char reaction kinetics. As the table shows, the range is
three or four orders of magnitude, and the primary factor responsible is the particle size as indicated by Fig. 2 (4). In the
range 100 µm and above, the combustion time is determined largely
by the rate of oxygen diffusion through a boundary layer; and the
time is proportional to the square of the particle diameter with a
relatively low sensitivity to temperature. Prediction in this
region is respectably accurate. Below 100 µm, however, the range of
times becomes very much greater, more sensitive to temperature, and
essentially unpredictable a priori at this time.

Combustion time is only a significant factor in engineering
design in the entrained bed (pulverized) and some fixed bed systems.
In pulverized coal firing, the combustion time must not exceed the

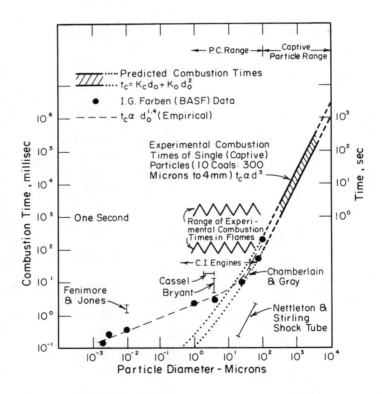

Fig. 2. Variation of combustion time with particle diameter from
various experimental and theoretical sources. (Source-Ref. 4).

gas residence time for there would otherwise be loss of combustible from the boiler and/or flame impingement in the convection banks, and this requirement is invariably satisfied.

In moving-grate systems (a sub-set of fixed beds), the coal must burn out before the particles are rejected to the ash pit.

The comparative magnitudes of Table I provide some basis for engineering selection of one system over another. Grate firing systems rarely exceed about 250,000 lb H_2O/hr as there seems to be a practical engineering limit when grate areas reach about 1000 sq. ft. (roughly 30 ft x 30 ft). Heat rates in excess of 10^8 Btu/hr, therefore, have been met by pulverized coal firing (entrained bed). This has been largely satisfactory with the two chief drawbacks being unit size (up to 150 ft high) and a highly variable heat flux over the boiler tube surfaces. The variable heat flux greatly aggravates design, since it creates a potential for hot spots that are very difficult to predict and that can cause tube failure.

The fluid bed combustor was originally proposed in some quarters as a device to reduce the unit size, and, hence, capital cost for a given duty. Later, it was realized that this could also be the answer to the SO_2 emissions control problem; and it had the further advantage of much more uniform heat flux over the heat exchange surface which could eliminate the hot-spot problem. It should also have advantages in reducing the NO_x emissions. However, commercial FBC boilers are only just appearing at the 50,000 lb H_2O/hr rate; scale-up to utility sizes is still some years away.

4. SELECTED PROBLEMS IN COAL COMBUSTION

In spite of the complexity of the coal combustion (and gasification) systems, this has been no deterrent to modeling. Most has been of the deductive, first-principles variety. Most of this modeling shows no attempt at validation by comparison with experiment so that its value is marginal or zero. Where validation has been attempted, it has usually been obtained at the cost of multiparameter adjustment to obtain agreement. This opens the question of uniqueness: sufficient parameters to adjust the agreement obtained may be spurious and the apparent validation becomes questionable (see also Review, 5). Such first-principles modeling is still the final target of all modeling, however; and it still has an immediate value in helping to define more precisely what is well known and well established, what is arguable and needs better definition, and what are the relative magnitudes of different mechanisms so that research priorities can be set (see eg. 6).

The alternative to deductive modeling is inductive modeling, also known as experimentally based modeling as it relies heavily for certain information on rather specific experimental results. It may also utilize or may be known as the "assigned parameter" approach. Its use is illustrated in three examples here.

4.1 <u>Heat Flux Profile Prediction in a Boiler</u>. This is a problem of major practical concern to boiler manufacturers. In attacking this problem, Bueters, et al (7) decided to attempt what could be done rather than what, for the long term, they would like to have done. The analysis focusses on the radiative heat transfer. In considering the complexities of unknown three-dimensional mixing, this was reduced to a manageable problem by assuming overall, one-dimensional flow up the furnace so it could then be **divided** up into a number of finite-element slices, as illustrated in Fig. 3 (7). Figure 3 shows that the boiler is also divided into three regions: a lower, recirculation-zone region representing the known penetration of the flow into the ash hopper; a middle, heat-release zone (incorporating backmix) whose location in the computer program was allowed to float; and a top, plug-flow zone that carried the products of combustion out of the furnace. A number of acceptable assumptions were incorporated which closed the flow pattern, and this completed the treatment of the aerodynamics. The reaction rate was reduced to the assumption that all heat was released uniformly in the heat release zone whose size and location was then determined by fitting.

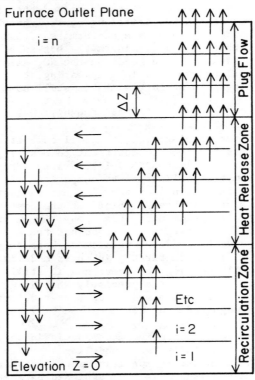

Lower Furnace Program Sample
Gas Circulation Pattern

Figure 3. Source-Ref. 7

321

The convected heat-flow was then set by the input heat release assumptions and the flow model.

Only the radiation was treated with any level of detail, utilizing a finite element form of the Equation of Transfer. The model was then used to compute the heat flux to the wall, and the heat release zone location was adjusted to obtain best fit between the calculated and experimental heat flux profile on the walls. A typical result is shown in Fig. 4: the circles are the calculated points and the triangles are experimental.

There is an element of arbitrariness that allows the fitting by way of adjusting the location and size of the heat release zone, but the model results were then tested in three important ways. First, the computer location of the heat release zone was compared with the physical location of the burners; in all cases, the one bracketed the other, as illustrated in Fig. 4. Then the furnace

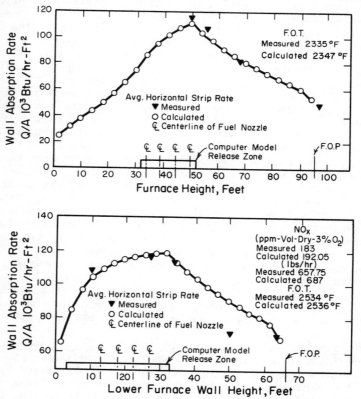

Fig. 4. Wall absorption rate vs lower furnace wall height. Predicted vs measured values for utility unit B: 415-Mw oil fired; test 7; 0 degree tilt; 10.7% EA; 0% gr; NHI=4172.2 (10.6) Btu/hr; release zone ht=30 ft; release zone centroid=18 ft; F_E=1.44; NHI/PA = 2.52. (Source-Ref. 7).

outlet temperature (FOT) and NO_x exit levels were calculated and
also compared with measurement. Both generally showed excellent
agreement, thus providing additional important support for the model.

In the original model development, the model was also used in
an inductive mode. Initial flame emissivity assumptions were found
to underestimate the radiative transfer. Some simple theoretical
considerations led to the conclusion that there were grounds for
adding a correction factor, the parameter F_E so that the nonluminous
emissivity ϵ_H would be transformed to the effective value, $\bar{\epsilon}$, given
by $\bar{\epsilon} = (1 + \epsilon_H - 1/F_E)$. Experimental evaluation of the numerical
F_E values for gas and oil flames showed that the values had little
range for each fuel, at about 1.20 ± 0.02 for gas and 1.475 ± 0.075
for oil.

The potential for criticism of this type of approach is quite
evident; nevertheless, calibration of the model on one set of boiler
measurements has permitted prediction of performance of other
boilers with surprising, but entirely acceptable, engineering
accuracy (7). A number of questions come to mind. The two most
important relate to fuel reactivity and to boiler/burner design.
Regarding the first, the model is "calibrated" against a range of
bituminous coals. What, therefore, will happen if the fuel is
athracite or lignite? There is no published answer to this. However, noting that the fuel is simply assumed to burn, the effect of
reactivity could be expected to control the width of the heat
release zone, and this width is shown to control the peakiness of the
wall flux. Faster reaction--say, using lignite--might, therefore,
increase the possibility of local tube burnout by heat-flux hot
spots. This suggests that the relation between heat release width
and fuel type and firing conditions needs investigation.

Regarding the influence of boiler design, the model should be
equally applicable to front and opposed-wall firing systems if the
approximated flow pattern of Fig. 3 is still applicable. The
"calibration" could then be expected to turn up some significant
differences, such as the heat release zone width; but the potential
value of the model suggests that this should now be put to the test.
What do we learn from this? Primarily that scientific and engineering insight can be effectively used as the basis for relatively
simple, but valid, models of operating engineering systems. Moreover, the models also provide quantitative estimates of different
components in the model that can then be used for setting research
priorities.

4.2 <u>Flame Propagation in One-Dimensional Pulverized Char Flames</u>. In
a one-dimensional flame, the mixing and aerodynamic problem is
eliminated by premixing the fuel and air supplies and by designing
out the backmix by stabilizing a one-dimensional flame against a
water-cooled tube bank (8,9). In modeling the system, it was again
accepted that the basis of the model had to be radiation (10,11),
but there was clear experimental evidence that char reactivity
played a most significant role, with measured ignition times varying
from 0.1 sec for high reactivity char and coal up to 1.0 for low
reactivity char.

Inclusion of reactivity in the model thus presented a problem that was as much a matter of policy as one of science. In the first-principles approach to modeling, others have included in their models complex expressions for the reaction rate, with selection of kinetic constants (such as activation energy) taken from the literature. Rather curiously, however, I do not recall ever seeing any carbon combustion modeling by itself that attempted to model the complete lifetime of a carbon or char particle in a flame. There is no evidence, therefore, that the published complex rate expressions (or simple ones for that matter) will actually predict realistic particle life histories. Indeed, the informed expectation is, in general, that they will not since the complex rate expressions (that I have seen) intended to model life histories universally lack ability to model the change in internal area as pores open out and ultimately coalesce.

On that account, it was decided instead to use available experimental data on char combustion reduced to an empirical expression, rather than the "fundamental" expressions. This could be said to inject a component of "experimental modeling" into the overall model. The model was, therefore, constructed as a multidimensional radiation model in a one-dimensional flame system, with the local char reaction rate called out by an empirical expression that represented the experimental data obtained by Field (12). In addition, to take into account the internal surface changes already referred to, another empirical expression obtained by Field to describe his experimental results was used. For a particle of radius a and density σ, Field found that he could describe the variation by

$$(\sigma/\sigma_o) = (a/a_o)^n \qquad (1)$$

where σ_o and a_o were the initial density and radius, and n was an empirical constant of value 3.

The value of Eq. 1 is that it describes the variation of σ and a without invoking any -- potentially arguable or restrictive -- mechanisms. With $n = o$, then $\sigma = \sigma_o$, and the particles burn at constant density (shrinking core behavior). With $n = \infty$, then $a = a_o$, and the particles burn internally (falling σ) at constant radius. At all other values of n between o and ∞, the particles burn in both modes.

The success of the model using the radiation equations and the experimentally-based empirical expressions for the char reaction is illustrated in Fig. 5. The model was also able to predict flame speeds and experimentally verified burning times. Most important of all, the relationship between reactivity and flame speed was established. The link was found to be the influence of particle size on flame emissivity. Flame propagation was governed by radiative heat transfer from the flame to the approaching cold cloud. The magnitude of the radiative flux crossing the flame front was determined by the flame temperature, which was about the same for all chars, and by the flame emissivity; and the flame emissivity

324

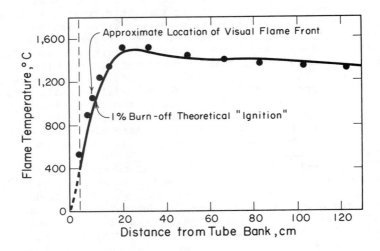

Fig. 5. Comparison of prediction and experiment for flame temperature profiles. Modeling was for <u>char</u>; experimental data for bituminous coal (data source; Howard [14] Run 14). Rosin-Rummler mean dia, of 50 μm. Mixed mechanism, Case 3, model. (Source-Ref. 11.)

was determined by the particle size. The high reactivity chars were found to burn as much internally as externally, while the low reactivity chars only burned externally. Thus, the effective particle size and hence the flame emissivity was higher in the former case. Consequently, the coupling between flame speed and reactivity was due to the influence of burning mode on particle size which controlled the flame emissivity and, thus, the rate of heat exchange between flame and unburned cloud.

What we learn from this example is the value of using empirical expressions to represent real behavior where the fundamental equations are still essentially incomplete (lacking factors to account for internal surface changes). This has provided us with the necessary conceptual insight that explains the nature of the coupling between reactivity and flame speed. It has also given us good grounds for using the empirical expression of Eq. 1 in a wider range of flame applications and it has also identified for us the most important next task at the more fundamental level, which is to develop a char-burning model that will essentially predict the behavior described by Eq. 1. When we succeed at that, we will then know we have a reliable basis for calculating the changes in internal surface.

4.3 <u>Fixed Bed Gasification</u>. The difference between combustion and low-Btu gasification in air (Producer Gas) is that the product gas is burned above the bed in combustion, and a deeper bed (up to 10 to 12 ft) is used in gasification to provide adequate heat exchange and to optimize combustible gas production. In modeling such a fixed bed, using combustion pot data (13) as a verification (validation) source, Barriga (14) used the assigned parameter method for virtually identical reasons to Cogoli's use (9) of experimental data sources for char reactivity.

In fixed bed gasification (combustion), air generally enters the coke or coal column through the grate at the bottom. As the air flows up the bed, the oxygen first burns to generate CO_2 at a high temperature in the lower part of the bed. The hot CO_2 is then reduced to CO by reaction with more coke until the temperature drops to a level where the C/CO_2 reaction is too slow to measure. Above that, any additional coke column acts as a counter-flow, packed-bed, heat exchanger. If the fuel comes in as coal, this heat will first dry it and then pyrolyze it.

In modeling the system, it is also found to be dominated by radiative heat transfer (which can be treated as a radiative conductivity), aided by a strong convective flow. Once again, however, there is a question of how to treat the coke reactivity. As before, there is the potential for reaction at either extreme of reaction, at the surface only (shrinking core behavior) or internal reaction with no change of radius -- or a combination of the two. In particular there is here a strong likelihood of partial penetration of the reaction, possibly to a depth of a few milimeters, in particles that are initially several centimeters diameter. This would only utilize a fraction of the total available internal surface (that in cokes can run from 100 to 500 m^2/g), but this could still be an order of magnitude greater than the superficial or external surface of a (nominally) spherical particle.

In this instance (14) the problem was first simplified by appeal to experiment (13) in which it had been found that the particles reacted with diminishing diameters at almost constant density. The almost constant density, however, did not rule out the possibility of partial penetration into the particle to an extent that the surface of attack was significantly increased, but the density change was so small as to be obscured by experimental scatter.

The approach was, therefore, to define a parameter, S_v, the internal reactive surface area per unit volume of the bed. Coupled initially with the assumption of the shrinking sphere, the value of S_v was equated to the external surface of spherical particles at a number density set by burnoff and appropriate close packing assumptions. An additional surface area multiplier was then introduced to allow for possible partial penetration, and this was given values of 1, 10, and 50 in successive calculations. The multiplier thus acted as a single fitting parameter and as a sensitivity analysis parameter. It was not necessary, of course, to have made the initial identification between S_v and the particle diameters; the quantity S_v could have been used instead as the fitting parameter, and this would have been more appropriate if the shrinking sphere behavior was unknown. Be that as it may, the procedure was found to be quite effective, and a typical calculation is illustrated in Fig. 6. The full procedure is described in the publication (14); it should be noted that the agreement shown was not possible by arbitrary adjustment of the kinetic coefficients as this was found to improve fit, for example, to the gas analyses at the expense of no fit at all to the temperature profile, and vice versa.

What we see from this is very much what was said about the one-dimensional p.c. flame, except for the lack, as yet, of any definitive predictive or experimental differences of behavior

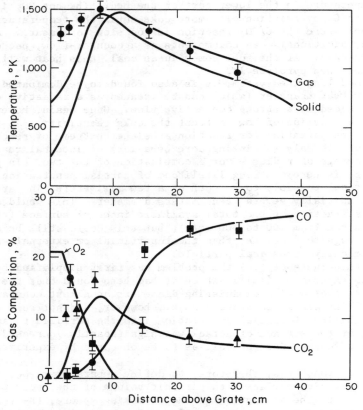

Fig. 6. Comparison of prediction and experiment for gas and solid temperatures, and gas composition (O_2, CO_2, CO), with variation through bed. (Source-Ref. 14.)

between high and low reactivity cokes or coals. This then identifies a principal target for future research.
4.4 <u>The Single Particle</u>. Reviewing what has been described above,
what we see is a need for decreasing accuracy in knowledge of the
carbon and coal kinetics as the complexity of the flame system
increases. However, we can see some need for better understanding
of the actual combustion mechanism -- explicitly, to be able to
explain Eq. 1, for example -- and we can also develop a feel for
the priority of the information in relation to the other behavior
involved (heat transfer, mixing, and so forth).

In fact, the qualitative behavior, as already outlines in
Sec. 2, is now fairly clear -- or more accurately, the possibilities.
What has still not been accurately predicted by any useful modeling
is a real, lifetime behavior, except for particles above 100 μm
where the reaction is dominated or controlled by diffusion (see
Fig. 2).

Amplifying the descriptions of Secs. 2 and 4.2, the behavior
identified is that of oxygen diffusing through a boundary layer
and into a particle to react at internal surfaces. The internal
surface, however, is provided by pores of different sizes, and
reaction takes place to different extents in the pores, depending
on the degree of accessibility. Conventional descriptions identify
macropores as greater than 200 Å and micropores less than 20 Å, with
transitional pores between these two limits. Typically, the majority
of the internal surface is in the micropores (as high as 95%), but
access also depends critically on the macropores.

In the case of the different chars used by Cogoli (9), all four
had about the same internal surface (measured by N_2), but two had
much greater macropore volumes. This accounts for the significant
difference in reactivity and flame speed.

At the same time, measurements of low ($500^{\circ}C$) temperature
reactivity on the same chars by TGA showed essentially the same
reactivity. From this we can, therefore, construct the following
qualitative picture.

At low temperatures (below $500^{\circ}C$) reaction is so slow that the
extent of oxygen consumption is small and there is oxygen penetration to the center of a particle, even if there are no macropores
for transportation. At the same time, reaction is slow relative to
the rate of internal diffusion, and all available (N_2 and CO_2)
surface is participating in the reaction. As the temperature rises,
the reaction rate increases and, for uniform accesibility to all
surface (monopore size), the oxygen consumption rate as it diffuses
in is finally too great to be in excess at the center of the particle. The penetration depth starts to fall, and the reaction zone
becomes thinner, steadily withdrawing, by the classical description,
from penetration to the particle center, until it is finally confined
essentially to the external surface. The particle will then burn
out in the shrinking core mode at essentially constant density.

If there is a significant pore size distribution, however, such
that the large pores still provide necessary access to the particle
interior, there can still be a steady reduction in the area par-

ticipating in the reaction since we can expect that the depth of penetration into the local micropores branching off the macropores will steadily drop. The fraction of micropores nominally accessible, however, will not drop so fast, or even remain constant, so that there is effectively continuing uniform volumetric reaction throughout the particle. This will be accompanied, of course, by the pore growth and coalescence that will also strongly affect the area involved in reaction.

In modeling these two options, the first has been modeled many times, though only once (15) with growth included. The second option does not seem to have been modeled. Most of the modeling would appear to be worthless, particularly when no account has been taken at all of internal reaction. The problem is extremely complex, however, and presents serious difficulties to model realistically. The changes in active surface area are what must be focussed on. In summary, the surface area changes on two accounts: first, there is reduced penetration of the reaction into a single micropore as the temperature rises, either because the reaction at the mouth is speeding up, or because the pore is being starved (no macropores); and second, the areas of the reacting pores are changing because of reaction and coalescence.

At this time, first-principles modeling does not look particularly attractive: there are too many options for the initial formulation of the problem, and too many options for assumptions regarding the progress of the reaction, definition of coalescence, and so forth. The question of uniqueness then becomes quite critical. What is needed at this time is sufficient experimental data that regular patterns -- if any -- of surface area development during reaction can be identified: this is saying that the problem is still mainly in the data gathering stage (6). Analysis will best be accomplished by inductive procedures using assigned parameters to codify the surface area data and their development.

DISCUSSION AND CONCLUSIONS

We are now in a position to consider more directly the question posed by the title of this paper. The opening point made in the Introduction is that fundamental combustion research has not or does not appear to have contributed much or anything in the past to aid boiler, or furnace, or burner design. It must then be an article of faith that what fundamental research has failed to do in the past, it will not fail to do in the future.

But can we be certain of this? Clearly not; but as the examples cited in Sec. 4 can show, there are certainly very favorable omens. There are those, of course, who enthusiastically endorse fundamental research as a matter of researcher's right, if only he can think of some problem to be solved, relevant or not. This, however, is not a good reason for supporting applied research, even at the fundamental level.

There is a further school of thought that now believes that everything can be done by computer modeling. This is a question of

much greater substance. What we must recognize first of all is
that the complexity of the interim and final problems is such that
the only way to handle the interacting factors is by computer modeling. However, this often results in what I have heard described as
"kitchen sink" modeling in which everything conceivable is included
(including the kitchen sink). One can take issue with many such
modeling advocates, however, on two counts:

- First, it is arguable that greater detail in the model
necessarily results in greater accuracy in the predictions. This
was the background to the point being made in Sec. 4 -- that
judicious simplifications or use of experimental sources of data
may be effective where more complex "fundamental" models would
either fail or be open to attack on grounds of lack of demonstrated
uniqueness. Such partial modeling is, of course, very much more of
an art, and requires more physical insight and intuition to be
effective, which could be a serious discouragement to those posess-
ing only technique and lacking scientific imagination. However,
such partial modeling would appear to be much more in the main-
stream of classical research that invokes the parsimony of minimum
hypotheses.

- Second, the more complex the model, the more difficult it is
to validate the model. This validation requires not only demonstration of the effectiveness of the model, but also the validity and
uniqueness of the model components.

Validation of models, however, requires experimental data, and
this is both costly and difficult to obtain, particularly as scale
increases. This requires something else: proper financial support
of appropriate groups on a continuing basis. To be sure, this
is the eternal cry; but if fundamental research is to make any contribution, it must be properly supported. Lack of adequate support
is finally too discouraging.

What then are my conclusions?

First, that there is need and scope for fundamental research,
with the proviso or caveat that the word "fundamental" should be
interpreted liberally rather than conversatively. This would allow
or require work up to the scale of one-dimensional flames and small
three-dimensional flames. The scope of such work should cover all
the individual topics mentioned in Sec. 2 and illustrated in Fig. 1,
individually and in combination. Blueprints for research projects
can readily be constructed from the diagram and the relevant
literature. They are not provided in this paper as it would be less
than courteous to assume that readers lack the necessary imagination
to construct their own proposals (but see Ref. 6).

Second, the research should be both experimental and theoretical,
with close coupling between the two. This would inevitable involve
the development of increasingly complex computer programs; but for
models to be convincing, validation against adequate experiments
is ultimately essential. There is certainly scope for purely
computer studies of alternative models or mechanisms, particularly
if these can be formulated on an _experimentum crucis_ basis for
ultimate experimental test. However, it is still astonishing to me

330

how many modeling studies of even pure carbon combustion are published without any attempt to check the predictions against available experimental behavior.

Third, preference should be given to parsimoneous modeling rather than to the all-encompassing modeling that is currently favored, it would seem, by most investigators at this time.

Fourth, to demonstrate relevence or utility, particular efforts should be made to assemble the best possible experimental data base of performance data of existing, full-scale boilers and furnaces in actual use. This should be coupled with development of parsimoneous models, such as the Bueters model (7), to predict such performance and that would then be validated or calibrated by comparison with the developed data base.

As a final word, it is now very much my impression from some of the recent successes in coal research that the problem is finally yielding to analysis and investigation. It is still possible to squander the research resources, however; and if we develop unverifiable models and do stupid experiments, we shall fail to achieve anything worthwhile and provide support for those who believe that fundamental research is, indeed, a waste of time. This provides excellent incentive to show that properly planned and executed fundamental research does, indeed, have utility.

REFERENCES

1. Wagoner, C. L. and Weingartner, E. C., "Further Development of the Burning Profile", ASME Paper No. 72WA/FU-1 (1972).
2. Suuberg, E. M., Peters, W. A., and Howard, J. B., Proc. 17th Symp. (Internat.) on Combustion, pp. 117-128, The Comb. Inst., Pitts., Pa. (1979).
3. Solomon, P. R. and Colket, M. B., Proc. 17th Symp. (Internat.) on Combustion, pp. 131-140, The Comb. Inst., Pitts., Pa., (1979).
4. Essenhigh, R. H., "Coal Combustion", Ch. 3 in Coal Conversion Technology (C. Y. Wen and E. S. Lee Eds.), Addison-Wesley Publ. Co. (1979).
5. Essenhigh, R. H., "Fundamentals of Coal Combustion", Ch. 19 in Chemistry of Coal Utilization (M. Elliott Ed.), John Wiley (in press).
6. Essenhigh, R. H., "Research Opportunities for Universities in Combustion of Coal", in Proc. Workshop on Research in Coal Technology--The University's Role, OCR/NSF-RANN Workshop (SUNY-Buffalo), Oct. 1974.
7. Bueters, K. A., Cogoli, J. G., and Habelt, W. W., Proc. 15th Symp. (Internat.) on Combustion, pp. 1245-1260, Comb. Inst., Pitts., Pa. (1975).
8. Howard, J. B. and Essenhigh, R. H., Ind. Eng. Chem. Proc. Des. & Devel. 6, 74-84 (1967); Proc. 11th Symp. (Internat.) on Combustion, pp. 399-408 (1967).
9. Cogoli, J. G., Gray, D., and Essenhigh, R. H., Comb. Sci. and Technol. 16, 165-176 (1977).

10. Cogoli, J. G. and Essenhigh, R. H., Comb. Sci. and TEchnol. <u>16</u>, 177-185 (1977).

11. Xieu, D., Masuda, T., Cogoli, J. G., and Essenhigh, R. H., Paper accepted for presentation at 18th Symp. (Internat.) on Combustion, Univ. Waterloo, Canada, August 1980.

12. Field, M. A., Comb. and Flame, <u>13</u>, 237-252 (1969); <u>14</u>, 237-248 (1970).

13. Eapen, T., Blackadar, R., and Essenhigh, R. H., Proc. 16th Symp. (Internat.) on Combustion, pp. 515-522, Comb. Inst., Pitts., Pa. (1977).

14. Barriga, A. and Essenhigh, R. H., "A Mathematical Model of a Combustion Pot", Proc. Spring Meeting, Western States Sec. of the Combustion Institute, April 1979.

15. Simons, G. A., and Finson, M. L., Comb. Sci. and Technol. <u>19</u>, 217-226 (1979); Simons, G. A. Comb. Sci. and Technol. <u>19</u>, 227-236 (1979).

16. Essenhigh, R. H., Proc. 16th Symp. (Internat.) on Combustion, pp. 353-374, The Comb. Inst., Pitts., Pa. (1977).

17. Smoot, L. D. and Horton, M. D., Prog. Energy Combust. Sci. <u>3</u>, 235-258 (1977).

18. Laurendeau, N. M., Prog. Energy Combust. Sci. <u>4</u>, 221-270, (1978).

19. Krazinski, J. L., Buckinus, R. O., and Krier, H., Prog. Energy Comb. Sci. <u>5</u>, 31-71 (1979).

20. Smoot, L. D. and Pratt, D. T., "Pulverized Coal Combustion and Gasification", Plenum Press, NY (1979).

ADDENDUM

I have refrained from including detailed mathematical analyses in this paper on the grounds that those who know them don't need them; and those who need, but don't know them, won't follow them in any possible summary form. Readers requiring such background are referred to the following Review articles, in particular, in the References following, as starting points for their investigations: References 4, 5, 6, and 15-20.

HEAT CAPACITY AND NMR STUDIES OF WATER IN COAL PORES

S. C. Mraw and B. G. Silbernagel
Corporate Research-Science Laboratories
Exxon Research and Engineering Co., Linden, NJ 07036

ABSTRACT

Heat capacity and NMR studies examine the freezing phenomenon and molecular dynamics for water in Illinois bituminous and Wyoming sub-bituminous coals as well as an Arkansas lignite. For high water content a broad maximum in heat capacity is observed below 273 K. For low water content a continuous transition from liquid-like to solid-like properties is observed. This latter deviation from the properties of bulk water is attributed to water-coal surface interactions. Wideline and transient NMR observations on coals containing H_2O and D_2O suggest a continuous change in characteristic times of motion with temperature on a microscopic scale. Observed activation energies for diffusion, $\Delta \sim 0.17$ eV, are comparable to those for bulk water.

INTRODUCTION

Coal consists of three components: an organic matrix, inclusions of inorganic mineral matter, and an extensive network of pores. Porosity is a major feature which distinguishes coal from other hydrocarbons curenly employed or contemplated as energy sources, such as petroleum, tars an oil shales. Low rank coals can have surface areas exceeding 100 m^2/gm (based on CO_2 chemisorption). Access to this surface is a prime consider tion in any scheme for coal conversion.

The pore network usually contains water in "as mined" coals and the amount of included water provides a measure of the pore volume. Data for three coals of different ranks are shown in Table I. Throughout the pape water content is expressed as grams per gram of dry coal (g/g). A four-f variation in water content is observed in going from an Illinois bitumino coal to an Arkansas lignite. Molar and valence bases provide complementa perspectives of the water amount. The volume of coal occupied by water c be very large: for the Arkansas lignite it approaches 50% of the total volume of the coal. Data for the two subbituminous coals examined, Rawhi and Wyodak, were comparable.

The present paper describes heat capacity and NMR observations of molecular motion and the freezing process of water in coals of varying rank. The first objective of the work is to understand the ways in which inclusion in the coal structure modifies water's properties from those observed in the bulk. In particular, pore sizes and molecule-surface interactions of both physical and chemical types can modify these properties, which should be reflected in changes in the rate and type of molecular motion. The second, and ultimate objective is to obtain information about the pore structure itself,

Table I Water Content of "As-Mined" Coals
(Expressed per gram of dry coal)

	g-H_2O	m Moles	cm^3
Illinois Bituminous	0.16	8.8	0.16
Rawhide (Wyo.) Sub-bituminous	0.48	26.6	0.48
Arkansas Lignite	0.67	37.2	0.67

such as surface area and pore size distributions. Heat capacity and NMR observations are an appropriate combination because they provide complementary information about the system. Heat capacity measurements determine the long-time, macroscopic average of the water behavior while NMR provides a picture of motion on the molecular scale at short times (10^{-5}-10^{-10} s). Related observations have been discussed previously.[1,2] While other NMR studies have appeared in the literature,[3,4,5] they have been cursory and of limited scope.

The present heat capacity data show several inequivalent forms of water behavior in coal pores: a monotonic increase in heat capacity upon heating from 121 K - 295 K for low water content and a broad heat capacity maximum occurring below 273 K for coals of higher water content. We attribute this latter peak to a "freezable" component of the water in the coal pores. NMR studies do not show an appreciable loss in proton NMR intensity between 200 K and 300 K, suggesting that water is not being selectively frozen out in pores of given sizes upon cooling. A monotonic variation in correlation times for molecular motion and activation energies comparable to those for bulk water suggest a continuous transition from the liquid to the solid state-at least on a microscopic scale.

SAMPLE PREPARATION AND EXPERIMENTAL DETAILS

Samples with different water content were prepared by exposing 10x20 mesh coals to a low humidity environment at ambient temperatures for varying periods of time. Ceramic alumina analogs and samples with included D_2O were prepared by exposing the dried samples to H_2O or D_2O, respectively. Water content was determined by sample weighing and by determining the integrated intensities of ^{1}H and [2]D NMR signals. Heat capacity measurements were performed with a Perkin-Elmer DSC-2 differential scanning calorimeter operating between 100 K and 300 K. Transient NMR measurements were made with a Bruker SXP spectrometer at frequencies of 12 MHz and 30 MHz. Wideline NMR measurements were made with a Varian WL-112 spectrometer at 12 MHz and 15 MHz. Temperatures were varied between 140 K and 300 K on both NMR systems using chilled gases.

334

HEAT CAPACITY MEASUREMENTS

Heat capacity data for a Wyodak subbituminous coal with varying water content are shown in Figure 1a.[1] The apparent water heat capacity

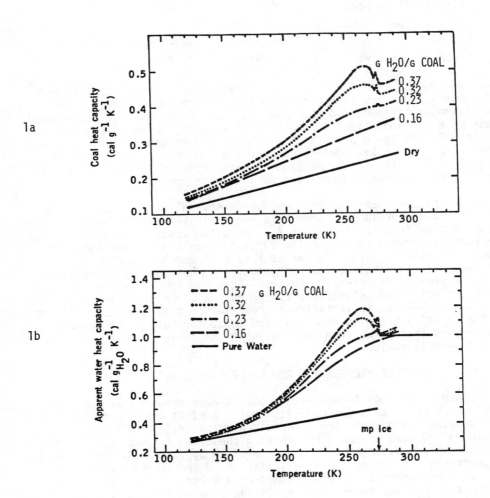

Figure 1. Total heat capacity and apparent water heat capacity for Wyodak coals of different water content

is obtained by subtracting the nearly linear contribution from the dry coal, as seen in Figure 1b. Results for bulk water are included for comparison. The heat capacity of bulk water is nearly independent of temperature above the melting point, has a discontinuity of ∿80 cal/g at

273 K and decreases linearly in the solid state from 273 K to ~120 K. Near 120 K and again at room temperature, the heat capacity of water in coal pores is similar to that in the bulk, but the behavior at intermediate temperatures is significantly different. No latent heat of fusion is observed. Furthermore, the temperature dependence of the heat capacity depends upon water content. At low water levels, such as 0.16 g/g for the Wyodak case in Figure 1b, the heat capacity rises monotonically from solid-like values near 125 K to liquid-like values at room temperature. At higher water content, such as 0.37 g/g and 0.31 g/g, a broad heat capacity maximum is observed, the maximum occurring significantly below 273 K. As mentioned above, the water fraction not exhibiting a heat capacity maximum will be called non-freezable, the remainder will be called freezable.

Measurements of water condensed on porous alumina ceramics show the same freezable/non-freezable dichotomy, as shown in Figure 2. Since the pore size distribution and surface area are known for these ceramics, the freezing process can be correlated with the extent of

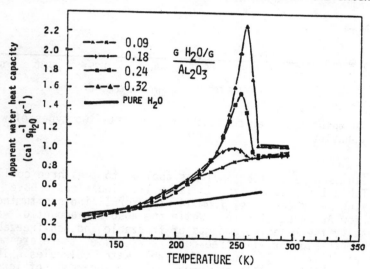

Figure 2. Analogous heat capacity behavior is seen for water on porous ceramics

molecular coverage. The total experimental enthalpy from 121 K - 295 K was determined by integrating the heat capacity of the included water over that temperature range. This total enthalpy is plotted as a function of water content in Figure 3. The data can be described in terms of two linear variations, as suggested in the figure. The linear slope is 158 cal/g-H_2O at high water content, the value expected for bulk water in this temperature range. The slope drops to 102 cal/g-H_2O at low water content. The change in slope occurs at near 0.17 g/g, which corresponds to roughly two monolayers of water on the surface of the ceramic if close packing of water molecules at their van der Waals radii is assumed.

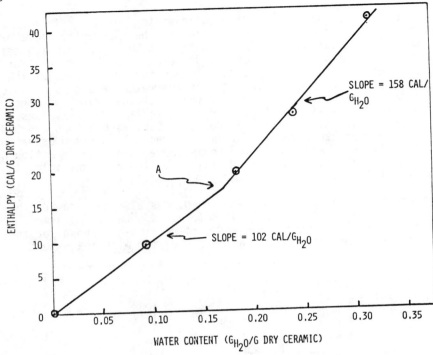

Figure 3.   An enthalpy-water content plot suggests two types of water
behavior

The results of a similar analysis applied to the three coals
listed in Table I are summarized in Table II.   These coals have
widely varying water contents, from 14 wt.% for Illinois bituminous
to 40 wt.% for Arkansas lignite.   While the absolute amount of water
varies greatly, the relative amounts of freezable and non-freezable
water are comparable for all three coal ranks - i.e. there are nearly
equal numbers of freezable and non-freezable water molecules.   The
difference between Illinois bituminous coal and the coals of lower
rank is in the absolute number of molecules: 0.08 g/g as compared
to 0.25 g/g for the other cases.   Arkansas lignite differs from Rawhide subbituminous coal in its large capacity for additional water at
high relative humidity.   Nearly 20% of the total water content in the
lignite case exists at relative humidities greater than 98% (as determined by equilibration with a K_2SO_4 solution), in spite of the fact
that the lignite in this state appears dry by visual inspection.   This
component has a much more "liquid-like" heat capacity - the width of
the heat capacity maximum is much narrower than seen in Figures 1b and
2.   This component is much less pronounced for the Rawhide coal and
not observed for the Illinois bituminous coal.

Table II Distribution of Water Types for Different Coal Ranks
(units-g-H_2O/g-dry coal)

	Illinois Bituminous	Rawhide Subbituminous	Arkansas Lignite
Total as Received	0.16 (14 wt.%)	0.48 (32 wt.%)	0.67 (40 wt.%)
Non-Freezable	0.07	0.25	0.25
Freezable (<98% RH)	0.09	0.21	0.27
(>98% RH)	0.00	∿0.03	∿0.15
Effective H_2O Surface Area (m^2/g)	130	460	460

Reasoning by analogy with the ceramic model systems, the amount of non-freezable water should be proportional to the surface area of the coal. Assuming that the first two water monolayers adjacent to a coal surface are non-freezable, surface areas of 130 m^2/g for Illinois and 460 m^2/g for Rawhide and Arkansas samples are obtained. While this is an <u>effective</u> surface area because of the model dependent assumptions involved in its estimation, the agreement with surface area determinations for Illinois coal by CO_2 absorption (128 m^2/gm) and small angle x-ray scattering (140 m^2/gm), as well as heats of absorption of methanol ∿130 m^2/gm is excellent.[6] By contrast, the very large surface areas inferred for the lower rank coals suggest that not all water resides in pores. Some must actually be included in the organic structure.

NMR STUDIES

Motion of the included water molecules strongly effects their NMR properties. Nuclear interactions are averaged by molecular motions occurring on time scales shorter than 10^{-5} - 10^{-6} s - a phenomenon known as motional narrowing. Motions on the 10^{-7} - 10^{-8} s time scale produce fluctuations which cause energy exchange between the nuclei and their environment - a process called spin-lattice relaxation. Wideline NMR measurements of the shape and width of the NMR absorption and transient NMR measurements of the spin lattice relaxation time (T_1) probe these motions.

The present observations were made on two types of samples: Wyodak coals with water contents of 0.31 and 0.18 and Wyodak and Illinois coals with 0.17 and 0.12 g/g, respectively, of included D_2O. The 0.8 g/g Wyodak sample contains the maximum amount of water exclusively in the non-freezable form, as inferred from the heat capacity data. By contrast the 0.31 g/g Wyodak sample has a pronounced freezable component. The D_2O samples are useful for two reasons. First there is very little background signal from the organic material, a problem which significantly complicates the wideline NMR measurements on protons. In addition the ^{2}D nucleus has very different NMR properties from those of the ^{1}H nucleus. Resonance properties of protons are determined by the interaction of their nuclear spin with local magnetic fields. These fields come from the other proton on the same molecule, protons on adjacent molecules, and paramagnetic impurities in the organic and inorganic components of the coal matrix. Resonance properties of deuterons are dominated by the interaction of the nuclear electric quadrupole moment with electric field inhomogeneities of the O-D bond in the molecule. This interaction is roughly ten times stronger than the proton's magnetic interaction and essentially intramolecular and therefore even more highly local. These different species provide complementary pictures of molecular motion, as illustrated below.

The temperature dependence of the proton linewidth in the two Wyodak + H_2O samples, shown in Figure 4, provides an example of motional narrowing. The residual linewidth at room temperature, ~ 0.1 gauss, arises from inhomogeneities in the applied magnetic field and the inhomogeneous magnetic properties of the coal. The NMR line broadens to a width of approximately 8 gauss at low temperatures with the transition centered around 180 K. This implies that the characteristic times for molecular motion exceed 10^{-5} s in this temperature range. Subtracting the background proton signal arising from the coal matrix reveals the characteristic Pake doublet powder spectrum at low temperature which is expected for water molecules.[7] The most striking feature of Figure 4 is the fact that the 0.18 and 0.31 g/g samples behave identically - i.e. there is no obvious distinction between freezable and non-freezable water at molecular distances on a 10 μs time scale. The relative intensity of the NMR signals from both samples remains constant from 200 K - 300 K, suggesting that one component of the sample is not being selectively "frozen out" as the temperature is lowered.

A measure of the activation energy for molecular diffusion is obtained using a simple expression for the variation of linewidth with changing correlation time:[8]

$$\delta\omega^2 = \delta\omega_0^2 \, (2/\pi) \, \tan^{-1} \, (\alpha\delta\omega\tau_c) \qquad (1)$$

where $\delta\omega$ is the resonance linewidth at a given temperature, $\delta\omega_0$ is the low temperature width, τ_c is the correlation time for motion, and α is a model-dependent constant of order unity.

The resulting τ_c's appear thermally activated

$$\tau_c^{-1} = \tau_0^{-1} \, \exp \, (\Delta/kT) \qquad (2)$$

with an activation energy of $\Delta = 0.2$ ev = 4.6 Kcal/mole.

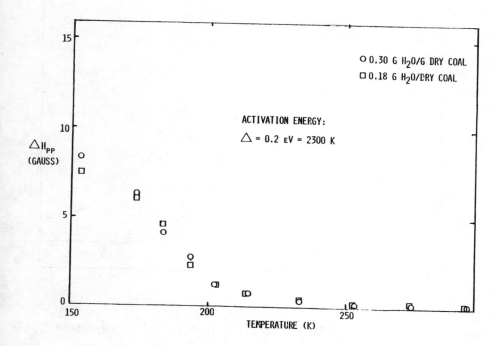

Figure 4. Proton linebroadening is similar for Wyodak coals of different water content

Similar linebroadening behavior is found for the D_2O samples, shown in Figure 5. The broadening occurs at higher temperatures as a consequence of the greater strength of the interaction. The broadening occurs at higher temperatures for the Wyodak case - suggesting slower D_2O motion in that sample. The activation energy for the Wyodak coal, determined in an analogous way to that discussed above for H_2O, yields an activation energy of Δ = 0.15 eV = 3.4 Kcal/mole.

Transient NMR measurements of D_2O spin lattice relaxation show an exponential magnetization recovery, which suggests an equivalent energy exchange process for all observable nuclei. The exponential increase of relaxation times at low temperatures (Figure 6) suggests thermally activated diffusion. The onset of broadening near 250 K makes measurements at lower temperatures extremely difficult. In this temperature regime $1/T_1 \propto \tau_c$ and the activation energy extracted from the slope of the curve of Figure 6 is 0.17 eV = 4.0 Kcal/mole.

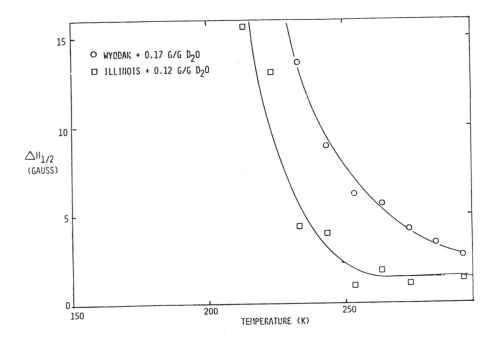

Figure 5. Different D_2O line broadening is observed for Wyodak +
0.17 g/g D_2O and Illinois + 0.12 g/g D_2O

Transient NMR measurements of the H_2O spin lattice relaxation
for the two Wyodak coals exhibit much more complex behavior. The
magnetization recovery is non-exponential suggesting a broad range of
inequivalent relaxation behaviors for the H_2O molecules. Although it
is clearly an oversimplification in this case, the magnetization recovery was decomposed into two components, long and a short relaxation
processes, for a preliminary analysis. The resulting relaxation times
$T_1{}^L$ and $T_1{}^S$ are plotted as a function of temperature for observations
on the 0.18 g/g Wyodak sample at 12 MHz in Figure 7. A pronounced
minimum in both times is observed for temperatures near 250 K. Comparable behavior is seen for the 0.31 g/g sample. A very weak increase
in the minimum T_1 values is seen in comparing observations at 12 MHz
and 30 MHz.

The magnitudes of T_1 and the frequency independence of the minimum T_1 values suggest that relaxation is occurring by the coupling
of the nuclei with paramagnetic impurities in the sample. In the
absence of impurity-related relaxation effects, the minimum T_1
values should be approximately 15 ms and vary inversely with the
resonance frequency. Similar impurity relaxation effects are observed for water condensed on high surface area charcoals[9] and

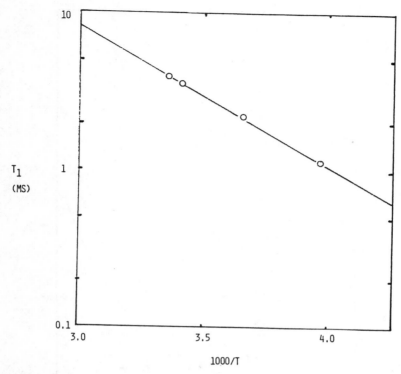

Figure 6. D$_2$O spin lattice relaxation appears thermally activated
for Illinois +0.12 g/g D$_2$O sample

mineral matter.[10] The low temperature increase of T$_1$, which should be
proportional to τ_c^{-1}, appears close to the 4 Kcal/mole value, as seen
by comparison with the dashed line on the figure. These paramagnetic
relaxation effects do not occur in the D$_2$O observations because the
quadrupole interaction is so much larger and the impurity relaxation
process is much weaker. The relaxation rate is proportional to the
square of the gyromagnetic factor. Since the gyromagnetic factor
of the deuteron is less than 1/6 of that of the proton, this process
is less than 1/40 of what it would be for the proton case. Thus the
D$_2$O observations provide an opportunity to study the molecular dynamics
of included molecule without interference from the paramagnetic impurities.

CONCLUSIONS AND COMMENTS

The present discussion is intended to provide some measure of
the scope of applicability of heat capacity and NMR techniques to the
study of coals. The heat capacity observations are rapid and can be
performed on as-mined coals without the need of pre-drying. They indicate a difference in the freezing process with different levels of

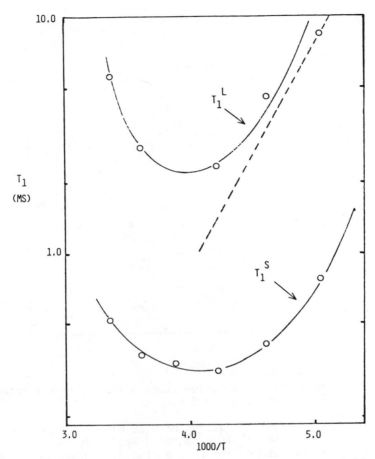

Figure 7. Proton spin lattice relaxation for Wyodak +0.18 g/g
H$_2$O is dominated by paramagnetic relaxation

water content and varying amounts of "freezable" and "non-freezable"
water are seen in coals of different ranks. NMR studies suggest that
these distinctions are not obvious on the molecular scale and that
the process of "freezing" does not necessarily involve discrete
phase transitions so much as a continuous decrease in rates of
molecular motion as the temperature is lowered. Activation energies
in the range of 0.15 - 0.2 eV are comparable to those observed in
bulk water. These observations suggest that physical interactions
with the surface of the pores and the small pore sizes are responsible for the broadening of the transitions.

A survey of related literature suggests that this freezing be-[11]
havior is found in a range of systems as diverse as water in α-Al$_2$O$_3$,
and N$_2$ on rutile.[12] "Non-freezable" behavior has been observed for

monolayers of water on materials as diverse as collagen, sheep's wool keratin and endellite. Further work in this area and a careful analysis of analogous systems should lead to a more quantitative picture of this phenomenon.

ACKNOWEDGMENTS

Wyodak coal samples were prepared and the heat capacity observations were performed by D. F. O'Rourke. D_2O coal samples were provided by R. B. Long and F. A. Caruso. NMR measurements were made by A. R. Garcia and L. A. Gebhard.

REFERENCES

1. S. C. Mraw and D. F. Naas-O'Rourke, Science 205, 901 (1979).
2. B. G. Silbernagel, L. B. Ebert, R. B. Long, R. H. Schlosberg, ACS Advances in Chemistry Series 92 , ___ (1980).
3. W. R. Ladner and R. Wheatley, J. Inst. Fuel 280 (1966).
4. A. D. Alekseev, R. M. Krivitskaya, B. V. Pestryakov and N. N. Serebova, Khim. Tver. Topliva 11, 94 (1977).
5. L. J. Lynch and D. S. Webster, Fuel 58, 429 (1979).
6. W. R. Grimes, A.I.P. Conference Proceedings , (1980).
7. G. E. Pake, J. Chem. Phys. 16, 327 (1948).
8. N. Bloembergen, E. M. Purcell and R. V. Pound, Phys. Rev. 73, 679 (1948).
9. H. A. Resing, J. K. Thompson and J. J. Krebs, J. Phys. Chem. 68, 1621 (1964).
10. A. Tinus, J. Petrol. Tech. 775 (1969).
11. K. Dransfeld, H. L. Frisch and E. A. Wood, J. Chem. Phys. 36, 1574 (1962).
12. J. A. Morrison, L. E. Drain and J. S. Dugdale, Can. J. Chem. 30, 890 (1952).

SURFACE CHEMICAL PROBLEMS IN COAL FLOTATION

By

S. R. Taylor, K. J. Miller, and A. W. Deurbrouck
Coal Preparation Division, PMTC, U.S. Department of Energy,
Pittsburgh, PA 15236

ABSTRACT

As the use of coal increases and more fine material is produced by mining and processing, the need for improved methods of coal beneficiation increases. While flotation techniques can help meet these needs, the technique is beset with many problems. These problems involve surface chemical and interfacial properties of the coal-mineral-water slurry systems used in coal flotation.

The problems associated with coal flotation include non-selectivity, inefficient reagent utilization, and excessive variability of results. These problems can be broadly classified as a lack of predictability. The present knowledge of coal flotation is not sufficient, in terms of surface chemical parameters, to allow prediction of the flotation response of a given coal. In this paper, some of the surface chemical properties of coal and coal minerals that need to be defined will be discussed in terms of the problems noted above and their impact on coal cleaning.

INTRODUCTION

In the United States, it is estimated that 10-15 million tons of fine coal per year is beneficiated by flotation.[1] The other, coarser 250-300 million tons of wet-cleaned coal is prepared by various specific gravity methods. Flotation, therefore, is the only process that utilizes the surface characteristics of coal and its impurities, rather than a difference in specific gravity, to make a separation.

Although coal particles are normally more hydrophobic than the associated clay or shale and pyrite or marcasite particles, there are many factors that can effect the degree of hydrophobicity. Among these are the oxidation of the coal surface, products of the oxidation of the pyrite or marcasite, various ionic species in the slurry (including those from coal-pyrite oxidation), porosity of the coal particle surface, and the presence of clay slimes in the slurry. Also, pH can be a factor, although this appears to be related to the type and abundance of ionic species in the slurry and their tendency to precipitate out as hydrophilic colloid to be adsorbed onto the coal.

In this paper, the various problems affecting the hydrophobicity and, therefore, the floatablity of coal will be discussed.

COAL FLOTATION PROCESS

Coal flotation is a physicochemical process for beneficiating fine coal. The process, in commercial use in the United States since the 1930's[2], utilizes a difference in the surface properties of coal and its associated material in an aqueous pulp to effect a separation. Hydrophobic or water-repellent particles (usually the coal) are floated to the surface by finely dispersed air bubbles to be collected as a froth concentrate. Other particles which are readily wetted by water do not stick to air bubbles and remain in suspension in the body of the pulp to be carried off as underflow.

Coal flotation is performed in machines or cells such as those shown in figure 1. The cells usually have mechanical agitators (figure 2) and froth scraper paddles, but some are pneumatic or combine pneumatic and mechanical methods.

Figure 1.   Bank of Denver flotation cells.

For flotation to be most effective, reagents must be added to the pulp. One type of reagent, the collector (usually a nonpolar oil), adsorbs on the particle surfaces, rendering them water-repellent or hydrophobic. Another type of reagent, the frother, facilitates the production of a transient froth capable of carrying the hydrophobic coal load until it can be removed from the flotation cell by the froth scraper paddles. Frothers in common use are methyl isobutyl carbinol or MIBC (a six-carbon branch chain alcohol), pine oil, and various water-soluble polyglycol types.

Coal flotation usually plays a secondary role to specific gravity separation processes for coarse coal cleaning. Because of this secondary or supplementary role played by flotation in coal washing, the portion of a coal preparation plant feed going to the flotation circuit is comparatively small, averaging probably 4 or 5 percent but sometimes ranging as high as 20 percent or more. The quantity depends on such factors as the mining methods, the degree of breaking or crushing to liberate impurities, and the tendency toward breakage during handling (friability) of the particular coal.

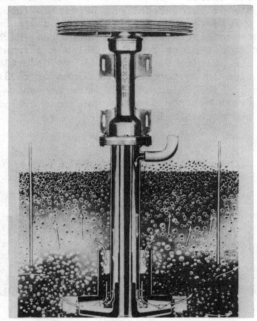

Denver cell—cut of mechanism.

Figure 2. Denver flotation cell mechanism.

Typically, feed particle size in coal flotation is minus 28 mesh (0.5mm) with perhaps 20 to 30 percent through 325 mesh (0.04mm). This material is in a slurry of two or three percent to 10 percent solids or more depending on the plant circuitry and the quantity of water used in coarse coal screening and classification.

Because in coal flotation the bulk of the material is floated (usually 70 percent or more), the froth created is very thick and abundant (figure 3). It is this abundance of froth and the rapid flotation of the hydrophobic coal that lead to some of the problems in selectivity; i.e., fine clay slimes and pyrites are frequently entrained in the froth. In order to get a cleaner product, flotation yield is sometimes sacrificed. The tailings or reject material, which should be largely clay, reports to a thickener or settling tank (figure 4) where the material is flocculated to form a sludge while

Figure 3.   Coal froth concentrate in industrial preparation plant.

Figure 4.   Coal preparation plant refuse thickener.

348

the clarified water is recirculated to the preparation plant. Sometimes, however, the tailings contain large amounts of fine coal that
floats on the water in the thickener. Such a condition, which is
shown in figure 5, is all too common and calls for better flotation
recovery.

Figure 5. Coal floating on refuse thickener.

SPECIFIC PROBLEMS

Coal Oxidation

Oxidation is a term sometimes used too freely to describe any
anomaly that diminishes coal flotation recovery. Although, as
pointed out earlier, there are many factors that will inhibit coal's
natural floatability, surface oxidation is indeed a significant one.

The chemistry of coal oxidation has been studied by many investigators. As Sun[3] points out, it is generally accepted that coal
oxidation can be divided into three stages. The first stage is a
superficial oxidation characterized by the formation of coal-oxygen
complexes with acidic properties. In the second stage, many of the
organic components of coal (hydrocarbons and carbon) are transformed
into hydroxy carboxylic or humic acids. Finally in stage three, the
humic acids are changed to water-soluble acids. A coal's resistance
to oxidation appears to increase with rank.

An attempt was made by Sun to determine whether the poor floatability of a mildly oxidized coal is caused by its water-insoluble
oxidized surface or by its water-soluble oxidation products. The
test results, table I, show that the water-insoluble surface of the

oxidized coal is chiefly responsible for its poor floatability. A comparison of tests 2 and 3 of table I shows that the floatability of the oxidized coal is slightly improved or at least not depressed by the presence of water-soluble oxidation products. Tests 3 and 4 show that once the coal surface is appreciably oxidized, it cannot be made floatable by washing with water or alkaline solutions.

Table I   Flotation of Pittsburgh bed bituminous coal under various conditions with 1.02 lb/ton purified kerosene and 0.64 lb/ton pine oil

Test No.	Oxida-tion	Coal treated before flotation	pH of pulp, 30 g coal and 125 cc water	pH of fil-trate of tailing	Recov-ery, Weight-percent
1	None	None	6.9	6.8	100.0
2	200°C	None	3.9	4.0	37.5
3	200°C	Washed 10 times with distilled water.	5.0	4.9	30.4
4	200°C	First conditioned in a warm NaOH solution of 1.0 N concentration for 10 min, then washed with distilled water 32 times	7.2	7.0	36.2

Oxidation Products and pH Control

Yancey and Taylor[4] showed that the oxidation products of coal-pyrite (ferrous and ferric sulfate) will depress pyrite flotation (table II). Although Yancey and Taylor did not discuss the effect of ferric sulfate on coal flotation, it is assumed that it would also depress coal.

Yancey and Taylor attributed the depressing action of ferric sulfate to the adsorption of either a ferric hydroxide sol or a basic ferric sulfate slime. The tendency for the depression to cease below about pH 5 is explained by the diminished hydroxyl-and ferric-ion concentrations and the increased hydrogen- and sulfate-ion concentrations. The lack of flotation depression on the basic side is explained by the fact that a flocculated ferric hydroxide is formed before contact with the particles takes place.

Baker and Miller[5] carried this work a step further by demonstrating that a number of other multivalent metal salts will depress pyrite, and coal as well, at specific pH levels and solution strengths. Their tests were conducted with $FeCl_3$, $AlCl_3$, $CrCl_3$, $CuSO_4$, and $CaCl_2$.

Table III, which illustrates the effect of a trivalent iron salt ($FeCl_3$) on the flotation of 100 by 200 mesh bituminous coal at different pH levels, shows that coal is strongly depressed at pH levels between 5.9 and 6.5 while it floats well at lower and higher values. At the lower pH levels, positively charged hydroxy complex ions [e.g.

$Fe(OH)^{++}$ or $Fe(OH)_2^+$] predominate, while in the pH range where the coal is depressed, positively charged $Fe(OH)_3$ colloid is quite likely present in higher concentration than the ionic hydroxy complex ions. Therefore, it is postulated that depression is most likely due to the adsorption of positively charged hydrophilic colloid on negatively charged coal in these specific pH ranges. At higher pH levels, precipitated colloidal $Fe(OH)_3$ is not positively charged, and the physical nature of the colloid changes; hence, the colloid does not adsorb on the coal at the higher pH values.

Table II Depression of coal pyrite with 0.02 gram of ferric sulphate

Sodium hydroxide, cm^3 of 1 N	Sulphuric acid, cm^3 of 1 N	pH	Yield, percent
4.00		10.4	20.3
1.50		10.3	27.3
1.00		9.1	23.2
.90		8.5	20.0
.80		8.0	12.1
.70		7.4	19.1
.50		7.1	5.8
.45		7.0	8.7
.40		7.0	.9
.25		6.7	.8
.20		5.1	.2
.15		5.0	.4
.10		4.4	1.0
	0.10	4.3	3.6
	.30	4.2	2.7
	.50	3.9	11.4
	.70	3.7	7.7
	1.00	3.5	8.5
	2.00	3.1	11.5
	3.00	3.0	16.0

Table III Flotation of 100 by 200 mesh coal in the presence of 5×10^{-5} mole/liter $FeCl_3$ at different pH levels -- pH adjusted with NaOH

pH	Recovery, weight-percent
4.5	86.0
5.1	87.7
5.9	47.9
6.5	39.7
6.8	79.2
7.0	86.8
7.4	89.1

The other multivalent metal salts had a similar effect on coal floatability at specific pH's and ionic solution strengths. $CaCl_2$, for example, depressed coal around pH 11.5-12.0; $CuSO_4$ between pH 6.5 and 7.5; and $AlCl_3$ between pH 5.5 and 7.0.

Coal-Pyrite vs. Ore-Pyrite

The floatability of coal-pyrite is somewhat different than that of ore-pyrite and presents some unique problems in coal desulfurization, both from the standpoint of selective pyrite depression and selective pyrite flotation.

Lyon[6] compared the flotation and adsorption characteristics of ore-pyrite with those of coal-pyrite and concluded that there are basic chemical and physical differences. The main cause of these differences appears to be the carbonaceous material in the coal-pyrite. The carbonaceous material impregnates the pyrite structure and lowers the apparent specific gravity and increases the porosity.

In Lyon's work, flotation of the ore-pyrite could be completely depressed by increasing the slurry pH to 10 (figure 6). This was not

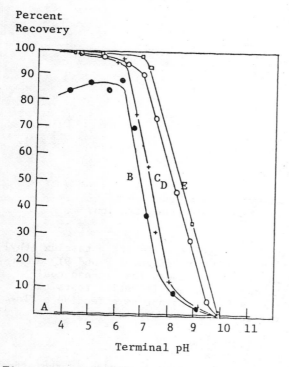

Percent Recovery

A - 0.00 pounds per ton
B - 0.02 pounds per ton
C - 0.04 pounds per ton
D - 0.08 pounds per ton
E - 0.15 pounds per ton

Terminal pH

Figure 6.   Flotation of ore-pyrite as a function of pH at five levels of xanthate concentration.

possible with the coal-pyrite--probably because of the carbonaceous material. At pH 10, about 20 percent of the coal-pyrite floated even without the addition of xanthate (figure 7). The flotation recovery of the coal-pyrite increased to 44 percent at pH 10 with the addition of 0.15 lb/ton of xanthate. The different flotation behavior of coal-pyrite was attributed to the carbonaceous material inherent in its surface.

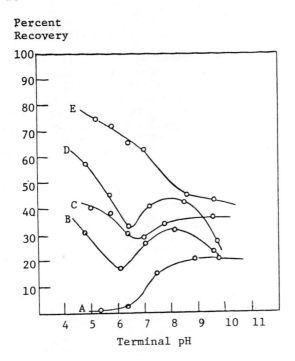

A - 0.00 pounds per ton
B - 0.02 pounds per ton
C - 0.04 pounds per ton
D - 0.08 pounds per ton
E - 0.15 pounds per ton

Figure 7. Flotation of coal-pyrite as a function of
pH at five levels of xanthate concentration.

The adsorption capacity of the coal-pyrite for potassium ethyl xanthate was twice that of the ore-pyrite (figures 8 and 9), which further reveals the basic differences between the ore-and coal-pyrites. The results of the adsorption and flotation tests taken together demonstrate that the xanthate was not as effective a flotation collector for coal-pyrite as for ore-pyrite.

Clay Slimes

Another factor to be considered in coal flotation is the effect of clay slimes or other hydrophilic colloidal material in the water. Table IV shows comparative flotation tests done with laboratory tap water and with clarified water from a commercial preparation plant.

353

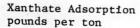

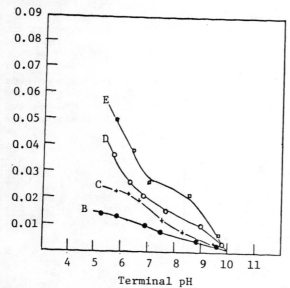

B - 0.02 pounds per ton
C - 0.04 pounds per ton
D - 0.08 pounds per ton
E - 0.15 pounds per ton

Figure 8. Adsorption of xanthate by ore-pyrite as a function
of pH at four levels of xanthate concentration.

Table IV   Flotation of Pittsburgh-Redstone bed coal
blend in laboratory tap water and in preparation plant clarified water

Product	Weight-percent	Ash, percent
Tap water		
Clean coal	68.7	7.0
Reject	31.3	33.0
Feed	100.0	15.1
Plant water		
Clean coal	48.0	6.6
Reject	52.0	23.1
Feed	100.0	15.2

Xanthate Adsorption
pounds per ton

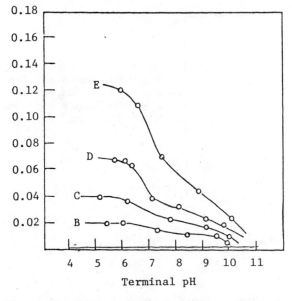

B - 0.02 pounds per ton
C - 0.04 pounds per ton
D - 0.08 pounds per ton
E - 0.15 pounds per ton

Terminal pH

Figure 9.   Adsorption of xanthate by coal-pyrite as a function
of pH at four levels of xanthate concentration.

Obviously the plant water had a depressant effect on coal flotation.
The reason for this was not determined, but since the plant water was
turbid, it probably contained clay slimes.

To help control the problem of clay slimes, it has been suggested that fine coal be deslimed or classified prior to flotation.
Also, the use of reagents (dispersants) has been attempted to prevent
slimes adsorption on coal. However, dispersion by chemical means has
not generally proven effective. This is an area in need of more
research.

The quantity of fine refuse associated with coal has a strong
influence on its amenability to washing by flotation. Table V illustrates the effect of different quantities of refuse in the coal.
Obviously, a high-ash feed results in a high-ash clean coal.

RECOMMENDATIONS

Probably the best assurance of maximum floatability of a given
coal is the use of freshly mined coals. Coals that are stored for
long periods of time prior to physical beneficiation become weathered
and subject to altered surface characteristics and dissolution of
various soluble substances into the slurry that can act as coal

depressants. Also, reagent requirements become difficult to judge because of the porous surface of the weathered coals and the partial coating of hydrophilic slimes.

Table V Effect of feed ash on quality of
froth concentrate

| Product | Product analysis, percent | |
	Weight	Ash
Clean coal	71.7	8.1
Tailings	28.3	77.1
Feed	100.0	27.6
Clean coal	40.4	13.6
Tailings	59.6	67.2
Feed	100.0	45.1
Clean coal	20.9	37.8
Tailings	79.1	74.0
Feed	100.0	66.4

After predictability is determined or established by controlled experiments with fresh or unweathered coals, the best assurance of reproducibility on a full scale is control of operating parameters such as reagent feed, water quality, pulp solids concentration, and slurry feed rate to the flotation cells. To insure that these variables are controlled, some instrumentation would likely be needed.

It would seem, therefore, that predictability in coal flotation is dependent on understanding and utilizing the surface chemistry of coal particles in aqueous slurry, subject to the control of operating parameters.

REFERENCES

1. Preparation Trends in North America. World Coal, May 1979, p. 24.
2. Burger, John R., Froth Flotation is on the Rise. Coal Age, March 1980, p. 99.
3. Sun, Shiou-Chuan, Effects of Oxidation of Coals on Their Flotation Properties. Mining Engineering, Transactions, AIME, April 1954, pp. 396-400.
4. Yancey, H. F. and J. A. Taylor, Froth Flotation of Coal: Sulfur and Ash Reduction. U.S. Bureau of Mines, Report of Investigations 3263, January 1935, 20 pp.
5. Baker, A. F. and K. J. Miller, Hydrolyzed Metal Ions as Pyrite Depressants in Coal Flotation: A Laboratory Study. U.S. Bureau of Mines, Report of Investigations 7518, May 1971, 21 pp.

356

6. Lyon, F. M. A Comparison of the Flotation and Adsorption Characteristics of Ore- and Coal-Pyrite with Ethylxanthate. M. Sc. Thesis, Michigan Technological University, Houghton, Michigan, 1969.

PHYSICAL AND CHEMICAL COAL CLEANING

T. D. Wheelock and R. Markuszewski
Ames Laboratory and Department of Chemical Engineering
Iowa State University, Ames, Iowa   50011

ABSTRACT

Coal is cleaned industrially by freeing the occluded mineral impurities and physically separating the coal and refuse particles on the basis of differences in density, settling characteristics, or surface properties. While physical methods are very effective and low in cost when applied to the separation of coarse particles, they are much less effective when applied to the separation of fine particles. Also they can not be used to remove impurities which are bound chemically to the coal. These deficiencies may be overcome in the future by chemical cleaning. Most of the chemical cleaning methods under development are designed primarily to remove sulfur from coal, but several methods also remove various trace elements and ash-forming minerals. Generally these methods will remove most of the sulfur associated with inorganic minerals, but only a few of the methods seem to remove organically bound sulfur. A number of the methods employ oxidizing agents such as air, oxygen, chlorine, nitrogen dioxide, or a ferric salt to oxidize the sulfur compounds to soluble sulfates which are then extracted with water. The sulfur in coal may also be solubilized by treatment with caustic. Also sulfur can be removed by reaction with hydrogen at high temperature. Furthermore, it is possible to transform the sulfur bearing minerals in coal to materials which are easily removed by magnetic separation.

INTRODUCTION

Coal, our most abundant fossil fuel, is frequently loaded with various impurities which cause problems of boiler fouling and corrosion, equipment wear, waste disposal, and environmental pollution. Sulfur which is of great concern can be present as an integral part of the organic structure and also as a component in discrete mineral phases dispersed throughout the coal. Nitrogen, on the other hand, is invariably part of the organic structure and not part of the mineral matter.

The distribution of mineral matter in coal is highly variable in amount, composition, state of aggregation, and particle size.[1] The bulk of the mineral matter of most coals includes clay, iron sulfides, carbonates, and silica.[2]

Western subbituminous coals and lignite may contain significant amounts of alkali and alkaline earth metals which are associated with the organic matter rather than the mineral matter.[3,4] These metal cations appear to be held by the carboxylic and phenolic groups present in the lower rank coals, which impart ion-exchange properties to the coal. Numerous trace elements may also be present in coal.[1,2] Some of these, such as germanium, beryllium and boron appear to be associated with the organic matter, whereas others such as cobalt, copper, nickel, chromium, and selenium are associated with the mineral matter.

The principal sulfur-bearing minerals in coal are pyrite and marcasite.[1,2] Other sulfide minerals, elemental sulfur, and calcium or iron sulfates may be present in minor amounts in un-weathered coal. Pyrite may be present in a wide variety of forms ranging from large sulfur balls to micron size crystallites.[1,5] The latter are particularly troublesome to remove since they may be finely disseminated throughout the coal. An appreciable portion of the sulfur content may be organic, probably as alkyl or aryl thiols, sulfides, and disulfides and as heterocyclic compounds of the thiophene type.[6,7,8]

Nitrogen seems to occur in heterocyclic rings of the pyridine or pyrrolic type which are part of the coal structure, but the information about nitrogen groups in coal is meager.[7,8]

Because of the diverse nature of various impurities in coal, some can be removed easily while others can be removed only with great difficulty, if at all. Generally, coarse discrete mineral particles which are freed from the coal matrix by crushing can be removed readily by physical separation methods. On the other hand, finely disseminated minerals and impurities which are bound chemically to the organic structure are difficult to remove. In some cases, finely disseminated minerals which can not be freed from the coal matrix except by grinding to a very fine size may be removed by chemical attack. On the other hand, impurities which are part of the organic structure can be removed only through chemical reaction.

Most industrial coal cleaning methods remove only the discrete mineral particles which are freed by relatively coarse crushing. The production of much fine-size material is usually avoided for economic reasons. Also the application of fine particle technology to coal cleaning is not as well developed as the application of coarse particle technology. Furthermore, chemical extraction methods have not yet reached the commercial plant stage of development. Consequently, even the best currently available technology may only achieve pyritic sulfur reductions of 43-80% and commensurate ash reductions.[9]

In the discussion of coal cleaning methods, the physical and chemical principles will be mentioned, especially of promising new methods. A discussion of technology as well as of methods which completely destroy the solid character of coal or which involve biological processes will be avoided.

PHYSICAL CLEANING METHODS

Physical coal cleaning always involves crushing or comminution of the coal followed by separation of the mineral phases from the organic phases. Although dry separation methods are sometimes used, wet methods are more common, particularly with smaller particle sizes. Wet methods entail dewatering and drying of the product which add substantially to the complexity and expense of cleaning.

The state of the art of physical cleaning has been reviewed thoroughly by Leonard[10] and new developments have been reviewed by Cavallaro and Deurbrouck[11] and by Liu.[12]

Coal Comminution

Mechanical equipment for crushing and grinding coal is highly developed and widely applied even though the scientific principles underlying particle size reduction are not completely understood, so that selection of equipment is based largely on empirical tests and operating experience. The equipment is quite varied with regard to operating principles, mechanical design, and what it will accomplish. Machines are available for a wide range of applications.[13] Unfortunately this type of equipment is notoriously inefficient with most of the energy supplied converted into heat.[14] Energy utilization is of greatest concern for fine grinding which requires a much larger input of energy per ton of coal than coarse crushing.[13] Furthermore, breakage induced by the sudden application of large mechanical forces tends to be nonselective.[15] Both coal and mineral matter are fragmented somewhat randomly with little tendency for breakage along boundaries separating organic and inorganic constituents. Consequently, excessive size reduction may be required to liberate mineral matter.

A promising technique known as chemical comminution results in selective breakage.[15-20] This technique employs specific chemical agents, such as anhydrous ammonia, which are strongly adsorbed by coal macerals but not by minerals. The coal separates along bedding planes and boundaries between organic and inorganic phases. Consequently, mineral matter is liberated without excessive size reduction.

Several theories have been proposed to explain the phenomenon. One relates the breakage to the swelling which takes place when coal is contacted with a comminuting agent.[21] The bulk volume of bituminous coal can increase 50% or more upon imbibing ammonia. Since the macerals swell but not the minerals, stresses develop along the interfacial boundaries separating these components, and fracturing occurs where the stresses are concentrated. Another possibility is that the comminuting agent disrupts the bonding between the major components of coal.[16] Thus ammonia which has the ability to hydrogen bond may interfere with the hydrogen bonding between coal macromolecules. Also, comminuting agents seem to have unshared electron pairs, and hence, some electron donor properties. This is a common characteristic of ammonia, alcohols, amines, acetone, glacial acetic acid, and perhaps liquid sulfur dioxide, all of which can comminute coal.[22-24] The best comminuting agents are small molecules, capable of diffusing into the microporous structure of coal. Hence, pyridine which is an electron donor and a good solvent for coal, is not as effective as ammonia for fragmenting coal.[18]

Coal is fragmented by anhydrous ammonia either in the liquid or the saturated vapor state, with similar results.[17-20] Ammonia does not react with the coal; less than 0.1% of the coal is dissolved by the liquid,[16-18] and virtually all of the ammonia can be removed

360

by desorption and washing with water.

The degree of fragmentation achieved with ammonia depends on several factors including the coal rank, the time of exposure, and treatment conditions.[15-18] Generally lower rank coals are fragmented to a greater extent than higher rank coals.

Coal can also be comminuted with aqua ammonia but with less effect than with anhydrous ammonia.[17,18] Davis[25] has proposed to facilitate comminution with a dilute solution of ammonia and sodium or potassium hydroxide by first pretreating the coal with a gaseous mixture of oxygen and nitrogen dioxide to increase the permeability of the coal. Stambaugh[26] has reported that it is also possible to comminute certain coals and lignite with superheated water at 275°C.

Float/Sink Separation

Since the density of the organic matter in coal is considerably lower than that of the mineral matter, the two materials are frequently separated by adding crushed coal to a liquid medium of intermediate density. The organic coal particles float because they are less dense, whereas the mineral particles sink because they are more dense than the medium. The separation can be relatively sharp for coal containing coarse, discrete mineral particles. However, the separation efficiency is lower when particles contain both organic and inorganic components and have a density close to that of the liquid medium.

The specific gravity is usually 1.3-1.7 for clean coal and 2.5-5.0 for mineral matter.[1] Single crystals of pure pyrite have a specific gravity of 5.0.[27] Although various liquid media have been employed for cleaning coal, the one in general use is a suspension of fine magnetite particles in water.[28,29] By varying the concentration of magnetite, the specific gravity of the suspension can be varied between 1.3 and 2.0. Even though the magnetite particles tend to settle, the presence of clay slimes reduces their settling rate and mild agitation keeps the particles in suspension. To effect separation, the refuse particles must have a significantly higher settling rate than the magnetite particles. Since the coal and refuse particles must be recovered from the magnetite suspension by screening and/or magnetic separation, these requirements limit the smallest size of coal which can be separated in a large relatively quiescent bath to about 6mm and in a heavy medium cyclone to 0.15-0.5 mm.[11,29]

A new method of cleaning smaller-size coal employs trichlorofluoromethane (Freon 11) as a heavy medium instead of a magnetite suspension.[30-32] This organic liquid has a specific gravity of 1.48 and viscosity of 0.4 centipoise at 20°C. Since this liquid is less viscous than a magnetite suspension, small particles of coal and refuse separate more rapidly than in a magnetite suspension enabling finer-size coal to be cleaned. Separations of even very fine-size material, e.g. 77 μm x 0 size, may be practical. An important feature of the method is the displacement of surface moisture from the coal by the nonpolar organic liquid. The moisture

and included slimes are transferred to the hydrophilic refuse
particles. Surface-active agents may be added to facilitate the
transfer of moisture. Because of the nonreactivity, low boiling
point (24°C), and small heat of vaporization (182 J/g) of Freon 11,
it is readily removed from both the coal and refuse particles so
that less than 100 ppm of the liquid is retained.

Hydraulic and Pneumatic Methods

Hydraulic methods which employ jigs, wet concentration tables,
hydrocyclones and other types of equipment are the most widely
used of all industrial methods for cleaning coal.[11] In most cases
these methods depend on the difference in settling rates; because
of their greater density, refuse particles sink faster than coal
particles of the same size. However, in the case of hydrocyclones
the principle of separation may be more like that of a dense medium
cyclone. Apparently, dense refuse particles accumulate in the
conical section of a hydrocyclone and form a dense medium which is
penetrated by other refuse particles but not by the lighter coal
particles.[29]

Generally hydraulic methods do not provide as clean a separation of coal and refuse particles as methods which employ a dense
medium. On the other hand, hydraulic methods require less ancillary
equipment and have lower capital and operating costs. Unfortunately,
none of these devices provide a good separation of very small
particles.

Although pneumatic methods have been used to clean coal for
a long time, the amount of coal cleaned by these methods has declined
steadily since 1965.[33] In a typical airflow cleaner, coal and
refuse particles are segregated into different layers while passing
down a sloping, perforated deck through which pulsating air flows.
These layers are divided and withdrawn separately. An airflow
cleaner is best suited for cleaning relatively dry coal (4% moisture
is ideal) in the size range of 0.5 to 19 mm, providing the coal is
amenable to separation at a specific gravity of 1.75-1.80. If
separation at a lower specific gravity is required, the efficiency
of separation will be less than optimum. Consequently the device
does not have the flexibility of a heavy media separator employing
magnetite where the specific gravity of the suspension can be controlled over a wide range. The need for a relatively dry feed has
also tended to limit the application of pneumatic separation.

Separations Based on Surface Properties

Several coal cleaning methods including froth flotation, oil
agglomeration, and solvent partitioning take advantage of the
difference in surface properties between the organic and inorganic
components of coal. Generally, the inorganic minerals except
pyrite tend to be quite hydrophilic or water avid, whereas the
organic macerals are either hydrophobic or at least less hydrophilic
than the minerals.

In the froth flotation method of separation, which has been used
industrially for many years, a suspension of fine coal particles in

water is aerated with numerous small air bubbles. The hydrophobic
coal particles become attached to these bubbles and are buoyed to
the surface of the suspension where they are recovered in a forth.
The hydrophilic minerals are left behind in the aqueous suspension.
A frothing agent such as methyl isobutyl carbinol (MIBC) or pine
oil is added to produce a stable froth and sometimes a nonpolar
collector such as kerosene or fuel oil to increase the hydrophobicity of the coal.

In the oil agglomeration method, a small amount of fuel oil
is added to an agitated suspension of coal particles in water to
coat the hydrophobic particles selectively. The oil-coated
particles stick together and form relatively large flocs or
agglomerates which can be separated from the unagglomerated mineral
particles by screening the suspension.

In the solvent partitioning method, a relatively large amount
of an organic solvent is mixed with a suspension of coal in water.
The organic particles are transferred to the solvent phase, while
the mineral particles remain in the aqueous phase. The liquids
separate into different layers which can be decanted.

In all of these methods, water has to be displaced partly or
completely from the surface of the coal particles (but not the
mineral particles) by either air, oil, or solvent. The ease or
difficulty with which this occurs will depend on the nature of the
coal surface and the nature of the water-displacing fluid. If the
surface is highly hydrophobic, water will be displaced more readily
than if the surface is less hydrophobic or somewhat hydrophilic.
Also water can be displaced more readily by certain solvents than by
various oils.

The extent to which a given fluid will displace water from the
surface of a coal particle may be expressed in terms of the angle
of contact between a bubble or droplet of the fluid and the solid
surface with which it is in contact. The angle will be small if
there is little tendency for water to be displaced and large if
there is greater tendency. A close relationship has been observed
between the angle of contact of an air bubble and the floatability
of various ranks of coal.[34-36] Low and medium volatile bituminous
coals with a carbon content of 88-90% on a moisture- and ash-free
basis exhibit the largest contact angles and are the most easily
floated. Coals of either higher or lower rank exhibit smaller
contact angles and are more difficult to float. Anthracite and
subbituminous coal require oily collectors in order to float, and
lignite is virtually nonfloatable.

Sun[36] proposed a surface components hypothesis to explain the
different floatabilities of various types of coal. According to
this theory coal is a mixture of hydrophobic and hydrophilic components; the overall wettability and, therefore, floatability is
determined by the balance between these components. Hydrocarbons
are regarded as highly floatable, carbon and graphite as moderately
floatable, and sulfur as slightly floatable. Mineral matter and
components rich in oxygen and moisture are regarded as nonfloatable.

Weathering or atmospheric oxidation can reduce the floatability of coal.[35] Air oxidation can affect coal floatability within a week or two at ambient temperature and within a few hours at 150-200°C.[37-39] Chemical oxidants such as potassium permanganate can increase the wettability of coal and reduce its floatability.[34,35] The higher oxygen content of lower rank coals and the presence of carbonyl and carboxyl groups may be responsible for the lower floatability of these coals.[40]

The floatability of coal can also be greatly reduced by the presence of slime coatings which render the surface hydrophilic.[36,41] Clay slimes appear to be held on the coal surface by electrostatic attraction. Such coatings have been removed by adding an ionic substance (e.g., disodium hydrogen phosphate) to change the charge on the colloidal particles.

Although various oils have been used to increase the hydrophobicity of coal, these oils do not spread spontaneously over a coal surface immersed in water.[34,42] The oils remain as discrete droplets forming a definite contact angle with the coal surface. While benzene and p-xylene spread more readily than either paraffin oil, n-decane, n-hexadecane, or a high-boiling-point petroleum fraction, benzene and p-xylene are much less effective in promoting the flotation of coal.[42] Interestingly, benzene and various other monoaromatic compounds are also less effective than paraffins or olefins of equivalent molecular weight for recovering coal by oil agglomeration.[43]

Other experimental evidence indicates that some of the same factors controlling the floatability of coal also control the oil agglomeration of coal. Thus Capes et al.[44] observed that sub-bituminous coal and lignite were more difficult to agglomerate than bituminous coal. Adams-Viola et al.[45] also observed that oleophilic (hydrophobic) coals were more easily agglomerated with paraffin oil than hydrophilic coals. A lower agitator speed and, therefore, less work was required to agglomerate a low volatile bituminous coal from Pocahontas Seam No. 3 (an oleophilic coal) than to agglomerate either a high volatile bituminous coal from Illinois Seam No. 6 or North Dakota lignite (both hydrophilic coals).

All of the cleaning methods based on surface properties have difficulty in separating coal and pyrite. Glembotskii et al.[40] noted that a fresh pyrite surface is wetted more readily by hydrocarbons than by water, and if pyrite particles are not heavily oxidized, they can be floated using only a frother. On the other hand, a wet pyrite surface is readily oxidized, particularly if alkali is present. The resulting film of hydrated ferric hydroxide will depress pyrite flotation. Moreover, the addition of an acid restores the floatability of the pyrite by seemingly removing the hydroxide film. Furthermore, pyrite is depressed by alkalis which furnish hydroxyl ions that are adsorbed on the pyrite surface.

The depressing action of various alkalis and of air oxidation in a warm alkaline medium on pyrite was confirmed by Wheelock and Ho.[46] In addition, Patterson et al.[47] showed that mild air oxidation of the surface of pyrite particles suspended in a warm alkaline medium prevented the particles from being oil agglomerated.

Not all forms of disseminated pyrite are equally difficult to remove. Capes et al.[48] found that it was much easier to remove flaky, imperfect polycrystalline pyrite from a Nova Scotia coal than to remove well-formed crystals of pyrite from a New Brunswick coal when both coals were treated similarly by fine grinding in water followed by oil agglomeration. It was hypothesized that the imperfect material was more susceptible to breakage and also to oxidation during grinding so that its surface was more hydrophilic.

Froth Flotation

Froth flotation, an increasingly popular method of removing ash-forming mineral matter from coal, is used in about 300 out of 400-500 coal preparation plants in this country; 10% of all the coal is cleaned by this method.[49] However, not all of the problems of coal flotation have been solved. The separation of coal and pyrite continues to be a serious problem. Also certain clays tend to float with the coal and the flotation of oxidized coal is still difficult.

A promising method of separating coal and pyrite involves two-stage flotation.[50] Coal is floated in the first stage using conventional reagents and subsequently depressed in the second stage by using a hydrophilic colloid. Some of the coarse pyrite and most of the high-ash refuse is rejected in the first stage. Much of the remaining pyrite is then floated in the second stage with a sulfhydryl collector (potassium amyl xanthate) while the coal remains in the water phase.

Another method of separating coal and pyrite which shows promise takes advantage of the difference in flotation kinetics of coal and pyrite.[51] It would be applicable to coals which float very rapidly. By using gentle operating conditions and a short residence time, much of the coal can be floated before an appreciable amount of the slower floating pyrite appears in the froth.

Further details of coal flotation principles and practice are presented in several excellent reviews.[40,50,52-54]

Selective Oil Agglomeration

Selective agglomeration of coal particles by oil added to an agitated suspension of fine-size coal and water was considered as a means of cleaning coal as early as World War I.[55] The process investigated at that time was known as the Trent process, and it was demonstrated in the laboratory by using a small motor driven churn. The Trent process used a large amount of fuel oil (30% of the dry coal weight) and required prolonged agitation. Pyrite tended to be agglomerated with the coal.

Another early oil agglomeration process, the Convertol process, was demonstrated in Germany following World War II.[56] This process utilized a high-speed phase inversion mill, a short mixing time, and smaller quantities of oil, only 2 to 7% of the weight of solids. It was subsequently tested in a pilot plant at Gary, W. Va., to recover metallurgical coal fines.[57]

Although not widely adopted, improved agglomeration methods are

being developed in various countries. Capes and his associates at
the National Research Council of Canada have been particularly
active in developing the principles of oil agglomeration.[44,58-60]
The method is capable of recovering coal particles which range in
size from a few microns to several millimeters across. The ability
to clean extremely fine particles is an important advantage where
coal must be reduced to a very fine size to release impurities.
Other methods including froth flotation are simply not capable of
cleaning extremely fine material. Another advantage of oil agglomeration is the ease with which the particles can be dewatered.
Large agglomerates can be dewatered adequately by draining on a
screen and small agglomerates by centrifuging, thus avoiding more
costly and wasteful thermal drying. The size of the agglomerates
and, therefore, the ease of dewatering depend on the amount of oil
used, the initial size of the coal particles, and the level of
agitation. With small concentrations of oil (e.g., less than 5%)
and intense agitation, microagglomerates are produced which are held
together by liquid bridges between the individual particles. Water
is retained in the interstices. With larger concentrations of oil
and less intensive agitation, larger agglomerates are produced and
the interstitial space may be partially or completely filled with
oil. Since the surface area per unit mass increases as the particle
size decreases, the ratio of oil to coal must be increased correspondingly. The type of oil affects coal recovery, agglomerate
size, and ash rejection. Lighter, more highly refined paraffinic
oils such as hexane, kerosene, and No. 2 fuel oil can be used to
agglomerate oleophilic coals with good ash rejection. More
viscous paraffin oils produce larger and stronger agglomerates and
provide good ash rejection but require greater agitation. Heavier,
more complex oils such as coal tars and crude petroleum can be used
to agglomerate hydrophilic coals but they do not produce as clean a
product.

Much of the work on oil agglomeration has been conducted with
high-speed mixing devices which produce large shearing forces.[43-45,
47, 56-60] Waring blenders and similar devices operating above
10,000 rpm with impeller tip speeds of 8 to 10 m/sec have been widely
used for laboratory batch tests in which adequate agglomeration is
achieved in 2-5 min.[58] Similarly, high-speed colloid mills with
peripheral speeds of 20-30 m/sec have been used for larger-scale,
continuous flow operations with short residence times (15-30 sec).
Such devices require very large power inputs. In addition, the large
shearing forces produced by such devices can result in attrition
of agglomerates.[45]

High-speed mixers are not always required for agglomeration.
Australian coking coals, which are probably hydrophobic, were
agglomerated successfully with slower speed turbine agitators in
laboratory and pilot plant tests.[61-64] However, these tests were
conducted with kerosene which is easily dispersed or with pre-
emulsified oil when more viscous oils were used. An experimental
study involving agglomeration of concentrated coal slurries (over
35%) with kerosene showed the effects of various process parameters

on the flocculation rate and agglomerate size.[61] The rate was
found to be a second order function of the solids concentration
which suggests that the rate is determined by the frequency of
binary collisions between oil-coated particles. The rate was
increased by increasing the agitator speed and by increasing the
amount of kerosene. On the other hand, the average agglomerate
size was not affected by the slurry concentration. The size
increased exponentially as more oil was used but decreased as
agitator speed rose. When diesel fuel was used for agglomerating
coal, it appeared that emulsification or dispersion of the fuel
into small droplets became an important rate-controlling factor.
By pre-emulsifying the fuel before adding it to the coal-water
suspension, the time required for agglomeration was greatly
reduced. The effect was even greater when heavy fuel oil was
used.[62]

Further laboratory tests with Australian coals showed that
similar levels of coal recovery and ash rejection could be achieved
with kerosene, diesel fuel, or heavy fuel oil when these materials
were pre-emulsified before use.[62] On the other hand, when the oils
were not pre-emulsified, the results obtained with heavy fuel oil
were noticeably inferior to those obtained with the lighter oils.

Since a good separation of coal and pyrite is difficult to
achieve, various methods have been proposed for improving the
separation. Capes et al.[65] showed that pyrite removal could be
increased by pretreating fine coal with iron-oxidizing bacteria for
1 to 3 days. The results indicated that the treatment oxidized the
surface of the pyrite, thereby making it hydrophilic and preventing
it from being agglomerated. Similarly Patterson et al.[47] found
that pyrite removal was increased by pretreating fine coal suspended
in a warm alkaline solution with air for several minutes to
oxidize the pyrite surface. The effectiveness of this treatment
was confirmed subsequently by Mezey et al.[66]. These investigators
also found that pyrite removal was improved by adding sodium
metaphosphate to the system before agglomeration. It was suggested
that the pyrite surface adsorbs phosphate ions to become more wet-
table. Recently Wheelock et al.[67] showed that pyrite was rejected
during oil agglomeration when an Iowa coal was suspended in a basic
solution (pH = 7 to 11) but not when it was suspended in an acidic
solution.

Solvent Partitioning

A solvent partitioning process under development by Dow
Chemical Company can recover selectively fine coal particles from an
aqueous suspension of coal and mineral particles.[68] The suspension
is contacted with an organic solvent such as perchloroethylene
which is capable of displacing water from the surface of the coal
particles but not the hydrophilic mineral particles. After mixing
the solvent with the aqueous suspension, the mixture is allowed to
settle to separate the liquids into an organic layer containing
coal and an aqueous layer containing the hydrophilic minerals.
These layers are separated, the particles recovered by filtration,

and residual solvent stripped with steam.

Perchloroethylene is particularly well-suited for this process because of its low solubility in water, relatively low boiling point (121°C), high specific gravity (1.62), and stability in the presence of water or steam. On the other hand, the treated coal may contain as much as 1% perchloroethylene.

Laboratory and pilot plant tests of the Dow process have shown that it can recover fine coal from preparation plant waste water, froth flotation tailings, and oil slurry pond wastes. For example, when the waste water from a coal preparation plant was treated in a pilot plant, 87% of the fine coal was recovered and 86% of the ash was rejected by the process. Moreover the ash content was reduced from 38% in the feed to 8.7% in the product. However, the sulfur content of the coal was not reduced noticeably. The moisture content of the product was about 13%.

Magnetic Cleaning

Since the organic material in coal is diamagnetic[69,70] whereas clay, shale, sandstone, and the ferruginous minerals are paramagnetic, the separation of these components by magnetic methods is theoretically possible. However, because the difference in magnetic susceptibility of coal and pyrite is very small, the separation is not satisfactory with conventional magnetic separators. Therefore, various treatments have been proposed for increasing the magnetic susceptibility of the pyrite particles.

One of the earlier treatments involved heating pulverized coal with steam and air to convert the surface of the pyrite particles to iron oxides and other compounds having higher magnetic susceptibilities.[71,72] The success achieved with this approach was limited.

Recently the use of microwave energy to enhance the magnetic properties of pyrite particles was demonstrated.[73] Also treating coal with iron pentacarbonyl at 170°C was shown to increase the magnetic properties of pyrite and other mineral matter.[74,75] The surface of the pyrite is converted to a pyrrhotite-like material and the surfaces of other mineral particles become a depository for iron crystallites. Consequently the affected particles can be removed by medium intensity induced-magnetic-roll separators. In one pilot plant test, 85% of the pyritic sulfur was removed from a Pennsylvania coal while 86% of the heat content was recovered.

The success of high gradient magnetic separation (HGMS) in removing weakly magnetic mineral impurities from kaolin has given an impetus to the development of similar cleaning methods for coal.[70] The method is based on trapping pyrite and other paramagnetic particles in a filamentary matrix of strongly magnetic material such as ferritic stainless steel wool which is placed in an intense magnetic field (e.g., 20 kOe).[70,76-78] The high field gradients (1-10 kOe/μm) produced in the vicinity of the filaments combine with the intense field to capture the slightly magnetic particles while the coal particles are carried through the matrix in a flowing stream of fluid. The importance of both a strong field and a high gradient is indicated by the following relation for the magnetic force F_m acting on a small particle of nonferromagnetic material in a magnetic field[77]:

$$F_m = XmH(dH/dx)$$

In this expression X is the mass susceptibility of the particle, m the particle mass, and H the field intensity acting in the x direction. In order to trap paramagnetic particles in the matrix, the magnetic force must be larger than the gravitational and hydrodynamic drag forces acting on the particles.

The technical feasibility of cleaning coal by means of HGMS has been demonstrated.[77-81] In most cases fine-size coal is introduced in a water slurry, but in some cases the coal is suspended in air or in an oil slurry.[82] Generally, results are better with water slurries than with air suspensions. One of the problems with air suspensions is the strong tendency for micron-size particles of dry coal to agglomerate which interferes with particle separation.[81] Also the particle retention time in the magnetic separator may be too short because of the high gas velocity required to suspend the coal particles. A fluidized bed magnetic separator is being developed to overcome these difficulties.[81,83] In a preliminary three-pass separation, the pyritic sulfur content of Upper Freeport coal (74 μm x 149 μm size) was reduced 87% and the ash content 52% while recovering 80% of the coal which was better than the results achieved with wet magnetic separation.

Electrostatic Separation

Electrostatic methods can be used to separate discrete particles of materials with different dielectric properties. Separation is achieved when charged particles are subjected to the action of an intense electrical field. Various methods employed to apply an electrical charge on the particles include inducing a charge by application of an electrical field and subjecting the particles to a corona discharge.

The separation of coal and pyrite with a conventional electrostatic separator was demonstrated by Abel et al.[84] When a synthetic mixture of closely-sized particles of precleaned coal and precleaned pyrite was treated by the device, the separation was very sharp. Cleaning Pittsburgh coal by a combination of air classification with a centrifugal separator and electrostatic separation removed 35-55% of the pyritic sulfur with an overall coal recovery of 80-90%.

Recently several novel electrostatic devices were applied to remove pyrite and ash from coal.[85] These devices utilize triboelectrification which occurs naturally when nonconducting particles are fluidized with air or are transported in an air stream to charge the particles.

CHEMICAL CLEANING METHODS

The present interest in chemical coal cleaning stems from a need to remove finely disseminated pyrite and organically bound sulfur. A number of methods are under development which are designed to remove one or both of these forms of sulfur. In addition, some of these methods will also remove other mineral matter and certain trace elements. However, there appears to be little work on methods which are not specifically designed to desulfurize. The removal of nitrogen, for example, seems to be a largely neglected area.

Oxidative Treatments
Many oxidizing agents are sufficiently strong to convert the pyritic sulfur in coal to forms that can be readily removed in a gaseous stream or aqueous solution. It appears that some of these reagents are also capable of converting part of the organic sulfur to extractable forms. However, the oxidation of organic sulfur compounds in coal is more speculative. First of all, knowledge of the content and nature of organic sulfur groups in coal is meager.[6,86] Secondly, the reactions of organic sulfur in coal are difficult to study. Finally, it is doubtful that the methods available for determining organic sulfur in coal are applicable to chemically treated coal.

Despite these problems, evidence is accumulating that a significant portion of the so-called organic sulfur can be removed from certain coals by specific processes. There seem to be two types of organic sulfur in coal, one removable by chemical means and another which resists chemical attack. The first type may include groups like mercaptans, sulfides, and disulfides which are fairly reactive while the second type may include heterocyclic structures like dibenzothiophene which are more stable. Indeed there is some evidence that about one-half of the organic sulfur in coal may be of the cyclic and thus unreactive type.[86-90] The differences in reactivity may also be related to the different origins of the organic sulfur. Some of it can be traced to the sulfur-containing amino acids present in the starting plant material, and some can be traced to secondary sources during coalification.[91-93]

Oxidation by Oxygen or Air
Four major processes under development for the removal of sulfur from coal employ oxygen or air in aqueous solution at elevated temperature and pressure. In the Ledgemont Oxygen Leaching (LOL) Process, ground coal is leached with a solution containing dissolved oxygen under 10-20 atm partial pressure at 130°C for 1-2 h, either under mildly acidic conditions or in the presence of 1-3 $\underline{M}$ ammonium hydroxide.[94-98] Under acidic conditions, about 90% of the pyritic sulfur is removed in 2 h but none of the organic sulfur. The pyrite is oxidized primarily to ferric sulfate and sulfuric acid, which is the basis for the acidic condition. The reaction can proceed stepwise but can be expressed overall as:

$$4 \ FeS_2 + 15 \ O_2 + 2 \ H_2O = 2 \ Fe_2(SO_4)_3 + 2 \ H_2SO_4$$

The leaching process is complex and depends on the temperature, pressure, concentration, and pH. Under alkaline conditions, about 80-85% of the pyrite is removed, but 30-40% of the organic sulfur is also extracted from certain coals. The reaction for the oxidation of pyrite is given by[97]:

$$FeS_2 + 4 \ NH_3 + 7/2 \ H_2O + 15/4 \ O_2 = 2 \ (NH_4)_2SO_4 + Fe(OH)_3$$

The consumption of oxygen, amounting to about 0.1 kg/kg bituminous coal, is approximately evenly divided into reaction with pyritic sulfur, reactions with organic coal components, and reactions producing carbon dioxide. Under alkaline conditions the carbon loss is about 10% and under acidic conditions somewhat less. In addition, the ash and heavy metal contents are substantially reduced.

In the Promoted Oxydesulfurization Process,[99-101] the leaching procedure is similar, using oxygen at about 20 atm and an acidic aqueous solution at 120°C for 1 h. However, an iron-complexing agent such as oxalic acid or its salt is added to act as a promoter. The precise mechanism by which it functions is not disclosed. Recently another promoter of a proprietary nature has been used. Also, a modification has been introduced in which the product of the first leaching step is heated to about 350°C for 1 h. Almost all of the pyritic sulfur, 94% of the iron, and 50% of the ash were removed by the one-step process, while up to 35% of the organic sulfur was removed by the two-step process. The heating value recovery was better than 95%.

The PETC Oxydesulfurization Process also uses acidic leaching conditions but at higher temperatures (140-200°C) and pressures (34-68 atm), with air as the oxidizing agent.[102-104] At these more severe conditions, the leaching rate is rapid, with almost all of the pyritic sulfur removed as sulfuric acid in 1 h or less. The overall reaction of pyrite under these conditions appears as follows:

$$2 \ FeS_2 + 7\tfrac{1}{2} \ O_2 + H_2O = Fe_2(SO_4)_3 + H_2SO_4$$

$$Fe_2(SO_4)_3 + n \ H_2O = Fe_2O_3 \cdot (n-3)H_2O + 3 \ H_2SO_4$$

In addition, up to 45% of the organic sulfur is removed from some coals, with fuel value losses usually less than 10%. While the ash content may be reduced by as much as 20%, the oxygen content is increased significantly. With increasing severity of leaching conditions, the amount of organic sulfur removed is increased in direct proportion to fuel loss.

The kinetics of the PETC process were investigated recently.[105] For the oxidation of pyritic sulfur, two alternate rate-controlling mechanisms appeared equally promising. In one case the particles of pyrite were assumed to be finely dispersed and uniformly distributed throughout the coal matrix with the rate of reaction being a second order function of the pyrite concentration. In the other case, pyrite was assumed to exist as discrete particles free from the coal with the rate being determined by diffusion through the product shell surrounding an unreacted core of pyrite. The oxidation of carbon was found to be insignificant at 150°C but significant at 190°C. Tentatively, it appeared that organic sulfur removal followed a zero order reaction with an activation energy of 78.9 kJ/mole.

The Ames Process[106,107] employs alkaline solutions (e.g., 0.2 $\underline{M}$ sodium carbonate) at 150°C and oxygen at about 14 atm to leach over 95% of the pyritic sulfur within 1 h from certain coals. The removal of organic sulfur varies from coal to coal but can approach 50%, especially if the first leaching step is followed by a second leaching step at 240°C or higher in a nitrogen atmosphere.[108] Under conditions of the Ames Process, the pyritic sulfur is extracted as soluble sulfate leaving a residue of hematite.[109,110] The reactions can be expressed as:

$$2 \text{ FeS}_2 + 7.5 \text{ O}_2 + 4 \text{ H}_2\text{O} = \text{Fe}_2\text{O}_3 + 4 \text{ H}_2\text{SO}_4$$

$$\text{H}_2\text{SO}_4 + 2 \text{ Na}_2\text{CO}_3 = \text{Na}_2\text{SO}_4 + 2 \text{ NaHCO}_3$$

The rate of oxidation of pyrite is controlled by the diffusion of dissolved oxygen through the shell of hematite which forms around a core of unreacted pyrite.[109,110] The apparent energy of activation is 22.6 kJ/mole.

The mechanism of organic sulfur removal is not well established. It is evident, however, that higher oxygen pressures favor greater removal of organic sulfur. Recent work[111,112] on model compounds shed some light on possible reactions of organosulfur compounds under conditions of the Ames Process. Thiophenol was oxidized, but the reaction apparently stopped at the phenyl disulfide. Benzylic sulfides, on the other hand, were quite reactive producing benzaldehyde, benzoic acid, and the corresponding sulfonic acid.

It has been suggested that the removal of organic sulfur from coal requires two steps.[103,111,113] In the first step the sulfur is converted to an oxidized form, such as a sulfone. The addition of oxygen to the sulfur atom weakens the carbon-sulfur bond so that it can be cleaved in a second step such as heating in the presence of base.

Oxidation by Chlorine

The powerful oxidizing ability of gaseous chlorine is exploited for desulfurizing coal in a chlorinolysis process being developed by the Jet Propulsion Laboratory of the California Institute of Technology.[114-117] A suspension of powdered coal, either in water or in an organic solvent like methylchloroform, is treated with chlorine gas for 1-2 h at 50-100°C to oxidize the sulfur to sulfuric and sulfonic acids. Chlorine also reacts with the organic matrix. In a subsequent step, the chlorinated coal is hydrolyzed to release hydrochloric acid, and the soluble sulfates, chlorides, and sulfonates are extracted with water at 50-100°C. The recovered coal is further dechlorinated by heating at 350-550°C in steam or an inert atmosphere. The method can be used to remove up to 70% of the organic sulfur and up to 90% of the pyritic sulfur. In addition, the content of trace elements such as lead, vanadium, phosphorus, lithium, beryllium, and arsenic can be reduced by 50-90%.[115,117] However, despite the hydrolysis and dechlorination steps, the residual chlorine levels can be up to 1% in the cleaned coal.[117]

The chemistry involved in the various steps of this process is very complex and not well established.[115] The overall reaction of pyrite seems to be as follows:

$$FeS_2 + 7\ Cl_2 + 8\ H_2O = FeCl_2 + 2\ H_2SO_4 + 12\ HCl$$

However, other work[118] suggests that the reaction may proceed all the way to ferric chloride. For the organic sulfur reactions, cleavage can occur at C-S bonds, as in sulfides

$$R-S-R' + Cl_2 \longrightarrow RSCl + R'Cl$$

or at S-S bonds, as in disulfides

$$R-S-S-R' + Cl_2 \longrightarrow RSCl + R'SCl$$

The sulfenyl chlorides are oxidized further, possibly through sulfonyl chloride intermediates, to sulfonic acids

$$RSCl + 2\ Cl_2 + 3\ H_2O \longrightarrow RSO_2Cl \longrightarrow RSO_3H + 5\ HCl$$

or to sulfuric acid

$$RSCl + 3\ Cl_2 + 4H_2O \longrightarrow H_2SO_4 + RCl + 6HCl$$

The above reactions are exothermic and proceed favorably at moderate temperatures.[115] The chlorine requirements are 3-4 moles of chlorine per mole of organic or inorganic sulfur (depending on the final conversion product). In addition, chlorine is expended on the chlorination of the organic coal matrix, which is considered a substitution reaction:

$$RH + Cl_2 \longrightarrow RCl + HCl.$$

In the hydrolysis and dechlorination steps, the chlorinated coal hydrocarbons release hydrochloric acid according to:

$$RCl + H_2O \longrightarrow ROH + HCl$$

or $RCl + R'H \longrightarrow R-R' + HCl$.

For coal chlorinated at moderate temperatures, these dechlorination reactions proceed to completion.

Coals treated by this process exhibit a loss in hydrogen content, while the nitrogen content seems unaffected. There is also apparently a change in the structure of the chlorinated coal since the coal becomes nonvolatile, noncaking, and nonswelling.[115] It may be inferred that aromatization, cross-linking, or other forms of condensation reactions take place.

A similar loss in coking properties was observed when Japanese bituminous coals were treated with chlorine gas in an aqueous suspension at room temperature.[119] The inorganic sulfur reduction was 70-95% and the organic sulfur reduction 13-20%.

Oxidation by Ferric Salts

The reaction of pyrite with ferric sulfate has been known for a long time[120] and can be expressed as a sum of two consecutive reactions:

$$FeS_2 + Fe_2(SO_4)_3 = 3\ FeSO_4 + 2\ S$$

$$2\ S + 6\ Fe_2(SO_4)_3 + 8\ H_2O = 12\ FeSO_4 + 8\ H_2SO_4$$

Yurovskii[71] proposed to use these reactions for removing pyritic sulfur from coal and they have become the basis for the Meyers Process[121-126] in which crushed coal is leached with an acidic solution of ferric sulfate at 100-130°C for several hours. Up to 95% of the pyritic sulfur can be extracted, provided the leaching time is long enough. Of course, none of the organic sulfur is affected, but the content of many trace elements, like arsenic, cadmium, chromium, lead, manganese, nickel, vanadium, and zinc, is reduced significantly.[125]

Although the Meyers Process uses ferric sulfate as the oxidant, ferric nitrate and ferric chloride can be equally effective.[121,122,127] When coal is leached, the ratio of sulfate to sulfur in the products is about 1.5 regardless of conditions.[124] The elemental sulfur which is finely dispersed throughout the coal can be extracted with a heated solvent like toluene or acetone, or it can be removed by vaporization. The ferrous sulfate in the spent leaching solution can be converted by air or oxygen to ferric sulfate which is then recycled for use in the oxidative leaching step. Alternatively, the regeneration can be performed concurrently with the leaching step.

Garrels and Thompson[128] found that the rate of reaction of ferric ion with pyrite is a function of the ferric-to-ferrous ion ratio. They postulated that the instantaneous rate, determined by the differential adsorption of ferric and ferrous ions, is proportional to the fraction of the surface of pyrite covered by ferric ions. This dependence on the competitive adsorption between ferric and ferrous ions was also observed by Sasmojo.[129] His data indicate that only uncomplexed ferric ions are active and that the rate-controlling step appears to be an electron transfer reaction between the adsorbed ferric ions and the pyrite surface.

Several attempts have been made to explain the slow kinetics of the oxidation of pyrite by ferric ions. Yurovskii[71] concluded that the elemental sulfur formed in the reaction is deposited on the surface of the pyrite particles and is the main cause of the retardation of the oxidation by passivating the pyrite surface. He found that both ferric sulfate and aluminum sulfate, leached from the mineral matter, are also adsorbed on the surface of the pyrite. The adverse effect of the adsorbed aluminum sulfate is due to the formation of a protective film which obstructs the diffusion of the oxidizing solution to the pyrite particles embedded in the coal. He proposed adding nitric acid to the leaching solution to eliminate these problems and speed up the reaction.

Betancourt and Hancock[130,131] improved the reaction rate by increasing the temperature and adding a second leaching step with a fresh solution of ferric sulfate. Passivation did not appear to be caused by a coating of elemental sulfur on the pyrite surface, since repeated washings with proven solvents for sulfur did not improve the rate. Several studies[132,133] have shown that the rate of oxidation is improved by introducing oxygen into the ferric sulfate leaching solution. The removal of pyritic sulfur can be also improved by using the acidic ferric sulfate solution as a dense medium, usually at specific gravity of 1.33, to preclean the coal before the chemical leaching step.[71,126]

Oxidation by Nitrogen Oxides

In the KVB process, the active oxidizing agent is nitrogen dioxide.[134] Dry, coarsely ground coal is heated for 0.5 to 1 h at 100°C under atmospheric pressure with a gas mixture containing oxygen, nitrogen dioxide, nitrogen monoxide, and nitrogen. Selective oxidation of the sulfur constituents in coal occurs, with about one-half of the pyritic sulfur being converted directly to sulfur dioxide. The rest of the oxidized pyrite is removed in a second step, in which hot water extracts soluble iron sulfites and sulfates. In the final step, the washed coal is treated for 0.1 to 1 h with hot caustic solution to remove the oxidized organic sulfur compounds as inorganic sulfites and sulfates. The overall treatment can remove almost all of the pyrite and up to 40% of the organic sulfur.

The mechanism of oxidation has not been fully established. It is postulated that pyrite can react with nitrogen dioxide as follows:

$$FeS_2 + 5\ NO_2 = FeSO_3 + SO_2 + 5\ NO$$

or $$FeS_2 + 6\ NO_2 = FeSO_4 + SO_2 + 6\ NO$$

Also nitrogen dioxide may react with moisture in the coal to form nitric acid which then attacks the sulfur. The nitrogen monoxide produced is reacted with oxygen to reform nitrogen dioxide which is returned to the process.

The reaction of nitrogen dioxide with organic sulfur groups in coal is more complicated. It is known[135] that sulfides can be oxidized to sulfoxides and sulfones according to:

$$R_1-S-R_2 + NO_2 \longrightarrow R_1-S(O)-R_2 + NO$$

and

$$R_1-S(O)-R_2 + NO_2 \longrightarrow R_1-S(O_2)-R_2 + NO$$

Model compounds which have been oxidized as indicated include dihexyl disulfide, benzyl phenyl sulfide, dibenzyl disulfide, and dibenzothiophene.[103,134] The oxidized sulfur compounds can be decomposed to sulfur dioxide and sulfur trioxide by further reaction with nitrogen dioxide,[135] or they may be removed by a selective solvent such as methanol.[136] In coal desulfurization, it appears more feasible to hydrolyze the oxidized sulfur compounds by caustic treatment.[134]

A question requiring further research is whether the coal is nitrated when subjected to the treatment described above.

Oxidation by Other Oxidizing Agents

The use of oxidizing agents other than the ones discussed above has not been popular for several reasons. Either the reagents are unduly expensive or difficult to produce and handle, or they are unselective in their reaction with coal, resulting in considerable degradation of the coal and an accompanying loss in heating value and wasteful consumption of reagents.

Yurovskii[71] proposed to use the well-known reaction of nitric acid with pyrite to desulfurize coal. Depending upon the final destined use of the coal, 12-27% nitric acid was used to leach the coal at 90°C for 2-15 min. The overall reaction is expressed as:

$$4 \ FeS_2 + 30 \ HNO_3 = 2 \ Fe_2(SO_4)_3 + 2 \ H_2SO_4 + 13 \ H_2O + 15 \ NO_2 + 15 \ NO.$$

Up to 86% of the total sulfur was thus removed. In addition, the ash content was reduced by 10-40%. The treatment also removed all of the calcium, iron, and magnesium and part of the aluminum from the ash. However, the fate of organic sulfur or the possible nitration of the coal were not mentioned.

In a preliminary study,[137] an attempt was made to remove sulfur, especially organic sulfur, from coal by using up to 1.2% ozone in oxygen at 25-100°C as the oxidizing agent. Since the effluent gas was enriched in sulfur compounds, some desulfurization took place. Sulfur dioxide was the major sulfur component at lower temperatures and carbon oxysulfide at higher temperatures. In other work,[119] coal was treated with 3% hydrogen peroxide in an aqueous solution, and almost all of the inorganic sulfur was removed.

Caustic Treatments

Coal can be desulfurized by treatment with alkali as in the

Battelle Hydrothermal Process[26,138-140] which involves heating an
aqueous slurry of finely ground coal and 10% sodium hydroxide
plus 2-3% calcium hydroxide. After the treatment at about 250-
300°C for 10-30 min, over 90% of the pyritic sulfur and up to 50%
of the organic sulfur can be removed. The heating value recovery
of 90-95% can be increased by recovering organic matter solubilized
by the alkali. The process can extract trace elements such as
arsenic, beryllium, boron, lead, thorium, and vanadium.[139] The ash
content can also be decreased substantially if the hydrothermally
treated coal is subsequently washed with dilute acid.

The sulfur remaining in the treated coal is not all emitted
into the atmosphere during combustion. Small amounts of sodium
and calcium retained by the coal serve as sulfur scavengers when
the coal is burned.

Generally, the extracted sulfur appears in the spent leaching
solution as sodium sulfide, if oxygen is excluded from the system.
The mechanism proposed[139] for extracting sulfur is:

$$FeS_2 + 2\ NaOH = Fe(OH)_2 + Na_2S_2$$

$$Na_2S_2 + Fe(OH)_2 = Fe_2O_3 + Na_2S$$

$$Na_2S_2 + coal = CO_2 + Na_2S$$

Thus sodium disulfide appears as an intermediate which is reduced
to sodium sulfide by coal or by ferrous hydroxide. Many other
reactions are also possible.

The mechanism for the extraction of organic sulfur is not
resolved as yet. In some of the gases evolved during desulfurization of coal by this process, dimethyl sulfide (CH_3-S-CH_3) has been
identified.[139] Therefore some carbon-carbon or carbon-sulfur bond
cleavage may occur. Also some acidic sulfur groups may be
solubilized as follows:

$$R-SH + OH^- = R-S^- + H_2O$$

and $$R-S-R' + OH^- = R-OH + R'-S^-$$

However, the chemistry of organic and inorganic sulfur removal from
coal by alkali treatment is highly speculative.

Earlier work on the use of a caustic solution to extract
sulfur from coal was carried out by Reggel et al.[141] Only pyritic
sulfur was removed consistently at 225°C by a 10% sodium hydroxide
solution. Work by Jangg[142] on the leaching of mineral pyrite with
2.5 M sodium hydroxide showed that 95% of the pyrite was converted
to a ferric oxide (α -$Fe_2O_3 \cdot H_2O$) and sodium sulfide at 400°C in
1-2 h, but none was reacted at 200°C. Burkin and Edwards[143] also
observed that mineral pyrite did not appear to react with a 10%
sodium hydroxide solution at 150°C.

Bunn[144] studied the reaction of coal-derived pyrite with
aqueous sodium hydroxide (0-10%) at 150-215°C in a packed bed
tubular reactor. The rate of pyrite conversion was chemical reaction controlled with an apparent activation energy of 21 kcal/mole
and linearly dependent on the concentration of sodium hydroxide.
When coal was subjected to the same treatment, it showed signs of
degradation, although some sulfur was removed.

Another caustic treatment process involves reacting 1 part of
coal with 4 parts of 1:1 melt of potassium and sodium hydroxides.[145]
Below 150°C, no reaction occurs. At about 225°C, pyrite reacts
vigorously with the caustic to produce sulfides which are soluble in
the melt. The reaction is apparently

$$8 \text{ FeS}_2 + 30 \text{ MOH} = 4 \text{ Fe}_2\text{O}_3 + 14 \text{ M}_2\text{S} + \text{M}_2\text{S}_2\text{O}_3 + 15 \text{ H}_2\text{O},$$

where M = Na or K.

The pyritic sulfur can be completely removed within 5 min.
Above 225°C some organic sulfur is also removed. After prolonged
treatment at 400°C, the sulfur content seems to increase, apparently
because of the formation of organic sulfur compounds by reaction of
coal with sulfides in the molten caustic. About 90-95% of the coal
is recovered after the treatment. For lower rank coals, severe
decomposition decreases the recovery to about 50% after a 30-min
treatment at 200°C.

A similar approach was utilized by Meyers and Hart[146] to remove
89% of the pyritic and 23% of the organic sulfur from bituminous
coal by a 30-min treatment at 370°C. When the coal was precleaned
by a float-sink method[126] at a specific gravity of 1.33 to remove
most of the pyritic sulfur and then subjected to the caustic
treatment, an average of 72% of the organic sulfur was removed. It
was proposed that the reaction displaced organic sulfide by hydroxide,
followed by condensation to a ring structure at the high temperature.

Another process[73] uses a sodium hydroxide solution to treat
powdered coal which is then dewatered and subjected to microwave
irradiation. The pyrite is apparently decomposed to pyrrhotite by
the selective microwave heating and extracted as a soluble sulfide.
This combined caustic-microwave treatment can remove up to 90% of
the pyritic and 50-70% of the organic sulfur from some coals.

Reductive Treatments

Hydrogen is the principal reducing agent which has been considered for desulfurizing coal. High temperatures are employed which
lead to the production of hydrogen sulfide.[147] Generally, pressures
near atmospheric are used when the main objective is to desulfurize
rather than to liquify or gasify.

In the IGT Hydrodesulfurization Process, up to 95% of the sulfur
can be removed at 800°C after 2 h.[148] There is also a substantial
decrease in the content of nitrogen and volatile matter. To prevent
caking, the coal is pretreated by oxidation in air at 400°C for
30 min. This pretreatment step eliminates 25-30% of the pyritic

sulfur as sulfur dioxide and aids in the subsequent removal of organic sulfur by hydrotreating.

Coal char produced by the Flash Pyrolysis Process of the Occidental Research Corporation can be hydrodesulfurized after first leaching with 6 N hydrochloric acid at 80°C to remove iron and calcium sulfides.[149] This pretreatment greatly reduces the hydrotreating time at 870°C and allows for higher hydrogen sulfide concentrations in the off-gas.

The reactions of hydrogen with sulfur compounds can be very complex.[147,150] Sulfur removal below 600°C occurs mainly by pyrolysis and devolatilization; above 600°C direct reaction of hydrogen with organic sulfur may occur.[151] The reaction of pyrite is believed to occur stepwise[150]:

$$FeS_2 + H_2 = FeS + H_2S$$

and $$FeS + H_2 = Fe + H_2S$$

The equilibrium constant for the second step, however, is only about 0.003 at 870°C; thus the presence of even small concentrations of hydrogen sulfide will prevent removal of sulfide sulfur.[149,150,152] The removal of pyritic or sulfide sulfur prior to hydrotreating is therefore quite beneficial. The equilibrium constants for typical reactions of organic sulfur with hydrogen are much greater. Thiols react to form hydrocarbons and hydrogen sulfide. Aliphatic sulfides and disulfides are also unstable, decomposing to hydrogen sulfide and olefins via a reversible reaction. Aromatic sulfides and heterocyclic compounds containing sulfur are difficult to desulfurize by hydrotreatment.

A complicating factor in hydrodesulfurization is that under certain conditions, the back-reaction of hydrogen sulfide with partially desulfurized coal becomes important. There is substantial evidence that inorganic sulfur can be reincorporated in the coal as organic sulfur, probably as a cyclic compound.[150,151] These problems can be minimized by removing inorganic minerals before hydrotreating or by introducing a sulfur scavenger such as calcium oxide or iron oxide which can be subsequently removed.

Miscellaneous Treatments

A potentially interesting method of removing organic sulfur from coal involves reaction with excess iron pentacarbonyl in an organic solvent such as benzene, toluene, or anthracene oil at 70-100°C.[153] In one experiment about 80% of the organic sulfur by weight was removed. It is claimed that the iron pentacarbonyl reacts with thiophenic sulfur, which is one of the most difficult forms to remove. The products may include carbon monoxide, carbonyl sulfide, and compounds containing iron and sulfur.

In another method, liquid sulfur dioxide was used to remove about 40% of the organic sulfur in bituminous coal.[154] The reaction was carried out on 1:2 coal/SO_2 mixtures in sealed borosilicate tubes

at 25-70°C for 1-2 h without stirring. After reaction the coal was
washed extensively with water, acetone, and dilute nitric acid to
remove most residual sulfur dioxide from the coal. Apparently, the
Lewis acid properties of liquid sulfur dioxide resulted in adduct
reactions of the following type:

$$R-S-R + SO_2 = R_2S:SO_2$$

It is uncertain by what mechanism or what bond breakage these
adducts were separated from the coal. At the same time, some
comminution of coal occurred, probably because of similar donor-
acceptor reactions between the sulfur dioxide and the coal. The
extracts, containing 5-10% of the original weight of the coal, were
highly colored, which is typical of such addition compounds. Pyrite
apparently did not react.

CONCLUSIONS

Due to the complexity of coal, the large diversity of impurities which occur in coal, and the great variation between coals, it
is not surprising that a number of coal cleaning methods are in
commercial use and an even larger number are under development. For
the most part, the methods in general use involve coarse crushing
and physical separation of the larger and intermediate-size coal
and refuse particles by techniques which are based on differences in
specific gravity or settling rate of the various particles. Finer-
size material may also be cleaned by methods based on similar
principles, or it may be cleaned by froth flotation which makes use
of the difference in surface properties of coal and its associated
mineral matter. While these methods can separate discrete particles
of coal and mineral matter, they are not designed to remove
chemically bound impurities and are inadequate when it comes to
removing finely disseminated microcrystals of pyrite of other
minerals.

Several new methods under development are designed to improve
the physical removal of mineral matter from coal. One method
utilizes anhydrous ammonia to unlock the minerals without resorting
to excessive size reduction. Another method utilizes a dense organic
liquid to effect a separation of the heavier and lighter components
of coal. Another selectively oil agglomerates fine coal particles
suspended in water while rejecting the mineral particles, and a
closely related method uses an organic solvent to selectively extract
coal particles suspended in water. Both wet and dry magnetic
separation methods show promise. These methods either utilize HGMS
or enhance the magnetic susceptibility of the pyrite and other
minerals by chemical treatment. In addition, electrostatic separation may be developed into a practical means of cleaning coal.

Various chemical processes are also under development for
extracting sulfur and/or other impurities from coal. Several of
these processes make use of an oxidizing agent such as air, oxygen,
nitrogen dioxide, chlorine or ferric sulfate to oxidize the sulfur

to water soluble sulfates which are removed by leaching. Other processes utilize hydrogen to reduce the sulfur and to extract it as hydrogen sulfide gas. Still other processes use caustic to solubilize and extract sulfur. Some of these processes remove appreciable amounts of other impurities besides sulfur.

ACKNOWLEDGEMENT

Ames Laboratory is operated for the U.S. Department of Energy by Iowa State University under Contract No. W-7405-Eng-82. This work was supported by the Assistant Secretary for Fossil Energy, Office of Mining, WPAS-AA-75-05-05.

LITERATURE CITED

1. J. D. McClung, M. R. Geer, and H. J. Gluskoter, in Coal Preparation, 4th ed., J. W. Leonard, ed., (AIME, New York, 1979), Chap. 1, pp. 53-79.

2. H. J. Gluskoter, in Trace Elements in Fuel, S. P. Babu, ed., Adv. in Chem. Series No. 141 (Am. Chem. Soc., Washington, DC, 1975), pp. 1-22.

3. E. A. Sondreal, P. H. Tufte, and W. Beckering, Combustion Science and Technology 16, 95-110 (1977).

4. L. E. Paulson and A. W. Fowkes, "Changes in Ash Composition of North Dakota Lignite Treated by Ion Exchange", Report of Investigations 7176, (U.S. Bureau of Mines, 1968).

5. R. T. Greer, in Coal Desulfurization: Chemical and Physical Methods, T. D. Wheelock, ed., ACS Symp. Series No. 64 (Am. Chem. Soc., Washington, DC, 1977), pp. 1-15.

6. P. H. Given and W. F. Wyss, British Coal Utilization Research Association Monthly Bulletin 25(5), 165-179, (1961).

7. D. W. Van Krevelen, Coal (Elsevier, New York, 1961), p. 171.

8. D.D. Whitehurst, in Organic Chemistry of Coal, J. W. Larsen, ed., ACS Symp. Series No. 71, (Am. Chem. Soc., Washington, DC, 1978), pp. 1-35.

9. E. H. Hall and G. E. Raines, Proc. Symp. on Coal Cleaning to Achieve Energy and Environmental Goals (Sept. 1978, Hollywood, FL), EPA-600/7-79-098a (April 1979), Vol. I, pp. 416-47.

10. J. W. Leonard, ed., Coal Preparation, 4th ed., (AIME, New York, 1979).

11. J. A. Cavallaro and A. W. Deurbrouck in Coal Desulfurization: Chemical and Physical Methods, T. D. Wheelock, ed., ACS Symp. Series No. 64, (Am. Chem. Soc., Washington, DC, 1977), pp. 35-57.

12. Y. A. Liu, ed., New Physical Methods for Cleaning Coal, (Marcel Dekker, Inc., New York, in press).

13. L. G. Austin and J. D. McClung, in Coal Preparation, 4th ed., J. W. Leonard, ed., (AIME, New York, 1979), Chap. 7.

14. J. D. McClung, in Coal Preparation, 3rd ed., J. W. Leonard and D. R. Mitchell, eds., (AIME, New York, 1968), Chap. 7.

15. R. G. Aldrich, "Pre-combustion Coal Cleaning using Chemical

Comminution: Research Report:, TR-73-564, (Syracuse Univ. Res. Corp., Syracuse, NY, Oct. 1973).

16. P. Howard, A. Hanchett, and R. G. Aldrich, presented at Institute of Gas Technology Symp. II, Clean fuels From Coal, Chicago, IL (June 23-27, 1975).

17. R. S. Datta, P. H. Howard, and A. Hanchett, "Feasibility Study of Pre-combustion Coal Cleaning using Chemical Comminution", Final Report, FE-1777-4, (Syracuse Research Corp., Syracuse, NY, Nov. 1976).

18. P. H. Howard and R. S. Datta, in Coal Desulfurization: Chemical and Physical Methods, T. D. Wheelock, ed., ACS Symp. Series No. 64, (Am. Chem. Soc., Washington, DC, 1977), pp. 58-69.

19. "Will Chemicals Replace Crushing?", Coal Age 83(2), 176-9, (1978).

20. V. C. Quakenbush, R. R. Maddocks, and G. W. Higginson, Coal Mining & Processing 16(5), 68-72 (1979).

21. D. V. Keller, Jr. and C. D. Smith, Fuel 55, 273-80 (1976).

22. R. G. Aldrich, U. S. Pat. 3,815,826 (June 11, 1974).

23. F. Fischer and W. Glund, Berichte 49, 1469-71 (1916).

24. W. K. T. Gleim, U. S. Pat. 4,120,664 (Oct. 17, 1978).

25. B. W. Davis, U. S. Pat. 4,312,448 (Jan. 2, 1979).

26. E. P. Stambaugh, presented at 2nd Conf. on Air Quality Management in the Electric Power Industry, Houston, TX (Jan. 22-25, 1980).

27. W. A. Deer, R. A. Howie, and J. Zussman, Rock-forming Minerals, Vol. 5, Non-silicates, (Longmans, Green and Co., Ltd., London, 1962).

28. E. R. Palowitch and A. W. Deurbrouck, in Coal Preparation, 4th ed., J. W. Leonard, ed., (AIME, New York, 1979), Chap. 9, pp. 1-36.

29. M. Sokaski, M. R. Geer, and W. L. McMorris III, in Coal Preparation, 4th ed., J. W. Leonard, ed., (AIME, New York, 1979), Chap. 10, pp. 1-39.

30. D. V. Keller, Jr., C. D. Smith, and E. F. Burch, presented at SME-AIME Conf., Atlanta, GA (March 7, 1977).

31. C. D. Smith, presented at Coal Age Conf., Louisville, KY (October 23, 1979).

32. D. V. Keller, Jr., in New Physical Methods for Cleaning Coal, Y. A. Liu, ed., (Marcel Dekker, Inc., New York, in Press).

33. R. L. Llewellyn, K. K. Humphreys, and J. W. Leonard, in Coal Preparation, 4th ed., J. W. Leonard, ed., (AIME, New York, 1979), Chap. 11.

34. G. A. Brady and A. W. Gauger, Ind. Eng. Chem. 32, 1599-604 (1940).

35. R. M. Horsely and H. G. Smith, Fuel 30(3), 54-63 (1951).

36. S.-C. Sun, Mining Eng. 6(1), 67-73 (1954).

37. S.-C. Sun, Mining Eng. 6(4), 396-401 (1954).

38. J. B. Gayle, W. H. Eddy, and R. Q. Shotts, "Laboratory Investigation of the Effect of Oxidation on Coal Flotation", Report of Investigations 6620 (U. S. Bureau of Mines, 1965).

39. J. Iskra and J. Laskowski, Fuel 46, 5-12 (1967).

40. V. A. Glembotskii, V. I. Klassen, and I. N. Plaksin, Flotation,

(Primary Sources, NY, 1972), pp. 406-24.

41. A. Jowett, H. El-Sinbawy, and H. G. Smith, Fuel 36, 303-9 (1956).

42. D. J. Brown, V. R. Gray, and A. W. Jackson, J. Appl. Chem. 8, 752-9 (1958).

43. S.-C. Sun and W. L. McMorris III, Mining Eng. 11(11), 1151-6 (1959).

44. C. E. Capes, A. E. McIlhinney, R. E. McKeever, and L. Messer, Proc. 7th Internat. Coal Preparation Congr., Sydney, Australia, (May 23-8, 1976).

45. M. Adams-Viola, G. D. Botsaris, and Yu. M. Glazman, presented at ACS meeting, Houston, TX (March 27, 1980).

46. T. D. Wheelock and T. K. Ho, presented at Society of Mining Engineers of AIME, New Orleans, LA (Feb. 18-22, 1979).

47. E. C. Patterson, H. V. Le, T. K. Ho, and T. D. Wheelock, in Coal Processing Technology, (Am. Inst. Chem. Engrs., NY, 1979), Vol. 5, pp. 171-7.

48. C. E. Capes, G. I. Sproule, and J. B. Taylor, Fuel Processing Technology 2, 323-9 (1979).

49. J. R. Burger, "Froth Flotation is on the Rise," Coal Age 85(3), 99-108 (1980).

50. K. J. Miller and A. W. Deurbrouck, in New Physical Methods for Cleaning Coal, Y. A. Liu, ed., (Marcel Dekker, Inc., New York, in press), Chap. 5.

51. F. F. Aplan, in Coal Desulfurization: Chemical and Physical Methods, T. D. Wheelock, ed., ACS Symp. Series 64, (Am. Chem. Soc., Washington, DC, 1977), pp. 70-82.

52. R. E. Zimmerman, in Coal Preparation, 4th ed., J. W. Leonard, ed., (AIME, New York, 1979), Chap. 10, pp. 82-104.

53. D. J. Brown, in Froth Flotation, 50th Anniversary Volume, D. W. Fuerstenau, ed., (AIME, New York, 1962), pp. 518-58.

54. F. F. Aplan, in Flotation, A. M. Gaudin Memorial Volume, M. C. Fuerstenau, ed., (AIME, New York, 1976), Vol. 2, pp. 1235-64.

55. G. St. J. Perrott and S.P. Kinney, Chem. and Metall. Eng. 25(5), 182-8 (1921).

56. T. Fraser, "Convertol Process of Coal-Slurry Treatment" Information Circular 7660, (U. S. Bureau of Mines, April 1953).

57. A. H. Brisse and W. L. McMorris, Jr., Trans. Society of Mining Engineers, AIME 211, 258-61 (1958).

58. C. E. Capes and R. J. Germain, in New Physical Methods for Cleaning Coal, Y. A. Liu, ed., (Marcel Dekker, Inc. New York, in press).

59. C. E. Capes, A. E. McIlhinney, and A. F. Sirianni, in Agglomeration 77, K. V. S. Sastry, ed., (Soc. of Mining Eng. of AIME, New York, 1977), Vol. 2, Chap. 54.

60. C. E. Capes, R. D. Coleman, A. E. Fouda, and W. L. Thayer, presented at 16th Biennial Conf. of the Institute for Briquetting and Agglomeration, San Diego, CA (Aug. 6-8, 1979).

61. A. R. Swanson, C. N. Bensley, and S. K. Nicol, in Agglomeration 77, K. V. S. Sastry, ed., (Soc. of Mining Eng. of AIME, New York, 1977), Vol. 2, Chap. 56.

62. C. N. Bensley, A. R. Swanson, and S. K. Nicol, Internat. J. of Mineral Processing, 4, 173–84 (1977).

63. S. K. Nicol and A. Brown, Proc. Australasian Inst. Min. Metall., No. 262, 49–55 (June 1977).

64. L. W. Armstrong, A. R. Swanson, and S. K. Nicol, CIM Bulletin, 72, 89–92 (Nov. 1979).

65. C. E. Capes, A. E. McIlhinney, A. F. Sirianni, and I. E. Puddington, Can. Mining and Metall. Bull. 66(739), 88–91 (Nov. 1973).

66. E. J. Mezey, S. Min, and D. Folsom, "Fuel Contaminants: Vol. 4. Application of Oil Agglomeration to Coal Wastes", EPA-600/7-70-025b (Jan. 1979).

67. T. D. Wheelock, R. Markuszewski, W. G. Leonard, and C. Han, presented at ACS Meeting, Houston, TX (March 27, 1980).

68. T. A. Vivian, presented at Chemical Engineering Department Seminar, Iowa State University, Ames, IA (April 9, 1980).

69. S. Ergun and E. H. Bean, "Magnetic Separation of Pyrite from Coals", Report of Investigations 7181, (U. S. Bureau of Mines, 1968).

70. R. R. Oder, IEEE Trans. on Magnetics Mag-12(5), 428–35 (Sept. 1976).

71. A. Z. Yurovskii, Sulfur in Coals, translated from Russian by Indian National Scientific Documentation Centre, New Delhi, India, 1974, (Acad. of Sciences of the USSR, Moscow, 1960).

72. W. M. Kester, Jr., Mining Eng., 17(5), 72–6 (1965).

73. P. D. Zavitsanos, J. A. Golden, K. W. Bleiler, and W. K. Kinkead, "Coal Desulfurization Using Microwave Energy", EPA-600/7-78-089, (June 1978).

74. C. R. Porter and D. N. Goens, Mining Eng. 31(2), 175–80 (1979).

75. J. K. Kindig, in Industrial Applications of Magnetic Separation, Y. A. Liu, ed., (Inst. Elec. Electron. Engrs., Inc., New York, 1979), pp. 99–104.

76. J. A. Oberteuffer, "High Gradient Magnetic Separation: Basic Principles, Devices and Applications", in Industrial Applications of Magnetic Separation, Y. A. Liu, ed., (Inst. of Elec. Electron Eng., Inc. New York, 1979), pp. 3–7.

77. S. C. Trindade, J. B. Howard, H. H. Kolm, and G. J. Powers, Fuel, 53, 178–81 (1974).

78. C. J. Lin and Y. A. Liu, in Coal Desulfurization: Chemical and Physical Methods, T. D. Wheelock, ed., ACS Symp. Series No. 64, (Am. Chem. Soc., Washington, DC, 1977), pp. 121–39.

79. H. H. Murray, IEEE Trans. on Magnetics, Mag-12(5), 498–502 (Sept. 1976).

80. H. H. Murray, in Coal Desulfurization: Chemical and Physical Methods, T. D. Wheelock ed., ACS Symp. Series No. 64, (Am. Chem. Soc., Washington, DC, 1977), pp. 112–20.

81. Y. A. Liu and C. J. Lin, Proc. Eng. Foundation Conf. on Clean Combustion of Coal, EPA-600/7-78-073 (April 1978), pp. 109–130.

82. R. E. Hucko, in Industrial Applications of Magnetic Separation, Y. A. Liu, ed., (Inst. Elec. Electron. Engrs., Inc., New York, 1979), pp. 77–82.

384

83. D. M. Eissenburg, E. C. Hise and M. D. Silverman, in Industrial Applications of Magnetic Separation, Y. A. Liu, ed., (Inst. Elec. Electron. Engrs., Inc., New York, 1979), pp. 91-4.

84. W. T. Abel, M. Zulkowski, G. A. Brady, and J. W. Eckerd, "Removing Pyrite From Coal by Dry-Separation Method", Report of Investigations 7732 (U. S. Bureau of Mines, 1973).

85. I. I. Inculet, M. A. Bergougnou and J. D. Brown, in New Physical Methods for Cleaning Coal, Y. A. Liu, ed., (Marcel Dekker, Inc., New York, in press), Chap. 3.

86. A. Attar and W. H. Corcoran, Ind. Eng. Chem., Prod. Res. Dev. 16(2), 168-70 (1977).

87. G. K. Angelova and K. I. Syskov, Izv. Akad, Nauk SSSR, Met. Topl., No. 5, 153-8 (1959).

88. A. Attar, in Analytical Methods for Coal and Coal Products, C. Karr, ed., (Academic Press, New York, 1979), Vol. 3, pp. 585-624.

89. R. Markuszewski, C.-K. Wei, and T. D. Wheelock, Am. Chem. Soc. Div. Fuel Chem. Preprints 25(2), 187-94 (1980).

90. R. Kavcic, Bull. Sci., Conseil Acad. RPF Yougoslav. 2(1), 12-13 (1954).

91. V. S. Kaminskii, Khim. Tverd. Topl., No. 4, 143-5 (1972).

92. D. J. Casagrande and L. Ng, Nature 282, 598-9 (Dec. 6, 1979).

93. D. J. Casagrande, G. Idowu, A. Friedman, P. Rickert, K. Siefert, and D. Schlenz, Nature 282, 599-600 (Dec. 6, 1979).

94. J. C. Agarwal, R. A. Giberti, P. F. Irminger, L. J. Petrovic, and S. S. Sareen, Mining Congr. J. 61(3), 40-3 (1975).

95. S. S. Sareen, R. A. Giberti, P. F. Irminger, and L. J. Petrovic, presented at A.I.Ch.E. Meeting, Boston, MA (Sept. 7-10, 1975).

96. J. C. Agarwal, R. A. Giberti, and L. J. Petrovic, U. S. Pat. 3,960,513 (June 1, 1976).

97. S. S. Sareen, in Coal Desulfurization: Chemical and Physical Methods, T. D. Wheelock, ed., ACS Symp. Series 64 (Am. Chem. Soc., Washington, DC, 1977), pp. 173-81.

98. R. A. Giberti, R. S. Opalanko, and J. R. Sinek, Proc. Symp. on Coal Cleaning to Achieve Energy and Environmental Goals (Sept. 1978, Hollywood, FL), EPA-600/7-79-098b (April 1979), Vol. II, pp. 1064-95.

99. E. H. Burk, J. S. Yoo, and J. A. Karch, U. S. Pat. 4,097,244 (June 27, 1978).

100. L. H. Beckberger, E. H. Burk, M. P. Grosboll, and J. S. Yoo, "Preliminary Evaluation of Chemical Coal Cleaning by Promoted Oxydesulfurization", Final Report, EPRI EM-1044 Project 833-1 (April 1979).

101. E. H. Burk, J. S. Yoo, and J. A. Karch, U. S. Pat. 4,158,548 (June 19, 1979).

102. S. Friedman and R. P. Warzinski, J. Eng. Power 99(3), 361-4 (1977).

103. S. Friedman, R. B. LaCount, and R. P. Warzinski, in Coal Desulfurization: Chemical and Physical Methods, T. D. Wheelock, ed., ACS Symp. Series 64 (Am. Chem. Soc., Washington, DC, 1977), pp. 164-72.

104. R. P. Warzinski, J. A. Ruether, S. Friedman, and F. W. Steffgen, Proc. Symp. on Coal Cleaning to Achieve Energy and Environmental Goals (Sept. 1978, Hollywood, FL), EPA-600/7-79-098b (April 1979), Vol. II, pp. 1016-38.

105. D. Slagle, Y. T. Shah, and J. B. Joshi, Ind. Eng. Chem., Proc. Des. Dev. 19, 294-300, (1980).

106. C. Y. Tai, G. V. Graves, and T. D. Wheelock, in Coal Desulfurization: Chemical and Physical Methods, T. D. Wheelock, ed., ACS Symp. Series 64 (Am. Chem. Soc., Washington, DC, 1977), pp. 182-97.

107. R. Markuszewski, K.-C. Chuang, and T. D. Wheelock, Proc. Symp. on Coal Cleaning to Achieve Energy and Environmental Goals (Sept. 1978, Hollywood, FL), EPA-600/7-79-098b (April 1979), Vol. II, pp. 1039-63.

108. T. D. Wheelock and R. Markuszewski, Fossil Energy Annual Report, Oct. 1, 1978 - Sept. 30, 1979, IS-4714, (Iowa State University, Ames, IA, Jan. 1980).

109. K.-C. Chuang, M.-C. Chen, R. T. Greer, R. Markuszewski, Y. Sun, and T. D. Wheelock, Chem. Eng. Commun. 7 (1-3), 79-94 (1980).

110. R. T. Greer, R. Markuszewski, and T. D. Wheelock, in Scanning Electron Microscopy/1980, O. Johari, ed., (SEM Inc., AMF O'Hare, Chicago, IL, 1980), Vol. 1, pp. 541-50.

111. T. G. Squires, L. W. Chang, C. G. Venier, and T. J. Barton, presented at 2nd Annual DOE and Environmental Symp., Reston, VA (March 17, 1980).

112. L. W. Chang, W. F. Goure, T. G. Squires, and T. J. Barton, Am. Chem. Soc. Div. Fuel Chem. Preprints 25(2), 165-70 (1980).

113. A. Attar and W. H. Corcoran, Ind. Eng. Chem., Prod. Res. Dev. 17(2), 102-9 (1978).

114. J. J. Kalvinskas and G. C. Hsu, Symp. on Coal Cleaning to Achieve Energy and Environmental Goals (Sept. 1978, Hollywood, FL), EPA-600/7-79-098b (April 1979), Vol. II, pp. 1096-140.

115. G. C. Hsu, J. J. Kalvinskas, P. S. Ganguli, and G. R. Gavalas, in Coal Desulfurization: Chemical and Physical Methods, T. D. Wheelock, ed., ACS Symp. Series 64 (Am. Chem. Soc., Washington, DC, 1977), pp. 206-17.

116. G. C. Hsu, G. R. Gavalas, P. S. Ganguli, and S. H. Kalfayan, U. S. Pat. 4,081,250 (Mar. 28, 1978).

117. J. J. Kalvinskas, K. Grohman, and N. Rohatgi, presented at Coal Age Conf., Louisville, KY (Oct. 23-25, 1979).

118. M. I. Sherman and J. D. H. Strickland, J. Metals 9, 1386-8 (Oct. 1957).

119. S. Mukai, Y. Araki, M. Konishi, and K. Otomura, Nenryo Kyokai-shi 48(512), 905-11 (1969).

120. H. N. Stokes, "On Pyrite and Marcasite", U.S. Geol. Survey Bull. No. 186, 1901.

121. J. W. Hamersma, E. P. Koutsoukos, M. L. Kraft, R. A. Meyers, G. J. Ogle, and L. J. Van Nice, "Chemical Desulfurization of Coal: Report on Bench-Scale Developments", Vol. 1(1), EPA-R2-73-173a (Washington, DC, 1973).

386

122. R. A. Meyers, U.S. Pat. 3,768,988 (Oct. 30, 1973).
123. R. A. Meyers, Hydrocarbon Processing 54(6), 93-5 (1975).
124. R. A. Meyers, Coal Desulfurization (Marcel Dekker, Inc., New York, 1977).
125. J. W. Hamersma, M. L. Kraft, and R. A. Meyers, in Coal Desulfurization: Chemical and Physical Methods, T. D. Wheelock, ed., ACS Symp. Series 64 (Am. Chem. Soc., Washington, DC, 1977), pp. 143-52.
126. R. A. Meyers, Hydrocarbon Processing, 58(6), 123-6 (1979).
127. W. E. King and D. D. Perlmutter, A.I.Ch.E. Journal 23(5), 679-85 (1977).
128. R. M. Garrels and M. E. Thompson, Am. J. Science 258-A, 57-67 (1960).
129. S. Sasmojo, Ph.D. Dissertation, Ohio State University, 1969.
130. T. Betancourt and H. A. Hancock, Proc. Coal and Coke Sessions, 28th Can. Chem. Eng. Conf., Halifax, N. S. (Oct. 22-25, 1978), pp. 157-67.
131. H. A. Hancock and T. Betancourt, Proc. Coal and Coke Sessions, 28th Can. Chem. Eng. Conf., Halifax, N. S. (Oct. 22-25, 1978), pp. 168-75.
132. W. E. King and J. A. Lewis, Ind. Eng. Chem., Proc. Des. Dev., in print.
133. D. A. Mixon and T. Vermeulen, "Oxydesulfurization of Coal by Acidic Sulfate Solutions", (Lawrence Berkeley Laboratory, Univ. of Calif., Berkeley, CA, Oct. 1979).
134. E. D. Guth, Proc. Symp. on Coal Cleaning to Achieve Energy and Environmental Goals (Sept. 1978, Hollywood, FL), EPA-600/7-79-098b (April 1979), Vol. II, pp. 1141-64.
135. A. F. Diaz and E. D. Guth, U. S. Pat. 3,909,211 (Sept. 30, 1975).
136. E. D. Guth and A. F. Diaz, U. S. Pat. 3,847,800 (Nov. 12, 1974).
137. M. Steinberg, R. T. Yang, T. K. Hom, and A. L. Berlad, Fuel 56, 227-8 (1977).
138. E. P. Stambaugh, in Coal Desulfurization: Chemical and Physical Methods, T. D. Wheelock, ed., ACS Symp. Series 64 (Am. Chem. Soc., Washington, DC, 1977), pp. 198-205.
139. E. P. Stambaugh, H. N. Conkle, J. F. Miller, E. J. Mezey, and B. C. Kim, Proc. Symp. on Coal Cleaning to Achieve Energy and Environmental Goals (Sept. 1978, Hollywood, FL), EPA-600/7-79-098b (April 1979), Vol. II, pp. 991-1015.
140. E. P. Stambaugh, J. F. Miller, S. S. Tam, S. P. Chauhan, H. F. Feldman, H. E. Carlton, J. F. Foster, H. Nack, and J. H. Oxley, Hydrocarbon Processing 54(7), 115-6 (1975).
141. L. Reggel, R. Raymond, I. Wender, and B. D. Blaustein, Am. Chem. Soc. Div. Fuel Chem. Preprints 17(1), 44-8 (1972).
142. G. Jangg, Erzmetall 16(10), 508-14 (1963).
143. A. R. Burkin and A. M. Edwards, in Mineral Processing, A. Roberts, ed. (Proc. 6th Internatl. Congr, Cannes, May 26-June 2, 1963), pp. 159-69.
144. R. L. Bunn, Ph.D. Dissertation, Iowa State University, 1977.

145. P. X. Masciantonio   Fuel 44(4), 269-75 (1965).

146. R. A. Meyers and W. D. Hart, presented at Symp. on Removal of Heteroatoms from Fuel, ACS Meeting, Houston, TX (March 24-28, 1980).

147. A. Attar, Fuel 57, 201-12 (1978).

148. D. K. Fleming, R. D. Smith, and M. Rosario Y. Aquino, in Coal Desulfurization: Chemical and Physical Methods, T. D. Wheelock, ed., ACS Symp. Series 64 (Am. Chem. Soc., Washington, DC, 1977), pp. 267-79.

149. A. B. Tipton, in Coal Desulfurization: Chemical and Physical Methods, T. D. Wheelock, ed., ACS Symp. Series 64 (Am. Chem. Soc., Washington, DC, 1977), pp. 280-9.

150. A. L. Yergey, F. W. Lampe, M. L. Vestal, A. G. Day, G. J. Ferguson, W. H. Johnston, J. S. Snyderman, R. H. Essenhigh, and J. E. Hudson, Ind. Eng. Chem., Proc. Des. Dev. 13(3), 233-40 (1974).

151. E. T. K. Huang and A. H. Pulsifer, in Coal Desulfurization: Chemical and Physical Methods, T. D. Wheelock ed., ACS Symp. Series 64 (Am. Chem. Soc., Washington, DC, 1977), pp. 290-304.

152. C. E. Hamrin, P. S. Maa, and C. R. Lewis, "Measurement of Inhibition Isotherms for Kentucky Coal", Annual Report, July 1, 1972-June 30, 1973, (Inst. for Mining and Minerals Res., Univ. of Ky., Lexington, KY, Oct. 1973).

153. G. C. Hsu, U. S. Pat. 4,146,367 (Mar. 27, 1979).

154. D. F. Burow and B. M. Glavincevski, Am. Chem. Soc. Div. Fuel Chem. Preprints 25(2), 153-64 (1980).

ELECTROCHEMICAL COAL GASIFICATION –
OPERATING CONDITIONS, VARIABLES, AND PRACTICAL IMPLICATIONS

J. Hickey, S. Lalvani and Robert W. Coughlin
Department of Chemical Engineering
The University of Connecticut, Storrs, Connecticut   06268

ABSTRACT

Subjecting slurries of a wide variety of coals and other carbonaceous fossil fuels to anodic oxidation in an aqueous electrolyte produces relatively clean gaseous carbon oxides, with clean hydrogen production at the cathode. Reaction stoichiometry, some preliminary aspects of proposed mechanism and chemistry are discussed along with kinetic and thermodynamic considerations. Rate data interpretations are considered in reference to potential scale-up to specific rates of industrial interest and importance. Reaction products (in addition to hydrogen), product compositions, possible process flow systems, and approximate process economics are also treated in preliminary fashion.

INTRODUCTION

Ever increasing demands upon energy sources provide incentives for investigating new ways to convert coals and other forms of solid carbonaceous fossil fuels into usable fluid synthetic fuels and important chemicals.

The degree of fluidity of such substitute fuels is strongly influenced by the hydrogen content, which ranges from 4:1 (hydrogen to carbon ratio) for methane (natural gas) up to almost all carbon for the hard coals.

Most coal conversion processes require at least some hydrogasification of the coal to produce needed hydrogen. This hydrogen can be generated in large amounts only by processes which split the water molecule. The chemistry and technology of the coal gasification process are complex and detailed discussions, including construction of hardware and gasification equipment, have been published.[1-5]

Conventional hydrothermal gasification of coal involves the steam-carbon reaction:

$$C_{(s)} + H_2O_{(g)} \rightarrow CO_{(g)} + H_{2(g)} \qquad (I)$$

partial
combustion

$$C_{(s)} + O_{2(g)} \rightarrow CO_{2(g)} \qquad (II)$$

and the water-gas shift reaction:

$$CO_{(g)} + H_2O_{(g)} \rightarrow CO_{2(g)} + H_{2(g)} \qquad (III)$$

Reaction I is highly endothermic, and its equilibrium is highly unfavorable at ordinary temperatures; the standard enthalpy and Gibbs free energy changes for I : $\Delta H^o = +31.4$ kcal/mole, $\Delta G^o = 21.8$ kcal/mole. In practice, coal is gasified by treating it with a mixture of steam and oxygen (Reaction II). In this way, the heat released by II ($\Delta H^o = -94.1$ kcal/mole and $\Delta G^o = -94.3$ kcal/mole) provides the thermal energy and assures the high temperatures (~800°C) required by reaction I.

The complex gaseous products from coal gasification must be scrubbed to remove impurities like tars, ash, CO_2 and sulfur compounds. Reaction III is often used to further adjust the hydrogen and carbon monoxide content of the gaseous product mixture. This is done separately to provide the desired CO/H_2 ratio needed for subsequent reactions, to produce the desired products (e.g., reaction of $CO + H_2$ to form methanol, methane, synthetic natural gas), or higher hydrocarbons (from Fischer Tropsch reactions). The water-gas shift reaction can be adjusted to produce H_2 almost exclusively; the hydrogen can then be used in the Haber process for NH_3 manufacture, in the Bergiees process for coal liquefaction, or for hydrodesulfurization of liquid fuels containing sulfur.

In this paper we summarize experimental results for a coal gasification process which electrolytically converts coal and water into two separate gaseous products, one essentially pure hydrogen and the other, gaseous carbon oxides. The process chemistry can take place at mild temperatures (including ambient temperatures) and the product gases are generally free of tars, ash and sulfur compounds. Electrochemical gasification involves the anodic oxidation of coal for which we can write a simple half-cell reaction:

$$C_{(s)} + 2H_2O_{(\ell)} \rightarrow CO_{2(g)} + 4H^+ + 4e^- \qquad \text{(IV)}$$

The corresponding reaction at the cathode is:

$$4H^+ + 4e^- \rightarrow 2H_{2(g)} \qquad \text{(V)}$$

The net sum of IV and V is the equation for the predominant reaction

$$C_{(s)} + 2H_2O_{(\ell)} \rightarrow 2H_{2(g)} + CO_{2(g)} \qquad \text{(VI)}$$

for the electrochemical gasification of coal, where relatively pure streams of carbon oxides and hydrogen are produced at the anode and cathode respectively. This contrasts with the complex, impure mixtures of H_2, CO, and CO_2 obtained by more conventional steam-carbon gasification.

The present process, as represented by equations IV, V and VI should be distinguished from conventional water electrolysis:

$$2H^+ + 2e^- \rightarrow H_2 \text{ (cathode)} \qquad \text{(VIIa)}$$

390

$$H_2O \rightarrow 1/2\ O_2 + 2H^+ + 2e^- \text{ (anode)} \qquad \text{(VIIb)}$$

$$H_2O \rightarrow 1/2\ O_2 + H_2 \text{ (net reaction)} \qquad \text{(VIIc)}$$

By thermodynamic considerations, electrochemical gasification of coal requires ~9.5 kcal/mole of H_2 of electrical energy ($\Delta F^O = -MFE^O$) which corresponds to a reversible potential of 0.21 V. Water electrolysis requires 56.7 kcal with a corresponding potential of 1.23 V, for the same amount of H_2. The present process provides the free energy required to drive the reaction (VI) by supplying electrons as an additional reagent at a potential of

$$E^O = \frac{-F^O}{nF} = \frac{19.1 \text{ kcal} \times 4.18 \text{ J/kcal}}{4 \text{ equiv} \times 96,500 \text{ C/equiv}} = -0.21V$$

This voltage is significantly lower than the potential of 1.23V for conventional water electrolysis shown in reaction VII.

This process, in which coal is used to depolarize the anode, has also been extended to the electrowinning of metals, such as copper, from solutions of their ions.

This paper deals with the ongoing investigations into the operating conditions and variables affecting the electrochemical coal gasification process. A table of notation used in this paper follows.

NOTATION TABLE

E   –   potential

E^O –   standard potential

F   –   Faraday's constant

G   –   Gibb's free energy

H   –   enthalpy

H_B –   enthalpy of combustion of carbon

i   –   current

M/e   –   mass-to-charge ratio

n   –   number of electrochemical equivalents

N   –   number of moles, e.g. N_C = number of moles of carbon

NDL   –   North Dakota Lignite

ROI   –   return on investment

ROIC   –   return on invested capital

S   –   entropy

t   –   time

EXPERIMENTAL

Investigations were conducted using stirred coal slurries in aqueous electrolyte within the anode compartments of cells such as those depicted in Figure 1.

The external EMF was applied by a potentiostat (Model PAR 371 and PAR 179A were used) made by Princeton Applied Research Corporation and the electrodes (both anode and cathode) were Pt mesh, gauge 52 (0.004 in. diameter wire) supplied by Matthey Bishop Inc. Graphite has also been successfully used as anode.

Gaseous products were analyzed by chromatography using a Spherocarb (100/120 mesh) separation column. Aqueous electrolytes were prepared with Baker analyzed reagent grade sulfuric acid and (for electrowinning) Baker $CuSO_4 \cdot 5H_2O$ crystals. Coal samples in pulverized form were obtained from the U. S. Department of Energy Pittsburgh Energy Research Center. For electrowinning, pulverized North Dakota lignite was used; metal analysis was carried out on a Perkin-Elmer 460 atomic absorption spectrophotometer.

Additional details of experimental apparatus[6,7] and ultimate coal analyses are given elsewhere.[7]

RESULTS AND DISCUSSION

Initial investigations into electrochemical coal gasification showed the process to be feasible even at room temperature. Several carbonaceous fossil fuels were used - three coals, one lignite and one char. Their analyses are given in Table I.

These coals were subjected to anodic oxidation in the apparatus of Figure 1a and significant currents (in comparison to currents of 0.05 mA at 40°C and IV after 15 minutes of operation in 4.13 M H_2SO_4 in the absence of coal) were measured, along with hydrogen production at the cathode, at potentials significantly lower than those required to electrolyze water. Table II shows the typical results.

As reported elsewhere,[6] the observed currents increased with increases in electrode area, ratio of coal/electrolyte, and temperature. Particle size had an effect, but the size effect is complex and is discussed below. After combusting Montana Rosebud in air, blank measurements were made using the ash. These blanks produced currents less than 0.5 mA at 0.875V even at 40°C; this is compared to 8.2 mA for the parent char at 0.78V and 23°C in Table II. Table II also shows the lignite to be two to three times more reactive than bituminous coals. A preliminary account[8] and further details[6] of this work appear elsewhere.

Coal slurries in the anolyte were held in suspension by agitation with a magnetic stirring bar. Oxidation of the coal requires contact of the particles with the anode; only background current is measured if a glass frit (permeable to dissolved species but not coal particles) is used to isolate the anode from the coal. Oxidation of the coal at the anode can be considered a three-dimensional electrode process, as has been reported for stirred slurry electrodes

392

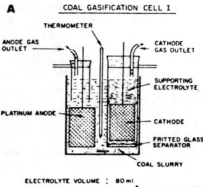

A   COAL GASIFICATION CELL I

THERMOMETER

ANODE GAS
OUTLET

CATHODE
GAS OUTLET

SUPPORTING
ELECTROLYTE

PLATINUM ANODE

CATHODE

FRITTED GLASS
SEPARATOR

COAL SLURRY

ELECTROLYTE VOLUME : 80 ml
ANODE   : 6.5 cm^2, PLATINUM GAUZE
CATHODE   : 9.0 cm^2, PLATINUM GAUZE
STIRRING   : MAGNETIC, 7/8" MAGNET

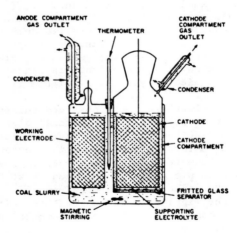

B   COAL GASIFICATION CELL II

ANODE COMPARTMENT
GAS OUTLET

THERMOMETER

CATHODE
COMPARTMENT
GAS
OUTLET

CONDENSER

CONDENSER

WORKING
ELECTRODE

CATHODE

CATHODE
COMPARTMENT

COAL SLURRY

FRITTED GLASS
SEPARATOR

MAGNETIC
STIRRING

SUPPORTING
ELECTROLYTE

ELECTROLYTE VOLUME : 475 ml
ANODE   : 96.5 cm^2, PT GAUZE, 52 Mesh/inch
CATHODE   : 156 cm^2
STIRRING   : MAGNETIC, 1 3/4 MAGNET AT 3
                    POSITION

Figure 1. Diagrams of coal gasification cells.

Table I. Characteristics of Coals Studied

	Pittsburgh Coal	Illinois no. 6 coal	North Dakota lignite	Montana Rosebud coal	Montana Rosebud char
			sample		
Analysis (As Received)					
moisture	2.2	6.2	18.9	7.3	1.6
volatile matter	36.9	37.6	35.5	36.1	6.1
fixed carbon	55.3	42.6	33.9	46.7	60.4
ash	5.6	13.6	11.7	9.9	31.9
hydrogen	5.3	4.9	5.2	4.9	1.3
carbon	77.5	63.2	50.3	61.5	63.0
nitrogen	1.5	1.2	0.8	0.9	0.4
oxygen	8.9	13.5	30.7	22.1	3.1
sulfur	1.2	3.6	1.3	0.8	0.2
free-swelling index no.	8.0	2.5	noncaking	noncaking	noncaking
heating value, Btu/lb	13740	11300	8238	10021	9583
Major Elements in Ash					
silica	43.5	44.3	12.3	36.7	44.3
Al2O3	21.9	15.8	11.5	18.9	21.5
Fe2O3	25.5	16.3	12.5	4.9	6.0
TiO2	1.3	1.1	0.3	1.0	1.0
CaO	1.4	9.2	23.0	16.8	18.2
MgO	1.6	1.1	8.9	4.4	4.9
Na2O	1.6	0.7	3.8	0.5	0.6
K2O	0.9	2.1	0.4	0.2	0.5
sulfites	0.6	6.8	25.4	13.2	2.3
Fusibility of Ash, °F					
initial deformation temperature	2630	1980	2030	2150	2150
softening temperature	2680	2100	2080	2180	2180
fluid temperature	2910	2180	2130	2210	2210

394

Table II. Comparison of Oxidation Rates of Different Coal Samples[α]

coal	potential of oxidation, V	coal concen, g/cm^3	oxidation rate, mA
Montana Rosebud char	0.78	0.475	8.20
North Dakota lignite	0.83	0.475	9.00
Pittsburgh coal	0.875	0.475	5.30
Illinois no. 6	0.875	0.475	3.00
Montana Rosebud coal	0.78	0.475	1.50

[α]Electrolyte: 3.7 M H_2SO_4; temperature: 23°C; electrode area: 6.5 cm^2; coal slurry concentration: 0.475 g/cm^3; particle size: 250μm and below; experimental apparatus: cell I.

(as shown in Figure 1), fluidized bed electrode,[9] packed bed electrodes,[10] percolating porous electrodes[11] or pumped slurry electrodes.[12]

Such electrodes are well adapted to continuous processes where the solid coal can be continuously added and products removed without operation interruption.

Particle Size Consideration

In one series of experiments, North Dakota lignite was seived into three different size ranges, and each was anodically oxidized at 39°C and 1.0V cell potential. Figure 2 shows the course of these oxidations plotted as a function of current vs. time. Initially, larger particles gave higher currents, but these relationships altered after 30 – 70 minutes of oxidation.

Electrode Surface Area

The electrodes were Pt gauge, and cathode and anode superficial areas were 9.0 cm^2 and 6.5 cm^2, respectively, for the Figure 1A cell (preliminary results); these were modified in the Figure 1B cell to 156 cm^2 and 96.5 cm^2 (cathode and anode, respectively). Current investigations show comparable results for graphite anodes and dilute HCl or H_3PO_4 electrolytes.

Activation Energy and the Rate Determining Step

At potentials of around 1V, North Dakota Lignite (NDL) gave apparent activation energies in the range of 10–12 kcal/mol. This low value suggests the rate determining step may not be a typical chemical reaction. Experiments currently in progress are investigating the influence of electrode potential and other conditions on the activation energy, and hopefully will better elucidate the process(es) connected with the low (10 – 12 kcal/mol°C) values observed to date.

Coal Concentrations

Arrhenius plots of log current vs. 1/T shown in Figures 3 and

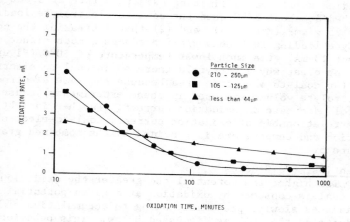

Figure 2.   Effect of particle size on the oxidation rate.

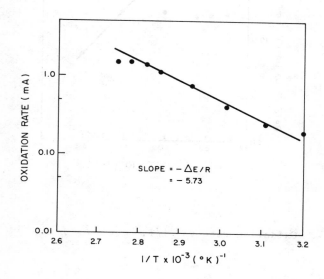

Figure 3.   Oxidation rate vs. 1/T.

4 indicate (1) that a limiting oxidation rate of about 1.5 mA is reached at about 85°C when the coal-to-electrolyte loading is about 0.00125 g/cm³ (Figure 3), and (2) that increasing the coal-to-electrolyte loading to 0.0625 g/cm³ produces a much higher current (about 13 mA) at a much lower temperature of 50°C (Figure 4). It is not yet clear why a fiftyfold increase in coal-to-electrolyte loading (or coal-to-electrode area because the same electrode area and electrolyte volume was used in these experiments) causes about a tenfold increase in oxidation current. Further insight into the interrelationship of oxidation current, coal-to-electrolyte concentration and temperature is provided by the combined graph shown in Figure 5.

Potentiostatic Studies

The higher the potential the greater the oxidation current; as the coal is consumed by oxidation at a given potential the current diminishes slowly, presumably owing to accumulation of reaction products on the coal as discussed below. This behavior is shown in Figure 6 where current is plotted vs. potential for various extents of coal consumption* as a parameter.

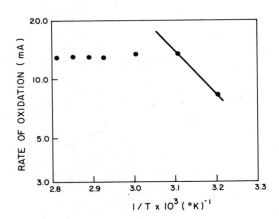

Figure 4. Oxidation rate vs. 1/T.

*Throughout this paper the extent of coal consumption is computed as the mass of carbon equivalent to total electrical charge passed during the experiment.

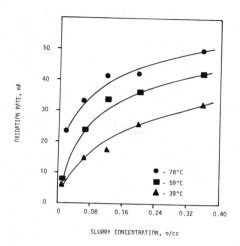

Figure 5. Effect of coal concentration on the oxidation rate.

Figure 7 shows similar information, but extended to very low potential for 1.75% carbon consumption. Figure 8 shows the observed reaction rates of three different coal samples and one activated charcoal at 1.0V, using slurry concentrations of 0.069 g/cc^3, plotted vs. cumulative coulombs passed. All of these results indicate that the rate of oxidation falls gradually as the reaction proceeds. But, only a portion (up to 18% for NDL) of the coal can be consumed before the oxidation rate falls steeply.

Previous workers [13-15] have reported that controlled oxidation of purer carbons, whether by electrochemical or chemical means, results in the formation of several surface oxides. The functional oxygen groups most often suggested[16] are:

1. carboxyl groups
2. phenolic hydroxyl groups
3. carbonyl groups
4. peroxide and ester groups

398

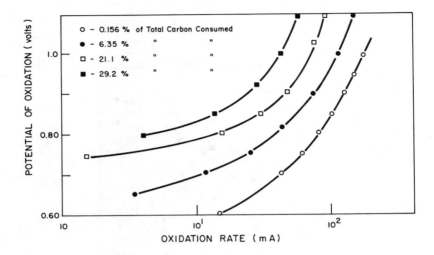

Figure 6. Effect of potential on the oxidation rate as the reaction proceeds.

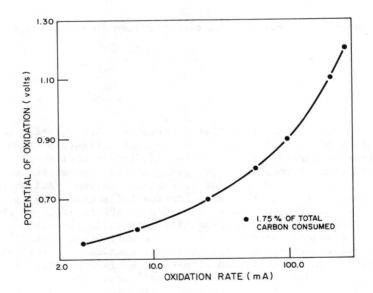

Figure 7. Effect of potential on the oxidation rate.

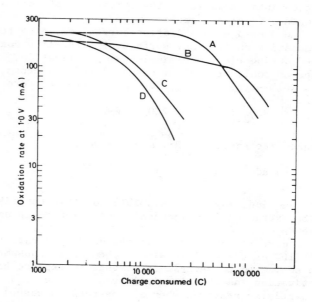

Figure 8. Oxidation rate of different samples at 1.0V as the reaction proceeds.

5. lactones
6. carboxylic acid anhydrides group
7. cyclic peroxide groups

It is our belief that such oxides also form during the anodic oxidation of coal, increasing in concentration on the coal particle surface as the reaction proceeds, thus lowering the oxidation current (increasing the oxidation potential) by making electron abstraction increasingly more difficult. Preliminary results from analysis of residual carbon values of partial electrochemically gasified Pittsburgh coal provides additional support. Based on the known reported[17] decomposition temperatures for such surface oxygen functionalities on carbon it seems feasible to maintain a higher steady oxidation rate at 200 - 600°C, a temperature range at which only minimal surface oxide concentrations would be expected to accumulate. It appears reasonable to consider surface oxides as

400

reaction intermediates in the conversion of coal and water to
hydrogen and gaseous carbon oxides.

The decrease in oxidation current may also result in part from
accumulation on the coal surface of a tar-like coating that is
formed during electrochemical coal gasification. Surface carboxyl
groups on coal, or smaller carboxylated fragments, might also react
via a Kolbe-type pathway, forming aliphatic hydrocarbons

$$RCO_2^- + R'CO_2^- \rightarrow 2CO_2 + R-R' + 2e^-$$

Surface free radicals, $R\cdot$, might also form

$$RCO_2^- \rightarrow CO_2 + R\cdot + 1e^-$$

Such reactions may be responsible for this tar-like coating that
can be extracted from anodically oxidized coals using hydrocarbon
solvents.

Figure 8 shows a drop in the oxidation rate by a factor of 7
after 16% of the coal is electrochemically consumed. This coal was
then subjected to several treatments to explore both the nature of
deactivation and the plausibility of regeneration. Original activity
was partially restored when a) the coal is washed with acetone, which
removed a tar-like coating with formation of a dark, possibly colloidal extract solution, and b) the coal is heated in air to 250°C,
presumably removing the oxygen containing surface functionalities
discussed above. In another experiment, NDL was anodically oxidized
to 26% consumption at 1.0V at 114°C. The remaining coal was
regenerated by filtration, acetone wash and heating in air to 250°C
for 2 hours. It was then returned to the anolyte and consumed an
additional 20%. The oxidation current for this regenerated coal
almost duplicated that for the virgin coal, and curve B of Figure 8
was essentially reproduced under the same conditions, with similar
product formation. This, then, suggests that conducting the
electrochemical coal gasification at temperatures above 200°C would
cause much greater coal consumption at still greater rates. Further,
experiments along this line are now in progress.

Galvanostatic Studies

NDL was electrochemically gasified under galvanostatic conditions. Fresh samples were oxidized using an oxidation current of
150 ma at first. Figure 9 shows a plot of the change in cell potential vs. % total coal consumption, which shows that to maintain the
desired reaction rate required a gradual increase of cell potential.
This is because the coal particles decrease in activity and concentration as the reaction progresses. From Figure 9 it is seen that
to maintain a constant current of 150 ma, the required potential rose
gradually to about 1.2V (after 97 hr), at which point it abruptly
jumped to 1.7V, which suggests the onset of a different reaction
mechanism. Immediately thereafter the current set point was lowered
to 100 ma and the reaction permitted to continue. The jump occurred

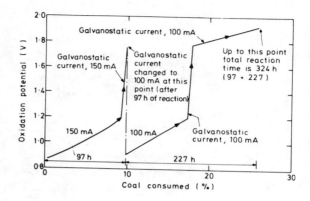

Figure 9.   Change of cell potential during galvanostatic study.

again at 100 ma.   Also of note is the fact that anode gas analyses
indicated that no O_2 was formed even when the potential rose to
1.98V.

Temperature Effects

It is suggested above that electrochemical coal gasification
at temperatures above 200ºC would consume greater quantities of
coal faster.   There are other benefits to be gained from operation
at higher temperatures:
    a) polarization potentials (overvoltages) are lowered
    b) reversible (thermodynamic) cell potentials are lowered
With respect to effect b), Figure 10 shows the influence of
temperature on the reversible cell potential derived from the thermodynamic ΔG and ΔH values of the electrochemical coal gasification
reaction.   The reaction cannot occur unless the cell potential at
least equals the reversible thermodynamic potential corresponding
to ΔG, the Gibbs free energy of reaction; as Figure 10 shows this
potential decreases with increasing temperature.   In principle,
electrolysis will occur at any potential above the reversible
thermodynamic value if $T\Delta S$ is supplied as heat from the surroundings.
In practice, however, larger potentials are required to give
reasonable reaction rates.
    Cell operation at or above the thermoneutral value, $(\frac{|\Delta H|}{nF})$,

402

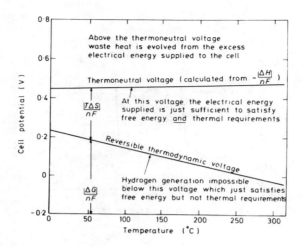

Figure 10. Effect of operating temperature on cell voltage.

supplies G and T S as electrical energy, with no additional energy
required because this electrical energy supplies the endothermic
heat of reaction. In fact, operation of the cell above the
thermoneutral value results in net heat reduction, requiring heat
removal.

Gaseous Products and Current Efficiency
     During oxidation of NDL, Pittsburgh coal, activated carbon and
Montana Rosebud char using cell potentials in the range 0.8 - 1.2V,
essentially pure H_2 gas evolved at the cathode. The current efficiency of hydrogen production may be expressed as

$$\text{current efficiency} = \frac{H_2 \ (\text{experimentally produced}) \times 100}{H_2 \ (\text{amount corresponding to coulombs passed})}$$

In each case the current efficiency of H_2 production was approximately 100% based on measured current integrated over time. The gas evolved in the anode compartment was almost pure CO_2 with small amounts of CO. This composition varied over the course of the reaction probably due to changes in surface oxide population on the coal. The volume ratio of gases collected at the cathode to those at the anode ranged from about 9.1 to 3.7, with the higher ratios obtained at the beginning of an experiment. From the stoichiometry of reaction VI this ratio should be 2, but ratios greater than 2 can be attributed to oxygen accumulation on the coal particles in the form of functional groups such as $-CO_2H$, $-CHO$, $-CH_2OH$ and the like. Also, higher relative amounts of H_2 may arise from the hydrogen content of coal, support for which comes from preliminary proximate analysis of the residual carbon after NDL gasification, which indicated preferential consumption of volatile components expected to be hydrogen-rich. Anode gas production is strongly related to the surface oxide concentration of coal. Binder et al reported[13] that a surface layer forms first on graphite and only then does CO_2 evolution commence. As the oxidation process advances, steady state concentrations of these surface oxides build up on the coal surface, giving a constant rate of gas evolution at the anode. Qualitative but sensitive mass spectrometric analyses of the anode and cathode gases showed no M/e values corresponding to SO_2 or H_2S in spite of the significant sulfur concentrations of the parent coals. H_2S is liberated, of course, when coal is first contacted with a strongly acidic electrolyte; but not during subsequent electrolysis. Presumably strongly anodic potential causes the oxidation of any residual sulfur compounds to sulfate which remains in the electrolyte.

Electrowinning of Metals

Extension of the electrochemical coal gasification process to metal electrowinning is accomplished by replacing the conventional half cell reaction of oxygen evolution at the anode with the coal consuming half cell reaction IV. The total overall cell potential of the resulting electrowinning process is lowered by about 1.10V with a corresponding reduction in electrical energy consumption. In one case Cu was deposited on a Pt mesh cathode (separated by a fritted glass barrier from the anode) at 60°C from an electrolyte consisting of 0.125 M $CuSO_4$ in 0.5 M H_2SO_4 using two different galvanostatic rates (5.9 mA and 12 mA). Figure 11 shows the corresponding changes in cell potential with time for both the conventional and coal-depolarized systems. For the conventional electrowinning process, the Cu was deposited at cell potentials of 1.65V and 1.73V. With anodic oxidation of a coal slurry identical conditions prevailed except that coal was simultaneously oxidized at the anode while copper was cathodically reduced and electrodeposited. Utilization of anodic coal oxidation lowered the overall cell potential by about 1.1V (compared to the conventional process, see Figure 11). The lowered cell potential means that the electrical energy consumption per mass of Cu deposited is lowered by about 2/3 due to the consumption of coal at the anode. When the efficiency (~35%) of producing

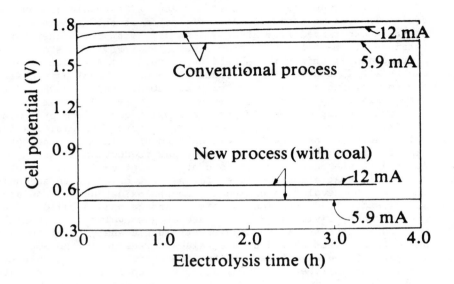

Figure 11.   Comparison of cell potentials of the copper electrowinning processes in galvanostatic conditions.

electricity from coal is considered, the net energy saving is about
50% for coal-depolarized electrodeposition as compared to the conventional process.   Details of efficiencies and economics have been
published elsewhere.[18]

TECHNOLOGICAL IMPLICATIONS

Hydrogen Production

It appears that production of hydrogen by the combined use of
fossil fuel and electricity as we report here has not been previously investigated or applied; there are several ways to view this
process.   In one respect, coal is gasified by reaction with water
using electrons as a third reagent, making the process thermodynamically favorable.   In another respect, the process electrolyzes
water to hydrogen, but the electrons in coal lower the operating
potential by depolarizing the anode.   The result is lowered consump-

tion of electricity and the production of CO_2 and H_2 instead of O_2 and H_2.

The following quantitative development gives a first order approximation of energy requirements for coal-consuming electrolysis vis-a-vis conventional water electrolysis. Conventional water electrolysis consumes energy at a potential of E_2 to produce N_{H_2} mol of H_2 in the amount of $2N_{H_2}FE_2$, while the current process under investigation, operating at a potential E uses the amount

$$E \int_0^t i dt + N_c (-\Delta H_B)$$

Substitution of the stoichiometric relationships $N_{H_2} = \int_0^t i dt/2F$ and $N_c = 1/2 \; N_{H_2}$ simplifies the expression to

$$2FN_{H_2} E + 1/2 \; N_{H_2} (-\Delta H_B)$$

Accordingly, the relative energy usage (REU) is

$$REU = \frac{\text{conventional electrolysis}}{\text{coal assisted electrolysis}}$$

$$= 2N_{H_2} FE_2/2FN_{H_2} E + 1/2 \; N_{H_2} |\Delta H_B|$$

$$= E_2/(E + |H_B|/4F)$$

Substituting $|\Delta H_B| = 94,100 \times 4.18$ J/mol and the value of F yields

$$REU = E_2/(E + 1.02)$$

Practical values of E_2 are approximately 2V, and E was observed in the present work to range between 0.8-1.0V at ambient temperatures. So, for the same rate of hydrogen generation, REU ≈ 1, and the total energy consumption for both conventional and coal-based electrolysis is about equal. The difference arises in that in the coal-based process, about half the required energy is obtained directly from coal. The electrical energy consumption of an electrolyte process is proportional to cell potential; the coal-assisted electrolysis operates at half the potential and therefore consumes half as much electricity. Higher-temperature operation should permit still lower cell potential and correspondingly lower electricity consumption.

Process and Pollution

No sulfur compounds were observed in the product gases formed during electrolysis in our laboratory investigations, and the dilute H_2SO_4 electrolyte is reusable with no apparent consumption of the acid. However, for a practical continuous process it is reasonable

to expect some consumption of H_2SO_4 due to reaction with the basic
mineral impurities of the coal. For example, H_2SO_4 reaction with
FeS_2 mineral impurities (pyrite, marcasite) produces H_2S. But,
anodic oxidation of the organic sulfides found in coal may produce
H_2SO_4. Figure 12 shows a preliminary conceptual flow diagram for
a continuous electrochemical coal gasification process. Experience,
however, dictates that a practical process would be far more complex. For example, the case of separating ash and unreacted coal
values by simple differential settling as shown here is open to
question. The problem of maintaining high coal concentrations in
the reactor with simultaneous continuous purging of the ash (with
minimal loss of unreacted coal) needs to be resolved. But the
purpose of this report is the presentation of initial findings and
early experiments.

Energy Efficiency

Thermodynamic efficiency depends on the quantities of coal and
electricity consumed per unit of hydrogen produced. These depend
in turn on operating potential, temperature and quantity of side-
products. Basing such efficiencies on the heating value of hydrogen,
and also accounting for the inefficiencies inherent in generating
electricity from coal, efficiencies estimated for electrochemical
coal gasification at potentials of approximately 1V are half those
cited for conventional gasification processes.[19]

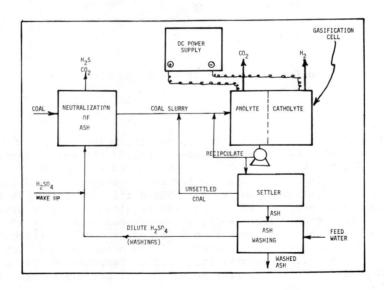

Figure 12. Flow diagram of a continuous electrochemical coal
gasification process.

The efficiencies for electrochemical coal gasification become much more favorable at lower operating potentials, as when operating at higher temperatures. Basing thermodynamic efficiency on the heating value of the gaseous product(s) does not account for energy required for product gas cleaning and shifting toward purer hydrogen. A fairer comparison could entail an economic estimate of hydrogen production costs for the different processes. A cursory and preliminary attempt at such an economic analysis follows.

Scale-Up Costs

An approximate extrapolation is based on the following data (see Figure 5) for a coal concentration of 0.36 g/cm^3 and an operating potential of 1V:

    a) 32.5 mA (39°C) $\cong$ 0.0036 g of C/hr

    b) 50 mA (78°C) $\cong$ 0.0056 g of C/hr

where the basis is 100 tons coal (as C) per hour (= 10^8 g of C/hr).

Assume a cell volume of 6.5 cm^3 (i.e., two flat electrodes each of 6.5 cm^2 area separated by 1 cm distance). Scale this up to 10^8 g of carbon/hour.

At 39°C,

$$\frac{10^8 \times 6.5 \text{ cm}^3}{0.0036} = 1.8 \times 10^{11} \text{ cm}^3$$

Thus, a volume of 1.8×10^5 m^3 (a cube $\doteq$ 56.36 m on a side) may be sufficient to gasify 100 tons/hr at 39°C.

At 78°C,

$$\frac{10^8 \times 6.5 \text{ cm}^3}{0.0056} = 1.17 \times 10^{11} \text{ cm}^3$$

Gasification of 100 tons/hr at 78°C may be accomplished in only a volume of 1.17×10^5 m^3, or a cube 48.81 on an edge.

The anodic rate of electrochemical coal gasification, i, is expressed as

$$i = i_o \exp (\alpha_a \eta F/RT)$$

where η is the overpotential, α_a the transfer coefficient, T the temperature, F the Faraday constant, and i_o a function of coal concentration. According to this equation, an increase in coal concentration and temperature would increase the rates of coal gasification, lowering the estimated volume requirements. More effective use of electrode area (i.e. higher current densities) could further reduce the required electrolysis volumes, but there may be limitations imposed by electrode and separator membrane areas. Additional experiments designed to probe current density upper limits and limitations are presently in progress.

The estimates above are approximate, and are presented here merely to indicate necessary avenues of investigation and development for possible practical utilization of electrochemical coal

408

gasification. It should be borne in mind that clean hydrogen is
produced directly in this process, rather than the contaminated
complex synthesis gas of conventional processes.

Economic Considerations
     Some very rough estimates of the cost of hydrogen produced
by electrochemical coal gasification can be made by consideration
of the production economics analysis for hydrogen for the period
1980 - 2000, performed under contract with Brookhaven National
Laboratory by Exxon Research and Engineering Company.[20] Table
III shows the derived costs of hydrogen in 1980 dollars; these
costs include a 20% return on invested capital. It is apparent
that conventional water electrolysis entails the greatest costs,
and cannot compete with other processes. Even the new General
Electric SPE (solid polymer electrolyte) technology cannot compete
economically with coal gasification if electricity costs are taken
at $0.027/kWh. If, however, off-peak electric power at $0.01/kWh
is used, then water electrolysis utilizing SPE could be economically
competitive. It is thus evident that the cost of electrolytic
hydrogen is roughly equally sensitive to the cost of electricity and
to capital costs (i.e. conventional vs. SPE technology).
     Because coal-based electrolysis of water is closely related
to the conventional water electrolysis process, an examination of
the details of the elecrolysis costs is in order. Table IV shows
such information, taken from the Exxon report, in slightly modified
form. SPE technology thus has its major effect on the economics,
by permitting much lower total capital investment, which would be
about 30% of that required for a conventional electrolysis plant.
There is a slight saving of electricity cost due to lowered cell
resistance, this along with savings in maintainance, plant overhead,
insurance, taxes, depreciation and ROI, each of which are computed

Table III. Cost of Hydrogen in 1980 Dollars from Several Processes
            (All Costs Include 20% ROIC)

Process	Cost in 1980 dollars/KSCF
methane reforming	1.97
partial oxidation of residuum	2.62
coal gasification	2.58-3.72 (range of 8 separate cases)
electroylsis (current technology)	6.25 (based on electricity @ $0.027/kWh)
electrolysis (G.E. SPE technology)	3.75 (based on electricity @ $0.027/kWh)
electrolysis (G.E. SPE technology)	1.65 (based on off-peak electricity @ $0.01/kWh)

Table IV. Comparison of Hydrogen Costs by Electrolysis, 1980$ (Basis: 33 x 10⁹ SCF/year; 7920 h/year Operation)

	current technology $/kW	current technology 10⁶$	normal power 10⁶$/year	off peak 10⁶$/year	new technology (SPE) $/kW	new technology (SPE) 10⁶$	normal power 10⁶$/year	off peak 10⁶$/year
onsites	650	255			167	65.6		
offsites	100	40			39	15.4		
total	750	295			206	81.0		
% contingency in investments	15	15			30	30		
working capital, 10⁶ $	18.9 (normal power) or 7.0 (off peak)				18.3 (normal power) or 6.9 (off peak)			
electricity @ 0.027 or $0.01/kWh			111.02	41.12			108.30	40.11
water and chemicals			1.25	1.25			0.74	0.74
labor and supervision			1.20	1.20			0.60	0.60
maintenance (4% of onsites)			10.20	10.20			2.62	2.62
plant overhead (2.6% of onsites)			6.63	6.63			1.71	1.71
insurance, property taxes (1.5% of total plant)			4.43	4.43			1.22	1.22
depreciation (10% of onsites, 4% of offsites)			27.50	27.50			7.33	7.33
interest on working capital (10%)			1.89	0.70			1.83	0.69
return on investment (20% of total)			59.00	59.00			16.20	16.20
total cost including return			223.12	152.03			140.55	71.22
oxygen credit ($24/ton)			16.70	16.70			16.70	16.70
net cost including return			206.41	135.33			133.85	54.52
hydrogen cost, $/KSCF (including return)			6.25	4.10			3.75	1.65

410

as percentages of capital investment.

The economic promise of SPE warrants a brief consideration of its technology, which utilizes a solid sulfonated fluoropolymer (e.g., DuPont's Nafion) as the sole electrolyte; a schematic diagram of an SPE electrolysis cell is presented in Figure 13. The technological basis is the design and use of the SPE, which permits much greater current densities requiring smaller cells; this greatly lowers capital investment. Lowered cell voltages also save in electricity costs. Table V presents a breakdown of projected 1985 SPE system costs.[21,22]

The savings and additional costs entailed in the introduction of coal into the water electrolysis are tabulated in Table VI. The cost of coal would be 8.28×10^6 \$/year assuming 0.276×10^6 tons/year @ \$30/ton. If the coal has a 10% ash content, and half of this ash is acid-consuming Na_2O, this would require 0.0218×10^6 tons H_2SO_4/yr x \$35/ton = 0.76×10^6 \$/year for acid. Capital investment

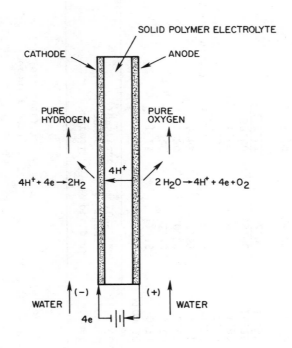

Figure 13. Schematic diagram of SPE electrolysis cell.

Table V. SPE System Costs ($/kW)[a]

electrolysis module	8.15
power conversion and switch gear	42.76
ancillary equipment	18.08
installation (not including land or buildings)	8.26
total	77.25

[a]1985 cost projection by G.E.

for equipment for dumping, unloading, storing, conveying, crushing, and pulverizing the coal is estimated at $9 million; ash handling, settling, and recycling equipment is estimated at $2 million.[23] If one assumes that electrolysis cell costs would double for an additional on-site investment of 8.15/77.26 x 65.6 = $6.92 million, the savings would also entail lowered electricity costs by half, reduced capital investment in electrical equipment by about half (= 1/2 x (42.76 + 18.18)/77.25 x 65.6 = $25.9 million), and reduced working capital by about half due to lowered electricity purchases.

The savings and costs tabulated (see Table VI) and discussed above are incorporated into cost analyses for coal-assisted electro-

Table VI. Costs and Savings Offered by Coal Gasification During SPE Electrolysis

Additional Costs

consumption of acid to neutralize ash (for acid electrolytes): $0.76 million/year
modified cells to permit flow of coal slurry: $6.92 million additional investment
coal preparation, handling and storage: $9 million additional investment
ash separation and handling: $2 million additional investment
loss of oxygen credits: $16.70 million/year
cost of coal: $8.28 million/year

Savings

reduced electricity cost (by about half)
reduced capital investment in power conversion and switch gear, in bus bars and circuit components: $25.9 million lower investment
reduced working capital due to greatly lowered electricity cost (i.e. lower by about half)

412

lysis for SPE technology presented in Table VII and for conventional
electrolysis technology as shown in Table VIII; the changes sug-

Table VII. Hydrogen Costs -- Coal Assisted Electrolysis -- SPE
Technology (Basis: 33 x 10^9 SCF/year; 7920 h/year
Operation)

investments	millions of $
onsites	57.6
offsites	14.4
total	72.0
working capital, $ million	9.2 (normal power) 3.4 (off peak)

operating costs $ million/year	normal power, 10^6 $/year	off peak power, 10^6 $/year
electricity @ $0.027 or $0.01/kWh (off peak)	54.15	20.06
cost of coal at $30/ton	8.28	8.28
water and chemicals	1.50	1.50
labor and supervision	0.60	0.60
maintenance (4% of onsites)	2.30	2.30
plant overhead (2.6% of onsites)	1.50	1.50
insurance, property tax (1.5% of total plant)	1.08	1.08
depreciation (10% of onsites, 4% of offsites)	5.82	5.82
interest on working capital (10%)	0.92	0.34
return on investment (20% of total)	14.40	14.40
oxygen credit	none	none
net cost (including return)	90.55	55.88
hydrogen cost, $/KSCF (including return)	2.74	1.69
savings over no-coal case, $/KSCF (%)	1.01 (27%)	none

Table VIII. Hydrogen Costs--Coal-Assisted Electrolysis--Conventional Technology (Basis: 33 x 10^9 SCF/year; 7920 h/year Operation)

investments	$ million
onsites	240
offsites	40
total	280
working capital, 9.5 (normal power)	3.5 (off peak)
$ millions	

operating costs, $ million/year	normal power, 10^6 $/year	offpeak power 10^6 $/year
electricity @ $0.027 or $0.01/kWh	55.51	20.56
coal at $30/ton	8.28	8.28
water and chemicals	2.01	2.01
labor and supervision	0.60	0.60
maintenance (4% of onsites)	9.60	9.60
plant overhead (2.6% of onsites)	6.24	6.24
insurance, property tax (1.5% of total plant)	4.20	4.20
depreciation (10% of onsites, 4% of offsites)	25.60	25.60
interest on working capital(10%)	0.95	0.35
oxygen credit	none	none
return on investment (20% of total)	56.00	56.00
total net cost (including return)	168.99	133.44
hydrogen cost, $/KSCF (incl. return)	5.12	4.04
savings over no-coal case, $/KSCF (%)	1.13 (18%)	0.06 (1.4%)

gested in Table VI are incorporated in each case. These analyses demonstrate that coal-assisted electrolysis has its most favorable effect when electricity costs are high, due to substitution of cheaper coal energy for more costly electrical energy. Its greatest effect (27% hydrogen cost reduction) is in application with SPE

414

technology at "normal" power costs of $0.027/kWh; the corresponding
conventional electrolysis cost reduction is 18%. No particular
advantage is gained, however, for direct incorporation of coal into
the electrolysis process for off-peak power at $0.01/kWh.

There are also tremendous scale effects to be considered when
comparing hydrogen costs for different processes, as demonstrated
in Figure 14.[24] Hydrogen production by steam-reforming becomes
prohibitively expensive for small plants, whereas for electrolysis
processes the effects are far smaller. The approximate economic
analysis presented above is very conservative in the projected
savings for coal-assisted water electrolysis, as it is based on
very large plant operation. As such, it is expected that for
smaller plants electrolysis would compete better with the other
processes in general, and the coal-assisted water electrolysis
should provide even greater proportionate savings in view of the
anticipated higher unit electricity costs at lower consumption rates
in smaller plants.

It must be emphasized that the foregoing economic development
is very approximate and may prove to be unreliable. In particular,
capital costs are based upon yet-nonexistent technology required
for coal slurry processing in electrolytic cells, and the estimates
and extrapolation of these costs presented here are necessarily

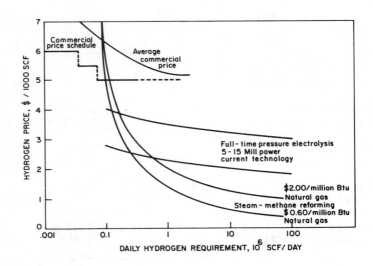

Figure 14. Comparison of hydrogen prices for various rates of
demand.

founded on judgement as much as on available data. The assumption is also made that the coal carbon is totally consumed, an uncertain premise, regardless of how reasonable it may seem to make use of only the most reactive coal fraction for electrolytic hydrogen production followed by subsequent burning of the unreacted portion for electric power generation. The hydrogen costs projected here, then, are only order-of-magnitude estimates. However, these estimates do provide some basis for judging the relative economic sensitivities of the various processing schemes, and some major economic impacts of capital and power costs. Capital investment may be reduced through innovative engineering of the electrolytic cell design. The interaction of coal cost and electric power cost and their influence on the price of the final product are also emphasized by these economic case studies.

The relative components of product hydrogen cost exhibit great potential incentives for increasing the substitution of coal for electricity by operating at lower cell potentials. The direction of this interplay is not unexpected, but the economic analysis suggests a strong cost-saving incentive.

SUMMARY

Our investigations have shown
1. The reaction rate for electrochemical coal gasification depends on coal concentration and particle size, operating temperature and potential. High temperature operation promises several advantages.
2. Activation energy falls in the range 37-54 kJ/mol, is independent of electrolyte concentration, but shows a weak dependence on coal concentration and particle size.
3. Contact of coal and anode is necessary to produce significant rates of electrochemical coal gasification. Different coals have different reactivities for this process, and in each case reactivity falls gradually as the reaction advances.
4. Original reactivity of the coal samples may be partially restored by washing with acetone wash or heating at temperatures between 200-600°C (or by a combination of these treatments).
5. Composition and generation rate of anode gas change as the reaction progresses. Two distinct coal oxidation mechanisms appear to occur at two correspondingly different stages of oxidation.
6. H_2SO_4 acts as a supporting electrolyte and does not appear to be consumed during the reaction.
7. Electrolyte hydrogen production and metal electrowinning from aqueous electrolytes are two processes in which electrochemical coal gasification may be applicable for lowering consumption of electrical energy.

Mass balance calculations, high temperature operation and use of anodes other than platinum are currently in progress.

REFERENCES

1. A. M. Squires, Science, 184, 340-46 (1974).
2. G. E. Klingman, R. P. Schaaf, Hydrocarbon Processing, pp. 97-101, (April 1972).
3. A. Verna, Chem. Tech., 372-81 (June 1978).
4. G. A. Mills, Chem. Tech., 418 (July 1972); Amer. Chem. Soc., Div. Fuel Chem. Prepr., 163rd National Meeting of the American Chemical Society, Boston, April 1972.
5. E. R. Kirk, D. F. Othmer, "Encyclopedia of Chemical Technology", Vol. 4, Interscience, NY 1966.
6. R. W. Coughlin, M. Farooque, Nature, 279, 303 (1979).
7. R. W. Coughlin, M. Farooque, I & EC (submitted).
8. M. Farooque, R. W. Coughlin, Fuel, 58, 705 (1979).
9. Goodridge, Electrochim Acta., 22, 929-933 (1977).
10. G. Kreysa, S. Pionteek, E. Heltz, J. Appl. Electrochem., 5, 305-312 (1975).
11. F. Coeuret, Electrochim Acta., 21, 185 (1976).
12. R. Dworak, H. Foess, H. Wendt, presented at the AICHE Meeting, Atlanta, Feb. 28 - Mar. 5, 1978.
13. H. Binder, A. Kohling, K. Richter, G. Sundstede, Electrochim. Acta., 9, 255 (1974).
14. R. E. Panzer, P. J. Elving, Electrochim. Acta., 20, 635 (1975).
15. R. W. Coughlin, Ind. Eng. Chem. Prod. R & D, 8, 12 (1969).
16. J. S. Mattron, H. B. Mark, Jr., "Activated Carbon", Chapter 3, Marcel Dekker, Inc., NY 1971.
17. Y. A. Zarifyanz, V. F. Kiseleve, N. N. Lezhnev, D. V. Nikitina, Carbon, 5, 127 (1967).
18. M. Farooque, R. W. Coughlin, Nature, 280, 666-668 (1979).
19. Dravo Corporation, "Handbook of Gasifiers", FE-1772-11, prepared by Dravo Corporation under DOE Contract E (49-18) 1772, published Feb. 1976.
20. H. G. Corneil, F. J. Heinzelmann, E. W. S. Nicholson, "Production Economics for Hydrogen, Ammonia and Methanol During the 1980-2000 Period", Exxon Research and Engineering Co., Linden, NJ 07036, BNL-50663, April 1977.
21. L. J. Nuttal, Int. J. Hydrogen Energy, 2, 395-403 (1977).
22. L. J. Nuttal, "Water Electrolysis Using SPE", presented at the Hydrogen for Energy Distribution Symposium, IGT, Chicago, July 1978.
23. Stone & Webster Engineering Co., 1979, private communication.
24. K. Darrow, N. Biedermann, N. Konopka, Int. J. Hydrogen Energy, 2, 175-187 (1977).

SELECTIVE MAGNETIC ENHANCEMENT OF PYRITE IN COAL
BY DIELECTRIC HEATING AT 27 AND 2450 MHZ

Delwyn D. Bluhm, Glenn E. Fanslow, and Stephen Beck-Montgomery
Ames Laboratory*, Iowa State University, Ames, Iowa 50011

Stuart O. Nelson
USDA, Richard B. Russell Agricultural Research Center
Athens, Georgia 30604

ABSTRACT

The objective of this project is to improve the magnetic separation of coal and pyrite by enhancing the magnetic susceptibility of pyrite in run-of-mine (ROM) coal through the use of selective pretreatments such as dielectric heating and induction heating. Separation of pyrite from coal would be facilitated by changing a portion of each pyrite particle to a more magnetic form.

The selective magnetic enhancement of pyrite in coal by dielectric heating was investigated fundamentally and experimentally. Fundamental treatment consisted of the measurement of the dielectric properties, at frequencies ranging from 1 MHz to 11.7 GHz, of a selection of coals and their pyritic fractions. These values were then used to calculate the theoretical heating that would be produced in these materials and their mixtures. The experimental treatment involved heating fractions and their mixtures at 27 and 2,450 MHz. Results showed limited agreement between theoretical and experimental heating and that it is possible to selectively heat pyrite in coal. Preliminary data indicate an increase in the apparent mass susceptibility of dielectrically heated pyrite samples that is roughly proportional to the time and power of pretreatment. Magnetic separation tests are being started.

INTRODUCTION

A. Background

The nation is once again looking to coal as the primary domestic resource to satisfy future energy needs. One of the shortcomings of presently developed coal beneficiation technology is that it cannot cope with the finely disseminated microcrystallites which constitute an appreciable part of the pyrite content of high sulfur coals. Development of an economical method for removing these microcrystallites which can be used in conjunction with the present beneficiation methods for removing coarse pyrites would greatly extend the possibilities for meeting air pollution control standards and would improve efficiencies of coal beneficiation techniques by allowing greater coal fines recovery. A potential method for removing pyrite microcrystallites is to grind the coal to a fine size and employ magnetic separation techniques to extract the pyrite; however, this is a difficult separation because of the small difference in the magnetic properties of coal and pyrite. By changing a part of each pyrite particle to a more magnetic form, the separation would be facilitated and the economic and technical feasibility of the method may be greatly improved.

It has been demonstrated that it is possible to separate pyrite from coal by high gradient magnetic separation (HGMS). However, the separation is difficult and requires powerful magnets and large field gradients. A typical HGMS unit is shown in Figure 1. By partially converting the pyrite in coal to a material with higher magnetic susceptibility, magnetic separation of coal and pyrite would be greatly facilitated. The grade and recovery of coal prepared by these methods would be increased and the required energy input for separation may be reduced. One possibility for increasing the susceptibility is to partially convert the pyrite to iron oxide. It has been shown in numerous experiments by Collison et al.[1] that pyrite can be readily converted to hematite by treating it with dissolved oxygen or air. Since the magnetic susceptibility of hematite is many times greater than that of pyrite, conversion of only a small portion of pyrite to hematite would significantly increase the magnetic susceptibility of an individual particle.

Another possibility is to partly convert the pyrite in the coal to a more magnetic iron-sulfur compound such as pyrrhotite by selective dielectric heating. Some preliminary investigations on the responses of coal to microwave heating have been reported by Beeson et al.[2] in 1975. Numerous studies have shown that the magnetic susceptibility of pyrite can be enhanced by heating; however, the problem in heating pyrite in coal is that of also heating the coal. This could result in pyrolysis of the coal in addition to being a waste of energy. A possible solution to this problem is the selective dielectric heating of pyrite in coal with a minimal heating of the coal. This may be accomplished if the dielectric properties of pyrite and coal differ sufficiently so the pyrite absorbs more electromagnetic energy and heats faster than the coal. The use of selective magnetic enhancement by dielectric heating may improve magnetic separation of pyrite from coal.

B. Previous Work

Physical separations of materials have been accomplished by magnetic means since before the 19th century. The mechanism that provides the basis for magnetic separation is the difference in the magnetic susceptibilities of the materials. In 1968, Kolm developed and patented a magnetic device[3] which proved to be of fundamental importance to high gradient magnetic separation (HGMS).

Very little work has been accomplished to couple high gradient magnetic separation for coal desulfurization with dielectric heating. Ergun and Bean[4] first proposed the use of radio frequency fields for selective heating of pyrite without affecting the coal. The amount of electromagnetic energy that a material exposed to electromagnetic radiation absorbs will depend on its dielectric properties. The local heating rate of the material depends on the power

419

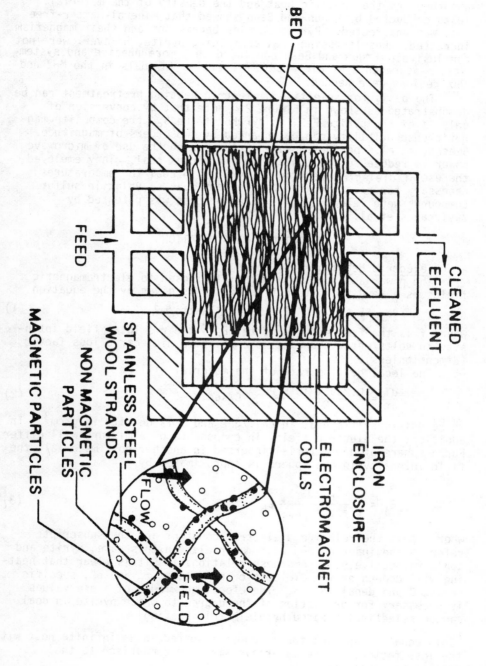

Figure 1. Vertical Section Through Center of High Gradient Magnetic Separator

absorbed and the specific heat and the density of the material. Tests conducted by Ergun and Bean showed that mineral-matter-free coal was unaffected. Pyrite samples became hot and their magnetism increased. Results using coal containing pyrite, however, were not conclusive. Ergun and Bean concluded that more precise and systematic measurements of dielectric properties in coals in the MHz and GHz regions were needed.

The basis for selective dielectric heating pretreatment can be demonstrated as shown in Figure 2. Note that by conversion of only 1% of the pyrite FeS_2 to pyrrhotite Fe_7S_8, the composite magnetic susceptibility can be enhanced by two orders of magnitude. Beeson, et al.[2] performed some early work on the use of microwave power to reduce sulfur content in bituminous coal. They employed the use of dielectric heating of pyrite in coal to temperatures necessary for reaction with a treatment gas so volatile sulfur compounds were released. A similar process was patented by Zavitsanos et al.[5]

THEORETICAL CONSIDERATIONS

A. Dielectric Heating

The power absorbed by a material exposed to electromagnetic fields as a result of dielectric losses is given by the equation

$$P = 55.63 \times 10^{-12} f E^2 \epsilon_r'' \quad \text{watts/m}^3 \tag{1}$$

where f is the frequency in hertz, E is the rms local field intensity in volts/meter and ϵ_r'' is the relative dielectric loss factor (dimensionless).

The local heating rate of the material will be

$$\frac{dT}{dt} = \frac{0.239 \times 10^{-6} P}{C \rho} \quad {}^\circ C/sec, \tag{2}$$

where C is specific heat in cal./g$^\circ$K and ρ is density in g/cm^3. In addition, the electric fields in components of a mixture will differ. For a sphere of one material inserted in another host material, the field intensity in the sphere is approximated by

$$E_i = \frac{3 \epsilon_{rh}'}{2 \epsilon_{rh}' + \epsilon_{ri}'} E_h \tag{3}*$$

where ϵ_r' is the relative dielectric constant and the subscripts refer to the inserted and host materials; in this case, pyrite and coal, respectively. From these relationships it is clear that heating will depend on the dielectric properties, ϵ_r' and ϵ_r'', specific heats, C and densities, ρ. Therefore, knowledge of these values is necessary for prediction of the degree to which pyrite in coal can be selectively heated by dielectric heating.

*This equation applies for a sphere inserted in an infinite host with the loss factors of the materials small in comparison to the dielectric constants.

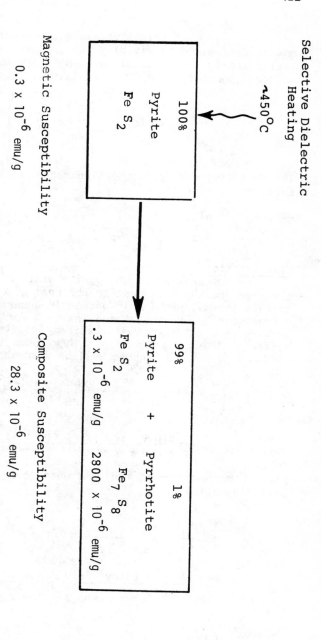

Figure 2. Selective Dielectric Heating Pretreatment

B. Selective Heating

The degree of selective dielectric heating of pyrite in coal may be determined by dividing the expected heating rate of pyrite by that of coal. When equations 1 and 3 are used, the ratio of the absorbed power for the two materials will be

$$\frac{P_p}{P_c} = \frac{P_{pyrite}}{P_{coal}} = \left[\frac{3\varepsilon_{r_c}'}{2\varepsilon_{r_c}' + \varepsilon_{r_p}'}\right]^2 \frac{\varepsilon_{r_p}''}{\varepsilon_{r_c}''} . \tag{4}$$

The selective dielectric heating ratio may then be determined from Equation 2, as follows:

$$\frac{dT_p}{dT_c} = \left[\frac{P_p}{P_c}\right] \frac{C_c}{C_p} \frac{\rho_c}{\rho_p} \tag{5}$$

Heat Transfer

The temperature of the pyrite will be a function of the power absorbed and the heat lost to the coal by conduction. An approach to this thermal problem is presently under study. It is based on a treatment by Goldenberg and Tranter[6] for a sphere, radius a, generating heat at a constant rate, A, inserted in an infinite medium of different thermal properties. They found that the temperature at radius r inside the sphere will be described by the equation

$$T_1 = \frac{a^2 A}{K_1} \left[\frac{1}{3} \frac{K_1}{K_2} + \frac{1}{6}\left(1 - \frac{r^2}{a^2}\right) - \frac{2ab}{r\pi} \int_0^\infty \frac{\exp(-y^2 t/\gamma)}{y^2} \right.$$

$$\left. \frac{(\sin y - y \cos y) \sin\left(\frac{ry}{a}\right) dy}{[(c \sin y - y \cos y)^2 + b^2 y^2 \sin^2 y]} \right] \tag{6}$$

and the temperature at radius r outside the sphere will be described by the equation

$$T_2 = \frac{a^3 A}{rK_1} \left[\frac{1}{3} \frac{K_1}{K_2} - \frac{2}{\pi} \int_0^\infty \frac{\exp(-y^2 t/\gamma)}{y^3} \right.$$

$$\left. \frac{(\sin y - y \cos y)[by \sin y \cos \gamma y - (c \sin y - y \cos y)\sin \sigma y] dy}{[(c \sin y - y \cos y)^2 + b^2 y^2 \sin^2 y]} \right] \tag{7}$$

where

$$b = \frac{K_2}{K_1}\sqrt{\frac{k_1}{k_2}}, \quad c = 1 - \frac{K_2}{K_1}, \quad \gamma = \frac{a^2}{K_1}, \quad \sigma = \left(\frac{r}{a} - 1\right)\sqrt{\frac{k_1}{k_2}}$$

In these equations, region 1 is within the sphere and region 2 is outside. The K's are thermal conductivities, the k's are thermal diffusivities and the T's are temperatures. An important factor to be considered in this problem is that the coal is being heated and, as its temperature increases, the heat loss from the pyrite will be reduced.

DIELECTRIC PROPERTIES MEASUREMENTS

While investigating the possibility of selective dielectric heating of pyrite in coal at 27 and 2450 MHz, it was decided to obtain dielectric properties measurements over a range of frequencies. In addition to providing the information needed for predicting heating at 27 and 2450 MHz, these measurements could assist in deciding on the possibility of an optimum frequency to be used for selective dielectric heating.

The dielectric properties of interest were the relative dielectric constant, ε_r' and the relative dielectric loss factor, ε_r''. Six different measuring systems were used. A Boonton 160-A Q-Meter with a special coaxial sample holder was used for measurements at 1, 5, 10, 20 and 50 MHz. Measurements at 150 MHz were taken with a Boonton 250-A RX Meter and coaxial-line sample holder. At 300 MHz, the measurements were obtained with a General Radio Admittance Meter and coaxial-line sample holder. The Roberts and von Hippel short-circuited-line technique was used for measurements at the microwave frequencies of 1, 2.45 and 11.7 GHz. Rohde and Schwarz power oscillators, standing wave meter and a 21 mm short-circuit, rigid air-line sample holder, were used in conjunction with a slotted line for measurements at 1 GHz and a non-slotted line at 2.45 GHz. A rectangular waveguide X-band system was used for measurements at 11.7 GHz. A computer program was used for calculating the dielectric properties from the short-circuited line and waveguide measurements. More complete descriptions of the equipment, methods and vertification of the reliability of the various measurements are given in references 7-15. Results are given in Table 1.

MATERIALS

The materials chosen for this study were a selection of run of mine (ROM) coals and fractions of these coals obtained by Deister table and heavy media gravitational separation. The fractions were 1.3 float and 2.0 sink. The 1.3 float was considered to be clean coal that was relatively pyrite free. The sink was treated as the pyritic portion of the coal because it had most of the coal removed and contained most of the pyrite from a given ROM coal. An additional coal fraction that was included was leached pyrite. This was obtained by leaching the sink portion with an acid that dissolved some of the other minerals and left the pyrite. Although none of the materials were pure pyrite or pure coal, they were derived from the coals and, therefore, it was assumed that they would provide the most accurate information on the heating of these materials. Recently, to assure removal of clay, shale and other minerals from the sink fractions of the coals, Bromoform was used as the heavy media for separations. Bromoform has a specific gravity of 2.876

424

Table 1. Dielectric Properties, Densities, and Fractional Porosities of Pulverized Coal Samples[a/]

Material	1 MHz ϵ_r'	ϵ_r''	5 MHz ϵ_r'	ϵ_r''	10 MHz ϵ_r'	ϵ_r''	20 MHz ϵ_r'	ϵ_r''	50 MHz ϵ_r'	ϵ_r''	150 MHz ϵ_r'	ϵ_r''	300 MHz ϵ_r'	ϵ_r''	1 GHz ϵ_r'	ϵ_r''	2.45 GHz ϵ_r'	ϵ_r''	11.7 GHz ϵ_r'	ϵ_r''	ρ_b g/cm³	ρ_p g/cm³	P_f
Dahm[b] ROH	3.06	0.27	2.81	0.27	2.79	0.26	2.72	0.26	2.65	0.27	2.48	0.15	2.51	0.10	2.38	0.11	2.31	0.10	2.21	0.12	0.625	1.79	0.64
Dahm[b] 1.30 Float	3.11	0.37	2.91	0.33	2.77	0.30	2.74	0.29	2.68	0.26	2.55	0.13	2.54	0.08	2.44	0.09	2.35	0.09	2.26	0.12	0.638	1.54	0.59
Dahm[b] 2.00 Sink	7.26	0.09	7.32	0.10	7.14	0.08	7.16	0.09	7.12	0.11	8.22	0.30	7.85	0.34	7.75	0.48	7.15	0.87	6.32	0.86	1.974	3.82	0.48
Illinois #6 ROM	3.29	0.34	3.03	0.31	2.92	0.31	2.90	0.32	2.79	0.31	2.66	0.19	2.62	0.11	2.53	0.11	2.43	0.10	2.36	0.12	0.767	1.59	0.52
Illinois #6 1.30 Float	2.71	0.10	2.59	0.11	2.53	0.11	2.57	0.11	2.56	0.13	2.46	0.07	2.52	0.06	2.44	0.06	2.34	0.07	2.29	0.08	0.730	1.38	0.47
Illinois #6 2.00 Sink	6.23	1.20	5.39	1.08	5.02	1.17	4.62	1.10	4.32	1.05	3.63	0.57	3.59	0.40	3.20	0.26	3.10	0.20	2.94	0.21	1.215	2.57	0.53
Kentucky #11 ROM	2.92	0.24	2.71	0.22	2.68	0.21	2.67	0.21	2.62	0.21	2.45	0.12	2.51	0.10	2.38	0.09	2.29	0.08	2.23	0.09	0.713	1.45	0.51
Kentucky #11 1.30 Float	2.59	0.12	2.50	0.12	2.46	0.11	2.43	0.12	2.43	0.12	2.39	0.07	2.44	0.08	2.36	0.07	2.26	0.06	2.19	0.06	0.728	1.36	0.46
Lignite[c] ROM	3.93	0.36	3.81	0.29	3.73	0.30	3.54	0.28	3.53	0.30	3.33	0.22	3.39	0.18	3.16	0.21	2.99	0.22	2.81	0.20	0.796	1.50	0.47
Pittsburgh #8 R.O.M.	2.63	0.16	2.56	0.18	2.50	0.17	2.44	0.18	2.41	0.18	2.32	0.09	2.44	0.04	2.28	0.06	2.21	0.05	2.17	0.05	0.745	1.55	0.52
Pittsburgh #8 1.30 Float	2.36	0.10	2.28	0.09	2.26	0.08	2.23	0.09	2.20	0.09	2.18	0.04	2.24	0.03	2.16	0.03	2.11	0.04	2.09	0.04	0.701	1.39	0.50
Pittsburgh #8 2.00 Sink	6.24	1.00	5.24	1.03	4.93	0.96	4.63	0.98	4.24	0.99	3.67	0.53	3.66	0.36	3.45	0.25	3.27	0.30	3.11	0.20	1.331	2.72	0.51

a ϵ_r' and ϵ_r'' - respectively, the dielectric constant and dielectric loss factor at indicated frequencies; ρ_b - bulk density, ρ_p - particle density, and P_f - fractional porosity.
b Dahm Mine, Mahaska County, Iowa.
c Beulah Mine, Mercer County, North Dakota.

(nominally 2.9). Previous sink material was prepared at a specific gravity of 2.0 as stated above. Chemical analyses, obtained by standard ASTM methods, are given in Table 2.

Table 2. Chemical Analyses of Coal Samples and Fractions

Material	Total Sulfur (%)	Pyritic Sulfur (%)	Ash (%)	Moisture (% w.b.)	Heating Value (BTU/lb)
Childers[1] ROM	8.1	3.2	14.5	2.9	10,813
Childers[1] 1.3 Float	4.4	1.2	4.9	4.5	13,252
Childers[1] 2.0 Sink	43.8	43.0	68.6	0.2	1,688
Dahm[1] ROM	3.0	1.5	11.2	5.0	12,103
Dahm[1] 1.30 Float	1.7	0.3	3.3	10.7	13,180
Dahm[1] 2.00 Sink	43.0	42.3	59.4	1.3	2,121
Illinois #6 ROM	4.1	3.0	19.9	3.1	11,082
Illinois #6 1.30 Float	1.8	0.4	3.2	3.2	13,783
Illinois #6 2.00 Sink	12.5	11.8	70.4	2.1	2,048
Kentucky #11 ROM	3.3	1.0	9.0	3.2	12,839
Kentucky #11 1.30 Float	2.7	0.5	3.4	3.1	13,740
Kentucky #11 2.00 Sink	8.3	6.6	71.7	2.4	1,777
Pittsburgh #8 ROM	2.8	1.3	16.3	1.1	12,456
Pittsburgh #8 1.30 Float	1.8	0.2	4.3	0.9	14,526
Pittsburgh #8 2.00 Sink	9.6	8.4	78.2	1.2	936
Leached Pyrite	51.7	47.7	65.5	0.1	875

[1] Iowa coal mine name is shown since there is no generally accepted bed name.

SELECTIVE HEATING AT 2450 MHZ

Predicted Results

The predicted values for the selective dielectric heating ratios of pyrite in coal were obtained using equations (4) and (5), the dielectric properties values from Table 1, and estimated specific heats and densities for coal and pyrite from the Chemical Engineering Handbook,[16] (C_c = 0.3, ρ_c = 1.3, C_p = 0.136, ρ_p = 5). The specific-heat and density values used for these calculations were not exactly those of the 2.00 sink material. However, the product term, $C_p \rho_p$, in Equation 5 for pyrite should be a good estimate because, although the density of the 2.00 sink material was somewhat less than 5, the

426

specific heat was somewhat greater than that for pure pyrite. Resulting values for the predicted selective dielectric heating ratios of each of the coals are given in Table 3.

Table 3. Selective Dielectric Heating Ratios for Pyrite in Coal at 2450 MHz

Coal	Predicted	Experimental
Dahm	2.0	2.3
Childers	3.3	2.6
Pittsburgh No. 8	3.1	1.9
Illinois No. 6	1.3	2.2

Experimental Heating

Heating at 2.45 GHz was performed on 25-gram samples that were approximately 1 cm high, 2 cm wide, and 30 cm long. Tests were performed in a rectangular cross-section WR-340 slotted waveguide. Power levels, as measured at the generator, were 2 kw forward and 100 W reflected. Energy not absorbed by the sample was dissipated in a water load.

The slotted waveguide was positioned with its axis horizontal and with the wide walls in a vertical orientation. Longitudinal slots, about 1 cm wide, along the center of the wide walls on both sides facilitated sample insertion and removal. Samples were positioned along the center of the waveguide axis and supported in a horizontal glass trough of semicircular cross-section. Average temperatures were obtained from observations made while drawing a thermocouple probe through the length of the sample upon removal of the glass trough and sample from the slotted waveguide.

Measured temperatures wnd treatment times for different "pyrite-coal sample" mixtures are given in Figure 3. In this Figure, the pyrite-to-sample ratios represent the percentage of the 2.00 sink and 1.30 float mixtures that is 2.00 sink material by weight. Since exact "pyrite-coal" mixtures were not known for ROM samples, their temperatures are shown on the ordinate.

The heating rate for each mixture was defined as the temperature increase from ambient (22°C) divided by the heating period. Assuming linear heating, the rates ranged from 3.97°C/sec. for the Dahm 2.00 sink to 1.30°C/sec. for the Pittsburgh No. 8 1.30 float as given in Table 4. The selective dielectric heating ratio was determined by dividing the heating rate of the 2.00 sink fraction of a given coal by the heating rate for the 1.30 float fraction of that same coal. These values ranged from 1.9 to 2.6 (Table 3).

Experimental heating rates reflect the temperature rise of the bulk sample of coal or pyrite, but not the rise in temperature of individual particles. Methods used for temperature measurements did not show actual peak temperatures reached during the dielectric heating. The sample temperature dropped from the peak temperature rapidly after heating stopped. This drop in temperature occurred in all samples even when arcing was observed. Most particles are smaller

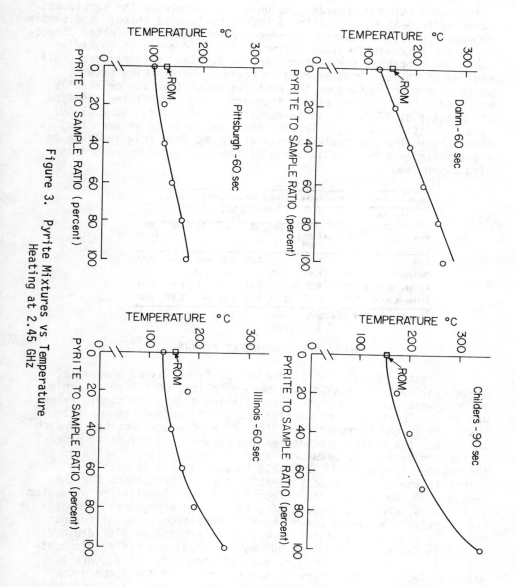

Figure 3. Pyrite Mixtures vs Temperature Heating at 2.45 GHz

than 125 microns (120 mesh) and lose heat rapidly during and after
heating. Since discrete particles cannot be dealt with, bulk temperatures are presented in the results. For heating at 2.45 GHz
using the slotted waveguide, the bulk temperature of the sample was
taken after heating by passing a thermocouple wire through the sample
immediately after heating. (Thermocouples cannot be inserted directly on the microwave heating region.) This did not give instantaneous and/or hotspot temperatures in the sample. Hotspots are small
regions of the samples, especially regions where arcing occurs,
where higher instantaneous temperatures occur than can be readily
measured at the present time. Attempts using an infrared gun-type
thermometer at 27 MHz or an infrared temperature scan at 2.45 GHz
have had little success. These techniques focus on an area that is
much larger than the heated sample area, and then only indicate an
average temperature. Calibration of the observed temperature is
also a problem.

Table 4. Sample Heating Rates (°C/sec) for 60-Second Treatment

Coal Source		Pyrite to Sample Ratio (%)						
Name	ROM	0 (1.30) Float	20	40	60	68	80	100 (2.00) Sink
Dahm	2.13	1.72	2.13	2.80	3.30	----	3.80	3.97
Childers*	1.42	1.42	1.70	1.98	----	2.26	----	3.64
Pittsburgh No.8	1.72	1.30	1.72	1.72	1.97	----	2.30	2.47
Illinois No.6	2.13	1.72	2.63	2.13	2.47	----	2.80	3.80

*Treatment time of 90 seconds.

SELECTIVE HEATING AT 27 MHZ

Dielectric heating at 27 MHz was performed using a parallel
plate applicator with a spacing between plates of 1.2 cm. Sample
holders were 1 x 10 cm glass crystallizing dishes and sample weight
was 25 g. Treatment time was limited to 100 seconds to minimize
heat loss to the surroundings. The 375 W of power absorbed by each
sample was the difference between the forward and reflected powers
at the generator. Temperatures were measured using an infrared
thermometer that viewed the sample through a 14 x 14 mesh wire
screen in the top plate of the applicator. (Corrections for the
presence of the screen ranged from 4 to 6°C.) Results are presented
in Figure 4. For the purpose of obtaining an approximate comparison of materials heated under the same conditions, two different
materials were heated at the same time. Selected results are presented in Table 5. Resulting values for predicted dielectric heating ratios, from equation 5, of each of the coals tested, are
given in Table 6. Experimental values are given for comparison.

Preliminary induction heating experiments were performed using
a 3-turn coil having an inner diameter of approximately 2.2 cm.
Sample holders were 16 x 150 mm test tubes and sample weight was

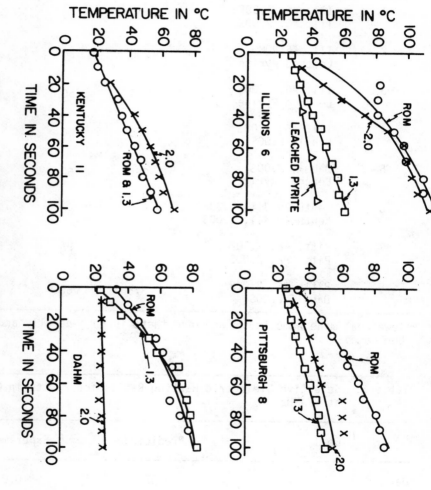

Figure 4. Heating of Coals at 27 MHz

Table 5. Heating at 27 MHz

Test No.		Power Absorption (Approx.) (Watts)	Time (Sec)	Temperature Rise (^{o}C)
1	Ill. #6, 1.30F Ill. #6, 2.00S	200	60	10 17
2	Ill. #6 ROM Leached Pyrite	200	60	30 8
3	Ill. #6, 2.00S Leached Pyrite	375	60	37
4	Ill. #6, 2.00S Dahm[1], 2.00S	365	60	44 4
5	Dahm[1], 2.00S Leached Pyrite	365	60	5 19
6	Kentucky #11, 1.30F Kentucky #11, 2.00S	365	60	25 30
7	Pitt. #8, 1.30F Pitt. #8, 2.00S	365	60	13 48
8	Pitt. #8, 2.00S Dahm[1], 2.00S	375	60	58 3

[1]Iowa coal mine name is shown since there is no generally accepted bed name.

Table 6. Selective Dielectric Heating Ratios for Pyrite in Coal at 27 MHz

Coal	Predicted	Experimental
Dahm	0.07	0.04
Illinois No. 6	3.56	2.18
Pittsburgh No. 8	3.78	1.12
Kentucky No. 11	0.26	1.33

15 g. Treatment time was 1 minute and power absorption by each sample, as indicated by the difference between the forward and reflected powers at the generator, was 500 W. Testing was limited to the leached pyrite and a coal having an ε_r' of 2.3, an ε_r'' of 0.19, 3.21% of pyritic sulfur, 8.13% of total sulfur and a moisture content of 2.9%, wet basis. Heating of the coal was negligible and the pyrite was warmed to a temperature of approximately 50°C.

DISCUSSION

Heating Results at 2450 and 27 MHz

The results demonstrated that pyrite can be heated faster than the coal. Further, they indicated that the degree of selective dielectric heating of pyrite in coal may be predictable if the physical properties are known. However, it is still not known if the susceptibility of pyrite in coal can be enhanced without undue loss of energy to the coal. To determine this, two pieces of information are required. They are the temperature required to produce the increase in susceptibility and the heat loss to the coal from the pyrite. Presently only the temperature measurements have been performed, and they have proven to be rather difficult to obtain. The problem is that, as it is heated, the pyrite undergoes an exothermic reaction that is hard to control. Additionally, it has been found through dielectric heating and through differential thermal analysis, DTA, that the exothermic reactions of the coal-derived pyrite were not the same as those predicted for pyrite in the literature. For the coal-derived pyrite, there were reactions at ~ 300° and 480°C. From the literature[17] one would expect the first reaction for normal pyrite to take place at ~ 480°C. However, additionally from more recent literature[18,19] reactions of "normal" pyrite at temperatures below 350°C are not uncommon. The reasons for the differences in initial reaction temperatures between coal derived pyrite and "normal" pyrite are being investigated.

Gas Chromatography and Mass Spectroscopy Analyses

In order to determine the products of reaction produced during dielectric heating, samples of leached pyrite were heated in a closed tube filled with air. The results presented in Table 7 show traces of SO_2, CS_2 and COS. A similar study using mass spectrometer analysis was made using samples covered by a back-fill of oxygen. The results are presented in Table 8.

X-ray Diffraction

Samples of heated and unheated pyrite have been analyzed by X-ray diffraction techniques. The unheated pyrite contained mostly pyrite and small amounts of marcasite. The heated samples of pyrite contained pyrite and varying amounts of pyrrhotite. A recent sample of the magnetic fraction of heated leached pyrite is shown in Figure 5. In this fraction, there is much more pyrrhotite than pyrite present. The nonmagnetic fraction of the heated leached

Table 7. Gas Chromatography Analysis

Material	Dielectric Heating		% vol/vol			
	Power Level KW	Time Seconds	H_2S	SO_2	CS_2	COS
Pyrite, leached, 2.9 sink	1.5	8	none	0.001	0.1	0.1
Same	1.5	6	none	0.001	none	none
Same	1.5	5	none	0.001	none	none
Same	1.5	7.7	none	0.001	none	none
Dahm Coal, 2.9 sink	1.5	8	none	none	detected	0.005
Same	1.5	10	none	none	0.05	0.3
Same	1.5	8	none	none	0.25	0.4
Same	1.5	12	none	none	0.025	0.25
Same	1.5	14.5	0.1	none	0.1	0.3
Same	1.5	16	0.1	none	0.5	0.35

Table 8. Mass Spectrometer Analysis

Mass Number of Ions	44^+	69^+	76^+ $(48^+ + 64^+)$	$(2 \cdot 15^+)$	$(26^+ + 27^+)$	(40^+) (41^+) (42^+) (43^+)	(55^+) (56^+) (57^+)
Atmosphere of O_2	CO_2^+	COS^+	CS_2^+	SO_2^+	ΣC_1^+	ΣC_2^+	ΣC_3^+ ΣC_4^+
0.9	620	19	1	1.2	16	22	12	1.2
0.8	1290	36	2	1.4	26	49	50	1.4
0.7	790	15	1	2.1	13	20	43	1.4
0.6	900	20	1	2.1	17	27	42	1.4
0.5	140	1	1	3.4	1	1	36	---
0.4	1520	57	6	2.2	120	82	76	8.3
0.3	2320	76	11	3.6	173	126	88	14.6
0.2	1350	22	7	4.1	8	32	43	2.7
0.2	1440	28	11	4.6	17	49	56	4.0
0.1	3170	73	39	7.3	105	190	82	13.4
0.0*	1380	28	3	40	96	88	?	?

*Air leak occurred, probably immediately.

433

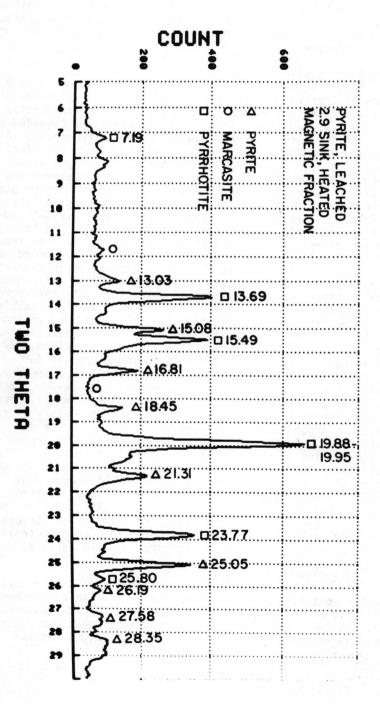

Figure 5. X-ray Powder Diffraction Analysis

pyrite contained mostly pyrite and marcasite with a small amount of
pyrrhotite. The X-ray diffraction procedure gives a qualitative
measure of the compounds present after heating, but not a quantitative measure.

Mass Susceptibility Measurements

Mass susceptibility measurements were made on the 2.9 sink fractions of leached Childers pyrite and Dahm Coal, mixtures of Dahm
Coal 1.3 float and pyrite 2.9 sink, and on the nonmagnetic fraction
of Dahm Coal 2.9 sink. This latter fraction was obtained by passing
the material through a low gradient magnetic separator. All samples
were measured for magnetic susceptibility before and after dielectric
heating at 2.45 GHz. The leached pyrite in Group I showed large
changes in susceptibility when arcing occurred between the pyrite
particles heated for 6 and 7.7 seconds. When no arcing occurred
during 5 seconds of heating, the change in susceptibility was negligible. In Group II, mixtures of Dahm Coal 1.3 float and pyrite 2.9
sink showed small increases in susceptibility as the heating time
was increased. These mixtures did not seem to heat as thoroughly as
the pyrite alone. It was assumed that the coal absorbed heat given
off by the pyrite and that the pyrite did not reach temperatures
necessary to increase susceptibility. The Group III Dahm Coal 2.9
sink fractions had a high initial susceptibility that decreased as
the heating time increased. Since this was an older sample of
coal, the iron pyrite could have partially oxidized to iron sulfate,
which has a higher susceptibility. All the samples in Group III
were analyzed by X-ray diffraction, and all contained only pyrite
and marcasite. No arcing occurred in these samples during heating,
and the susceptibility decreased slightly. The reason for the decrease in susceptibility cannot be explained; however, it could be
due to sample oxidation. Dahm Coal 2.9 sink material, Group IV, was
passed through a low gradient magnetic separator as a slurry and separated into nonmagnetic and magnetic fractions. The nonmagnetic
fraction was used for the heating experiments. No arcing occurred
during heating at the first 8, 10, and 12 second runs. The susceptibility of these samples increased by a factor of 20 to 40.
Arcing occurred in the second 8, 14.5 and 16 second runs. These
samples showed susceptibility increases of 50 to 80 times their
original values. The increase in susceptibility was due to the conversion of part of the pyrite to pyrrhotite. The largest increases
came when arcing occurred between particles, an uncontrolled reaction
at the present time. The results are shown in Table 9.

Frantz Isodynamic Separator

The Frantz Isodynamic Separator* consists of a vibrating chute
mounted centrally between the pole pieces of an electromagnet.
This unit may be inclined in any direction by the universal mount-
ing, and the electromagnet current is continuously adjustable
yielding magnetic field strengths of 0 to 15,000 gauss.

*S. G. Frantz, Registered Trademark.

Table 9.   Mass Susceptibility Measurements

Group	Material	Dielectric Heating at 2.45 GHz		Mass Susceptibility $x \times 10^{-6}$ emu/g	
		Power KW	Duration Seconds	Before	After
I	Pyrite, Leached, 2.9 Sink	1.5	6	0.67	79.66
	Same	1.5	5	0.67	0.77
	Same	1.5	7.7	0.67	12.15
II	Mix of 80% Dahm Coal 1.3 Float and 20% Pyrite 2.9 Sink	1.5	40	-0.52	-0.71
	Same	1.5	60	-0.52	-0.19
	Same	1.5	30	-0.52	-0.39
	Same	1.5	100	-0.52	-0.44
	Same	1.5	120	-0.52	+0.53
	Same	1.5	300	-0.52	+0.20
III	Dahm Coal, 2.9 Sink	1.5	60	35.19	31.13
	Same	1.5	90	35.19	29.50
	Same	1.5	120	35.19	25.19
	Same	1.5	150	35.19	25.75
	Same	1.5	180	35.19	21.13
	Same	1.5	210	35.19	21.54
	Same	1.5	240	35.19	21.79
IV	Dahm Coal, 2.9 Sink, Non-magnetic Fraction	1.5	8	3.24	132.91
	Same	1.5	10	3.24	101.07
	Same	1.5	8	3.24	255.62
	Same	1.5	12	3.24	68.24
	Same	1.5	14.5	3.24	184.36
	Same	1.5	16	3.24	236.09

436

Mineral separations are normally made by arbitrary adjustment of the current of the electromagnet and the transverse slope of the chute until the most suitable splitting is obtained. At any such setting the separator divides a sample into two fractions. The more magnetic fraction consists of particles each having a specific susceptibility above a value determined by the settings; the less magnetic fraction consists of particles of susceptibility below this value.

As dielectrically heated coal samples are produced, the Frantz separator is being used to determine the magnetic enhancement and magnetic separations that may be realized.

REFERENCES

1. Collison, G.F., T.D. Wheelock and R. Markuszewski, "High Gradient Magnetic Separation of Chemically Treated Pyrite," Unpublished Report, May 1979.
2. Beeson, J.L., G.E. Fanslow and T.S. King, "Using Microwave Power to Reduce the Sulfur Content in Iowa Coal," EMRRI, Iowa State University, Ames, Iowa 1975.
3. Kolm, H.H., "Magnetic Device", U.S. Patent No. 3,567,026. March 2, 1971.
4. Ergun, S. and E.H. Bean, "Magnetic Separation of Pyrite from Coals," USBM-RI-7181, 1968.
5. Zavitsanos, P.P. and K.W. Bleiler, "Process for Coal Desulfurization," U.S. Patent No. 4,076,607. Feb. 28, 1978.
6. Goldenberg, H. and C.J. Tranter, "Heat Flow in an Infinite Medium Heated by a Sphere," Brit. J. of Appl. Phys., v.3, p. 296-98, Sept. 1952.
7. Nelson, S.O., F.D. Yung, and L.H. Soderholm, "Determining the Dielectric Properties of Grain," Agricultural Engineering 34(9): 608-610, 1953.
8. Jorgensen, J.L., A.R. Edison, S.O. Nelson, and L.E. Stetson, "A Bridge Method for Dielectric Measurements of Grain and Seed," Trans. ASAE 13(1): 18-20, 24, 1970.
9. Stetson, L.E. and S.O. Nelson, "A Method for Determining Dielectric Properties of Grain and Seed in the 200- to 500-MHz Range," Trans. ASAE 13(4): 491-495, 1970.
10. Roberts, S. and A. von Hippel, "A New Method for Measuring Dielectric Constant and Loss in the Range of Centimeter Waves," J. Appl. Phys. 17(7):610-616, 1946.
11. Nelson, S.O., "Microwave Dielectric Properties of Grain and Seed," Trans. ASAE 16(5):902-905, 1973.
12. Nelson, S.O., "A System for Measuring Dielectric Properties at Frequencies from 8.2 to 12.4 GHz," Trans. ASAE 15(6):1094-1098, 1972.
13. Nelson, S.O., C.W. Schlaphoff, and L.E. Stetson, "Computer Program for Calculating Dielectric Properties of Low- or High-Loss Materials from Short-Circuited Waveguide Measurements," ARS-NC-4, Agr. Res. Serv., U.S. Dept. Agr., 1972.

14. Nelson, S.O., L.E. Stetson, and C.W. Schlaphoff, "A General Computer Program for Precise Calculation of Dielectric Properties from Short-Circuited-Waveguide Measurements," IEEE Trans. Instr. & Meas. IM-23(4):455-460, 1974.

15. Nelson, S.O., "Frequency Dependence of the Dielectric Properties of Wheat and the Rice Weevil," Ph.D. Diss., Iowa State University, Ames, Iowa, Pub. No. 72-19,997 Univ. Microfilms, Ann Arbor, Michigan, 1972.

16. Perry, R.H. and Chilton, C.D., Eds., Chem. Eng. Hndbk, 5th Ed., McGraw-Hill, New York, NY 1973.

17. The Differential Thermal Investigation of Clays. Edited by Mackenzie Mineral Society (Clay Minerals Group), London, 1957.

18. Craig, J.R. and S.D. Scott, "Sulfide Phase Equilibria," P.H. Ribbe, ed., Sulfide Mineralogy Washington, Mineralogical Society of America, 1974, short course notes, Vol. 1, pp C-1 to C-110.

19. Power, L.F. and H.A. Fine, "The Iron-Sulphur System, Part 1, The Structures and Physical Properties of the Compounds of the Low-Temperature Phase Fields," Minerals Sci. Engng., Vol. 8, No. 2, April 1976, pp. 106-128.

Panel Discussion - Interaction of Research
with Coal Technology Development

Panel Members

U. Merten
Gulf Research and Development Corporation
Pittsburgh, PA 15230

I. Wender
U.S. Dept. of Energy, Washington, D.C.   20545

W. H. Wiser
Dept. of Fuels Engineering, University of Utah,
Salt Lake City, UT 84112

D. E. Woodmansee
General Electric Corporate Research and Development
Schenectady, NY 12301

Chairperson of Panel and Compiler of Report:
B. R. Cooper
Dept. of Physics, West Virginia University
Morgantown, WV 26506

ABSTRACT

The concluding panel session was devoted to the interaction of
research with coal technology development. The four panel members
discussed this from the viewpoints of their respective experience in
the industrial, academic, and governmental sectors.

INTRODUCTION

The concluding panel session was intended to discuss the
bridging to technology development - how research can be of most
value to that development. The panel chairperson asked the panel
members to consider two types of contributions from research to the
technology of coal utilization and synthetic fuel production. The
first type of contribution might be characterized as "troubleshoot-
ing," helping to overcome obstacles in the presently envisioned line
of technology development. The second type is work leading to
genuinely new departures, such as a completely new technique for
coal liquefaction. The panel chairperson asked the panel members
to consider what they had heard during the three days of the conference, and to duscuss both what is being done that does provide the
desired bridging to technology, and also what is needed to make the
research atmosphere most productive for bridging to technology.
    The four panel members, drawn from the industrial, academic,
and governmental sectors, were: Dr. U. Merten, Vice-President,
Chemical and Minerals of the Gulf Research and Development Company;

Professor Wendell H. Wiser of the Department of Fuels Engineering of
the University of Utah; Dr. Donald E. Woodmansee, Liaison Scientist
from Corporate Research and Development to the Power Systems Sector
of the General Electric Company; and Dr. Irving Wender, Director of
the Office of Advanced Research and Technology in the U.S. Department
of Energy, Office of the Assistant Secretary for Fossil Energy. The
present report summarizes the initial remarks of each panel members.
The wide ranging discussion with the audience, which followed, is
not included here. The present report was compiled by the panel
chairperson, Professor Bernard R. Cooper of the Department of
Physics, West Virginia University, and has been reviewed by each of
the four panel members.

SUMMARY - REMARKS OF U. MERTEN

Several main points were made. Research does not have to make
coal utilization technology feasible. Such technology already is
feasible, although there are certainly problems that research can
help in solving. Research does not have to make coal utilization
economical. Present price levels for petroleum have already done
that. Getting broader utilization of coal "off the ground" is not
really research related. The main factors inhibiting wider coal
utilization are economic risks related to regulatory and other socio-
economic questions, rather than process economics as such.

Research does have a major contribution to make in cost reduction. The cost of much of the goods and equipment still in use in
our society is based on much lower energy prices than the present
ones. Only as such goods and equipment are replaced, will we pay the
full cost of present energy prices. Even marginal improvements in
the economics of coal conversion will help us in meeting this economic burden.

There are two kinds of research that are of interest. First is
research aiming at cost reduction in known processes. Among research
topics of this type not touched upon in the conference are questions
of reaction kinetics, of effects of particular coals, and questions
having to do with materials of construction. The second kind of
research is that leading to new processes, really different from
what we already have, and showing potential for improvement. For
example, underground gasification does offer the possibility for
doing something really different. An important aspect of any such
new processes is to be efficient.

In summary, we should recognize that the feasibility of coal
utilization does not have to be proven, and that cost reduction
should be our aim.

Summary - Remarks of W. H. Wiser

Dr. Wiser emphasized one problem specific to the academic
sector, growing out of the fact that much research is process
oriented. This problem is in dealing with the situation arising
from the necessity that the industrial sector carry out the commercial realization of the benefits of such research. This creates

difficulty in discussing the progress in such a research program
before completion, particularly in attempts to obtain funding for
further work. This can lead to an industrial company simply doing
the further work on its own, or perhaps even submitting a competing
proposal to a governmental agency.

The key point in this problem is how to take care of the transition from the university to industry, somewhere between the
laboratory and the pilot plant. There should be provision for
continuing useful imput from university people. The main operation-
al question is that of patents. There should be a way to have cooperation between university people and industrial companies on
going toward commercial scale operations; and the university people
should derive some of the financial benefits derived from commercialization.

Summary - Remarks of D. E. Woodmansee

As might be expected, the papers in the conference have concentrated on specific problems of the chemistry and physics involved in
coal utilization. Dr. Woodmansee commented that in practical utilization, and specifically in electrical power generation, one has to
consider all problems at the same time to insure feasibility of the
proposed approach under study.

To bring out this point, he discussed the major problems remaining in our overall efforts to utilize coal in electrical power
generation. This was done by referring to specific technologies:
conventional steam generator, combined power cycles involving gas
turbines, MHD, fuel cells. The conventional steam generator with
stack scrubber is the standard to which other power generating
cycles are compared. For the conventional generator, the major
problem is the removal of sulfur. In combined cycles involving gas
turbines, one problem is that one cannot simply substitute coal for
oil and gas. There are problems associated with corrosion by ash
and by alkali metals. One can go to an MHD process, that effectively combines the turbo expander and the generator into the MHD duct.
However, one then still has the problem of dust, compounded by the
presence of potassium seed. One can wash the gas from a gasifier
and use it in a gas turbine, since the washing can remove dust,
sulfur, and alkali metals.

For all power cycles, there are four main sources of problems:
(1)dust, (2)sulfur, (3)nitrogen, (4)alkali metals. In addition,
some processes, such as fuel cells, have chemical compositional
requirements.

There is an additional range of requirements and issues that
come in on considering liquefaction. These include several properties of the product: stability, octane rating, boiling range,
viscosity, toxicity (if product is for use by the general public).

In summary, it is important to keep the bigger picture in mind
while doing basic research on scientific problems in coal utilization. The researcher in one aspect should make sure someone is taking care of other aspects. This requires that some thought about
the pertinent process engineering aspects should be kept in mind at

all times.

Summary - Remarks of I. Wender

We are facing a new era in coal research. In the next year or so we will have a new situation. We will have in operation substantial pilot plants, e.g. using 250 tons of coal a day. We have to be ready to solve the problems that arise in the operation of these plants. We are not ready, as can be seen by some examples. Problems of construction materials are the main source of plant failure. Materials people dealing in such problems as corrosion and erosion have to be informed about specific process questions. For example, the zinc chloride process for making gasoline was made impractical by failure to deal with horrendous corrosion problems. With regard to corrosion, there are two ways to handle the problems: (1)to develop new materials to withstand corrosion, (2)to avoid the problem by treatment of the coal and/or a process modification.

Another problem area for needed work is instrumentation and control. For example, there is no way presently to really know what is happening in a pipe with two - or three - phase fluid flow going on inside. Another example is the need to know what happens to coal going into a gasifier.

There are research needs in both direct and indirect liquefaction. There are a number of problems for direct liquefaction. Effects that occur during short residence in the preheater should be investigated. Solid-liquid separation is a major problem. It would be worthwhile to know more about nitrogen in coal and its effects, although we probably now have a reasonably good picture of these effects. For both direct and indirect liquefaction, development of a successful gasifier is a key item. (In indirect liquefaction, the gasifier accounts for 70 or 80% of the capital cost; but even in indirect liquefaction - because of the need for hydrogen, it accounts for about a third of the capital cost.) In direct liquefaction, there is a need to gasify residues, which are quite difficult to deal with. Materials problems are very important in developing a reliable gasifier. Further work on molybdenum catalysts are warranted. The main research need in indirect liquefaction, other than work on the gasifier, is the development of more selective catalysts.

ABSTRACTS FOR CONTRIBUTIONS

PRESENTED AT POSTER SESSIONS

A total of 49 poster contributions were presented, divided between two sessions of about 90 minutes each. The abstracts for these poster contributions follow. The letter M preceding the abstract number indicates presentation on Monday, and T indicates presentation on Tuesday. Several abstracts were withdrawn, and the 49 abstracts printed here are only for those presentations actually given at the Conference.

M-1 CHEMICAL CHARACTERISTICS OF AMERICAN COALS

Dr. Swadesh Raj
Ebasco Services, Incorporated, 2 Rector Street, New York, NY 10006

ABSTRACT

Coal seams commonly consist of a number of superimposed layers, each of which may consist of different mixtures of macerals and minerals. The physical and chemical characteristics of these lithotypes are little understood. These lithotypes can possess significantly different processing characteristics. In this paper experimental results on some of the chemical characteristics of the lithotypes will be presented. The effect of these chemical characteristics will be corelated to the coal utilization industry. The results of the experimental investigation will also be corelated to the rank, age and petrographic composition of the coal.

M-2   A NEW PROCEDURE FOR THE SEPARATION OF COAL MACERALS*

G. R. Dyrkacz, C. A. A. Bloomquist, L. H. Fuchs and E. P. Horwitz
Argonne National Laboratory, Argonne, IL 60439

ABSTRACT

We are currently applying density gradient centrifugation (DCG) techniques to the separation of the maceral constituents of coal. The DGC technique has several advantages over the often used sink-float method. DGC permits rapid separation of 2-3 g of coal into the component maceral groups and provides a visual means for observing the maceral content and densities of the maceral species.
Prior to separating the macerals by DGC, the coal samples were ground to ~ 3μm average particle size using a fluid energy mill and nitrogen carrier gas. Maceral separation was performed by layering a slurry of the coal on a 1.6 L density gradient of CsCl, whose high and low density (1.0-1.5 g/cc) encompasses the densities of the macerals to be separated. After centrifuging, the density gradient was pumped into a fraction collector. The fractions were microscopically analyzed for maceral content.
We have separated two bituminous coals containing substantial amounts of exinites and inertinites. Many of the fractions were over 90% pure with respect to a single maceral group. Examination of the separated fractions revealed that exinites and inertinites have broad ranges in density and in H/C ratios in a single coal; thus care should be taken to carefully define the density range of macerals in this group for any chemical or physical properties to be examined.

*This work was performed under the auspices of the Office of Basic Energy Sciences, Division of Chemical Sciences, USDOE.

M-3   ENDOR STUDY OF BITUMINOUS COAL *

I. Miyagawa and C. Alexander
University of Alabama, University, AL   35486

ABSTRACT

This project is intended to develop an effective method for characterizing coal which will supplement the existing characterizing methods. We are using the new electron-nuclear-double-resonance (Negative ENDOR) technique which we have developed for the past six years.[1,2] The unique effectiveness of this method has already been confirmed:[3] three different ingredients of bituminous coal give entirely different ENDOR signals, although the electron spin resonance (ESR) signals from these samples are practically identical in both line shape and intensity. These ENDOR signals were further studied in the present investigation. It was found that the signal from the vitrain ingredient increased dramatically with increasing intensity of the radio frequency (RF) field. This result is quite different from the case of the radiation-produced radicals, for which the signal intensity saturates at a high RF field. It was also shown that the shape of the signal depended on microwave power. In the case of the clarain ingredient, the ENDOR signal suggested that no significant exchange interaction exists between the radicals. On the basis of these results, some characteristics of the radicals in each ingredient discussed have been determined, and will be presented.
*Research supported by a grant from The University of Alabama School of Mines and Energy Development

REFERENCES
1. I. Miyagawa et al., J. Magn. Reson. 10, 156 (1973).
2. Y. Hayashi and I. Miyagawa, J. Magn. Reson. 28, 351 (1977).
3. I. Miyagawa and C. Alexander, Nature 278, 40 (1979).

M-4   MATRIX ENDOR STUDIES OF THE CARBONIZATION OF
WESTERN CANADIAN COKING COALS

P. R. West and S. E. Cannon
Department of Chemistry, University of Victoria,
Victoria, B.C. Canada V8W 2Y2

ABSTRACT

Cretaceous bituminous coals of known rank ($\bar{R}_O$ max, vitrinite reflectance) have been examined by ESR (electron spin resonance)

and ENDOR (electron nuclear double resonance) techniques. Both
highly oxidised (outcrop) and unoxidised mine-run Balmer coal
from the Crowsnest field have been subjected to heat treatment
(200-900ºC), and the matrix proton ENDOR signal studied as a function of applied microwave and rf power. Changes in ENDOR line shape
and intensity are described with particular emphasis on the pre-
softening region of the unoxidised coal. A comparative study of
the carbonization of hvb and lvb coking coal from the Crowsnest is
reported.

M-5 SMALL ANGLE X-RAY SCATTERING STUDY OF THE POROSITY IN COALS*

P. W. Schmidt, M. Kalliat, and C. Y. Kwak
Physics Department, University of Missouri, Columbia, MO 65211

ABSTRACT

Small-angle scattering curves have been obtained for some
Pennsylvania State University PSOC coal samples and for several
other coals. The x-ray scattering data provide information about
the porosity in the coals and suggest that there are three classes
of pores, which have average dimensions of the order of 1000 Å,
30 Å, and less than 5 Å, corresponding to the macropores, transition
pores and micropores discussed by Dubinin. The principal factor
determining the form of the scattering curves has been found to be
the rank of the coal. In coals of all ranks, the specific surface
associated with the macropores is about 1 to 10 m^2/gm. The micropores are most highly developed in high-rank coals. Comparison of
the x-ray and adsorption results suggests that x-ray scattering and
nitrogen adsorption detect only the specific surface of the macropores and transition pores, while carbon dioxide adsorption measures
the total porosity from the micropores. Scattering data have also
been recorded for a series of coals which had been tested for their
suitability for conversion to liquid fuels. All the coals which
were well-suited for producing liquid fuels were found to have a
well-developed transition pore structure, while coals which were not
especially good for coal liquefaction processes had almost no
transition pores.

*Work supported by the U.S. Department of Energy.

M-6   X-RAY DIFFRACTION STUDIES OF FRAMBOIDS

A. H. Stiller, A. S. Pavlovic, and J. M. Cook
West Virginia University, Morgantown, WV 26506

ABSTRACT

Approximately one hundred framboids from coals, shales and

lake bottom sediments have been individually examined by x-ray diffraction (XRD) to determine their composition and their structure. Fresh, lake bottom framboids were largely composed of greigite accompanied with kaolinite. Aged, lake bottom framboids were mixtures of greigite and pyrite. Coal and shale framboids are predominately composed of iron pyrite either entirely or accompanied with other compounds. Occasionally, framboids were found entirely encased in quartz. Some framboids were composed of materials that could not be identified. The pattern of one such framboid was indexed with a tetragonal cell having the dimensions a = 4.1288 Å and c = 8.2400 Å. The Laue XRD patterns indicated that the microcrystallites of framboids were randomly oriented; no preferential orientation was found. It is concluded from these patterns which exhibit spots plus a continuous halo that the cementing material is a fine grain version of the same material of which the framboid is composed.

M-7   COAL CHARACTERIZATION BY X-RAY DIFFRACTION

A. S. Pavlovic, Jason M. Cook, J. J. Renton
West Virginia University, Morgantown, WV 26506

ABSTRACT

An attempt has been made to use the halo portion of the x-ray diffraction pattern of whole coals to characterize the constitution oc coals. After the mineral peaks and background portion of the x-ray pattern were subtracted from the patterns of coals of known properties, the Fourier transforms of the halos were calculated. The position of the transform peaks and the area under the transform peaks were correlated with known properties of the coals, such as % vitrinite, % mineral, % volatiles, Rank, BTU, % fixed-carbon. In general, it was found that the position of the Fourier transform peaks did not correlate with the characterization properties. However, the area under the Fourier peaks did give good correlations for some of the properties. These results will be given and discussed.

M-9   TRACE ELEMENT ANALYSIS OF COAL BY PROTON-INDUCED
X-RAY EMISSION (PIXE)

D. C. Buckle and G. C. Grant
Virginia Associated Research Campus, Newport News, VA 23606

ABSTRACT

PIXE provides simultaneous analysis of up to 80 elements with a single sample aliquot. An efficient hybrid statistical tech-

nique for interpreting the x-ray spectra has been developed. The
sensitivities vary smoothly with atomic numbers Z from .1 to 10
ng/cm^2 for elements above Z=20. Comparison of PIXE to Atomic
Absorption is made and good agreement is shown. Analysis results
for powdered and digested coal samples are presented along with
analysis of several USGS rock and sediment samples. PIXE is
found to be a cost-effective method of analysis, especially for
studies which require simultaneous detection of many trace
elements.

M-10   A RAPID, DIRECT METHOD FOR DETERMINING ORGANIC SULFUR
CONTENT IN COAL

T. D. Davies and R. Raymond, Jr.
Resource Characterization Group, Los Alamos Scientific Laboratory,
Los Alamos, NM 87545

ABSTRACT

A recently developed analytical method using the EPM (electron
probe microanalyzer) measures organic sulfur directly and avoids
the uncertainty of calculating organic sulfur content by difference. Analysis of 18 different coals (from 10 states, ranging
in rank from subbituminous C to low vol. bituminous, and in age
from Pennsylvanian to Paleocene) shows that organic sulfur contents
of coals can be calculated from the organic sulfur contents of the
vitrinite in the coals. This empirical relationship exists because
vitrinite is the dominant maceral type in most coals and generally
has a sulfur composition intermediate between exinites and
inertinites.

For EPM organic sulfur analysis, representative samples (-20
to -60 mesh) are potted in 1" diameter epoxy pellets, polished,
and carbon coated. Vitrinite grains are identified during analysis
by shape and texture, with results equivalent to oil immersion,
reflectance techniques. According to t-statistics, analyzing 15
vitrinite grains achieves a maximum variability of less than 0.20
wt % from the "true" organic sulfur content of coals containing less
than 2.00 wt % organic sulfur. Neither operator experience nor
variation in coal composition and rank appear to bias results.

A major problem with the EPM technique has been finding a
suitable EPM sulfur standard. Recently, in cooperation with L. A.
Harris, Oak Ridge National Laboratory, we have prepared a good
standard, derived from a petroleum coke, which is stable under an
electron beam and contains a uniform sulfur content.

M-11 A COMPARATIVE STUDY OF THE PROPERTIES OF MARCASITE AND PYRITE*

M. S. Seehra and M. S. Jagadeesh

Physics Department, West Virginia University, Morgantown, WV 26506

ABSTRACT

Marcasite (FeS_2) and pyrite (FeS_2) are major sources of Fe and S-bearing mineral matter in most coals. Recent studies have indicated[1] that these minerals may act beneficially as catalysts in coal conversion (of course, their negative role in pollution is well-known). In order to better understand the roles of these minerals in various technological processes, we have, in recent years, studied the various properties (magnetic, dielectric, optical) of pyrite[2] and similar studies on marcasite are currently underway. Magnetic susceptibility (χ) of a natural sample of marcasite, studied from 4.2 to 380 K[3], suggests low-spin configuration for Fe^{2+} and Van Vleck paramagnetism, somewhat similar to the case of pyrite. However, near 300 K, χ(marcasite) $\simeq 3\chi$(pyrite), suggesting large differences in the crystal-field splittings in the two cases. This is further confirmed by the recent electrical resistivity (77-400 K) measurements which yield the band gap $E_g \simeq 0.33$ eV (for pyrite $E_g \simeq 0.84$ eV[2]), the first such estimate for marcasite. This large difference in the values of E_g and χ for the two minerals, though consistent with each other, suggests different transformation properties in coal conversion processes in the two cases.

*Supported partly by the Energy Research Center, WVU.

1. P. H. Given et al, Fuel 54, 34 (1975); D. Gray, ibid 57, 213 (1978).
2. M. S. Seehra et al, Phys. Rev. B19, 6620 (1979) and references therein.
3. M. S. Seehra and M. S. Jagadeesh, Phys. Rev. B20, 3897 (1979).

M-12   COAL MINERAL ANALYSES AT THE COAL RESEARCH BUREAU
USING THE MULTIPLE ANALYSIS TECHNIQUE

W. Grady, D. Gierl, and T. Simonyi
Coal Research Bureau, College of Mineral and Energy Resources
West Virginia University, Morgantown, WV 26506

ABSTRACT

The Coal Research Bureau at West Virginia University is currently using a multiple analysis technique to examine minerals in coal. This technique is unique because it involves examining the

minerals in a coal sample utilizing five independent methods:
(1) X-ray powder diffraction, (2) infrared spectroscopy, (3) scanning electron microscopy with energy dispersive X-ray analysis, (4)
optical petrography, and (5) normative mineral calculations based on
elemental analyses. This technique of multiple analyses produced
both more accurate identifications and quantifications than would
have been possible with only one or two of the methods, and additionally important information on mineral compositions, associations,
and sizes and shapes.

Application of this multiple analysis technique to the
Pittsburgh coal of northern West Virginia revealed the presence of
the minerals illite, kaolinite, quartz, feldspars, muscovite,
calcite, dolomite, apatite, bassanite. gypsum, pyrite, marcasite,
hematite, and rutile. X-ray powder diffraction, infrared spectroscopy, and normative calculations were used to quantify most of these
minerals, while optical petrography and scanning electron microscopy
were used to identify minor minerals and to determine mineral
associations and morphologies.

M-13   ISOTHERMAL STUDIES OF THE PLASTIC STATE OF COAL

Linda Petrosky Yates, Henry E. Francis and William G. Lloyd
Kentucky Center for Energy Research Laboratory
P. O. Box 13015, Lexington, KY 40583

ABSTRACT

The plasticity of bituminous coals in the range 350-500°C is
of critical importance in thermomechanical fluidization such as is
required for coal pumping by heated screws, and in coal hydrogenly-
sis in the absence of added solvent.

Coal in its plastic state can be viewed as a suspension of
solids dispersed in a viscous liquid continuum. Equations defining
the rheology of solid suspensions are applied to the data obtained
from isothermal Gieseler plastometry.

These relationships are tested by measuring the effects of
the addition of stable liquid and solid phases upon the isothermal
plastic curves of bituminous coals.

In addition, we have applied a simple three-parameter model
to analyze the shapes of the isothermal plastic curves, providing
quantitative estimates of the effects of temperature and additives
upon the magnitudes of the melting and coking constants.

M-14 THE IMPERMEABILITY CHARACTERISTIC OF CAKING COALS UPON HEATING

Zhaoxiong Wang* and James K. Shou
Institute for Mining and Minerals Research
University of Kentucky, Lexington, Kentucky 40583

0094-243X/81/700449-01$1.50 Copyright 1981 American Institute of Physics

ABSTRACT

    The present study has set forth a hypothesis of impermeability
characteristics of caking coals. The impermeability of coal in
plastic state to gas is an extremely important property of caking
coals. The formation of impermeable plastic mass is a necessary
condition for the caking and coking process. The impermeability
creates a cage effect for physiochemical surface process which is
caused by and promotes complicated chemical interractions between
pyrolysis products. The determination of impermeability in plastic
state of coal along with other related parameters should be beneficial for making an overall evaluation of caking behavior. Based on
the penetrative plastometer, a modified experimental instrument was
designed and constructed. The unique advantage of measuring
impermeability of coal in its plastic state by this instrument is
providing a means to synthesize the impermeability with other data
such as volumetric shrinkage and coke quality. The parameters for
identifying impermeability of various coals have been recognized as
the maximum value of resistance, the initial temperature of intense
rising resistance and the temperature of maximum resistance. The
impermeability of coals in plastic state depends on their original
properties. Nevertheless, experimental results had shown that the
impermeability could be improved by way of regulating processing
conditions. This in turn may be helpful to broaden the marketability
of otherwise inferior caking coals.

M-15   INVESTIGATION OF STRUCTURAL DEFORMATION OF
                         COAL PARTICLE IN PYROLYSIS

                         M. Chiou and H. Levine
                       JAYCOR, Del Mar, CA   92014

ABSTRACT

    A model study of the rheological behavior of the coal melt is
proposed. Effects of intragranular heat and mass transfer, particle
size, porosity, deformation of viscous melt, etc., are considered,
so as to explain the striking phenomena of swelling and shrinking
deformations for a single particle at various heating rates. This
model gives both qualitative and quantative descriptions of the transient swelling characteristics, particularly the "cenosphere" formation during the pyrolysis of bituminous coal. The theoretical
predictions show reasonable agreement with the experimental observations of microscopic studies.

M-16   THE ELECTRICAL PROPERTIES OF COAL SLAG

Richard Pollina and Raymond Larsen
Montana State University,†
Physics Department, Bozeman, Montana 59717

ABSTRACT

The United States MHD Program has set as its goal, the development of a fullscale, coal-fired, MHD power plant some time in the
mid to late 1980's. Because of the severity of the MHD environment,
systematic materials characterization is necessary when studying
high-temperature materials, especially for use with coal. Experiments are underway to characterize Montana coal slags and the
ceramic materials they come into contact with. The inorganic constituents of coal remaining after high temperature combustion in an
MHD (Magnetohydrodynamic) power plant combustor form an iron-rich,
"dirty" glass whose electrical properties are important to the
operation of the MHD generator. In particular, alkali "seed"
(K_2CO_3) is added to enhance the conductivity of the plasma so the
slag layer which coats the walls and electrodes of the generator is
rich in K_2O. We present results of a systematic study of the
electrical conductivity of a Rosebud coal ash with graded amounts
of K_2CO_3 added. At high temperatures, the conductivity curves
are smooth with several ionic species contributing. At lower
temperatures the curves become more complex with the presence of
crystalline phases in the glass.

Supported by U.S.D.O.E. Contract

†ET-79-C-01-3087

M-18 ELECTRON BEAM IONIZATION FOR COAL FLY ASH PRECIPITATORS*

R. H. Davis, W. C. Finney and L. C. Thanh
Department of Physics, Florida State University
Tallahassee, Florida 32306

ABSTRACT

The continued importance of pulverized coal combustion will
depend in large measure on the economical improvement of particulate
control technology. Two attributes of electron beam ionization are
of major importance to the feasibility of electron beam precipitators for coal fly ash removal. First, copious ionization is produced by electron beams and ion current densities have been
measured which are 500 times larger than typical values for conventional precipitators. Second, electron beam technology provides
new flexibility in system design because of the wide choice which

can be made of the electron beam energy, beam current, and ionization zone geometry. The combustion of low sulfur coal produces a
fly ash of high resistivity which frequently impedes the charging
of fly ash particles and facilitates back corona in a particulate
layer. The features of electron beam ionization offer unique
approaches to the problem of high resistivity coal fly ash which
include new pre-charger deisgns.

*Supported in part by DOE Contract #ET-78-S-01-3199

M-19   OBSERVATION OF PARTICLE TRAJECTORIES AND INERTIAL
        EFFECTS IN A REGULAR LATTICE OF MAGNETIZED FIBERS

W. Lawson and R. Treat
Physics Dept., West Virginia University, Morgantown, WV 26506 and
Morgantown Energy Technology Center, USDOE, Morgantown, WV 26505
and F. Zeller, III,
Physics Dept., West Virginia University, Morgantown, WV 26506

ABSTRACT

   High gradient magnetic separation (HGMS) is a selective filter-
ing technique in which entrained magnetic particles are preferentially removed from a liquid or gaseous media by a ferromagnetic
filter mesh in an external magnetizing field. Among the applications proposed for this technique are pyrite removal from feed coal
and concentration of magnetic fractions of coal flyash for waste
recovery.
   The work presented here is part of an investigation of the
capture of small paramagnetic particles from a gaseous flow by a
regular square array of parallel ferromagnetic fibers. The trajectories of a 20-25 μm diameter MnO_2 particles are directly observed
in the fiber array via high speed cinephotomicrography. In the
geometry considered the fluid flow and external magnetizing fields
are parallel and the fiber array is perpendicular to both the flow
and magnetic fields. The Stokes number for particle-fluid interactions is greater than unity and inertial effects are seen to be
significant. Single fiber capture cross sections are measured in
the fiber array for varying flow velocities and magnetic field
strengths. The work also has implications for particle loading and
fiber shadow effects.

M-20   MAGNETIC SEPARATION, THERMO- AND MAGNETOCHEMICAL
        PROPERTIES OF COAL LIQUID RESIDUES*

E. Maxwell and D. R. Kelland
MIT-Francis Bitter National Magnet Laboratory, Cambridge, MA 02139**

453

I. S. Jacobs and Lionel M. Levinson
GE Corporate Research and Development, Schenectady, NY 12301

ABSTRACT

Iron sulfides in the form of ferrimagnetic pyrrhotite have been removed from solvent refined coal by the use of high gradient magnetic separation. The results of such separations have been correlated with those of studies of the thermochemical and magnetochemical properties of the undissolved solids present in the product of coal liquefaction. Complete inorganic desulfurization was achieved at the same temperatures at which the solids exhibit a maximum magnetization, at about 230 °C.

The preferred form for the highest magnetization is the monoclinic structure with a composition of $Fe_{0.875}S$. Where the sulfides do not occur in this form, it has been possible to convert them to the high magnetization state by exposure to a hydrogen sulfide atmosphere at an elevated temperature. To approximate the temperature cycle experienced by the liquid coal coming from the reactor toward the filter, samples were heated above the Curie point to destroy the magnetization and then cooled to about 260 °C for H_2S treatment. Measurements on the solids subsequently extracted showed a striking increase in the magnetization compared to the untreated material.

*Supported in part by Electric Power Research Institute.
**Supported by the National Science Foundation.

M-21   MICROSPECTROSCOPY OF COAL

J. S. Gethner
Corporate Research-Science Laboratories
Exxon Research and Engineering Co., Linden, NJ 07036

ABSTRACT

It has recently become popular to describe coal structure as a macromolecular polymer network. In order to quantitatively test this model and to study the influence of the physical structure of coal on liquifaction and gasification reactions, it is desirable to develop spectroscopic techniques which can be used to probe network and molecular properties. This has been difficult due to the high absorption of coal and heterogeneity of samples. We have used microspectrophometric techniques to measure absorption spectra in the near-UV and visible using a scanning microspectrophotometer and in the IR using a Fourier Transform IR spectrometer. Samples were prepared with thicknesses ranging from 500Å to 2μm and areas to 0.5 mm^2.

Broad featureless absorptions are observed in the near-UV and

454

visible and cannot be solely explained by either light scattering from the inhomogeneous coal network or the absorption of a heterogeneous misture of organic components.

The IR spectra show well resolved features and computer techniques can be used to eliminate background due to light scattering artifacts.

M-22 V^{4+} PARAMAGNETIC RESONANCE AS A PROBE OF THE STRUCTURE
AND DYNAMICS OF COAL RELATED FUELS

N. S. Dalal
Chemistry Department, West Virginia University, Morgantown, WV
26506

ABSTRACT

Vanadium is usually present as the VO^{2+} ion in fossil fuels. Because of its paramagnetic state (S = 1/2, I = 7/2), several electron paramagnetic resonance (EPR) studies of VO^{2+} in fossil fuels have been reported. One important conclusion mentioned in some studies is that EPR is not a good analytical technique for studying vanadium in fossil fuels. Another conclusion is that vanadium complexes are present in two structurally different forms, each dominating at different temperatures. On chemical grounds, however, such behavior for VO^{2+} complexes is unexpected. We have reexamined the literature data and find that several of the above discrepancies can be traced to the earlier analysis procedures. A critique of earlier vanadium EPR studies and details of our new results will be presented.

M-25   ON THE RAPID ESTIMATION OF % ASH IN COAL FROM SILICON
CONTENT OBTAINED VIA FNAA, XRF, OR SLURRY-INJECTION AA

D. G. Hicks, J. E. O'Reilly, and D. W. Kopenaal
Depts. of Chemistry, Georgia State University, Atlanta, GA   30303
and University of Kentucky, Lexington, KY   40506
and Institute for Mining & Minerals Research
Ironworks Pike, Lexington, KY   40583

ABSTRACT

The high-temperature ash levels of some typical U.S. bituminous coals can be rapidly estimated with a standard error (absolute) of 1.7 % ash from simple linear correlations between % Si and % ash. Si contents may be rapidly determined by Fast Neutron Activation Analysis (FNAA), X-Ray Flourescence (XRF), or Slurry-Injection Atomic Absorption Spectrometry (SIAAS) methods requiring little more sample preparation other than a brief grinding and weighing. These three instrumental methods showed roughly equivalent correlations. In contrast to earlier reports, logarithmic plots do not produce improved estimates. Also, correlations do not seem to be significantly improved by selecting coals within one rank or from one geographical area.

The slurry-injection atomic absorption method does not require significant modification of existing commercial AA instruments. The conventional method for determining ash content of coal, though simple and accurate, requires a considerable amount of time in the laboratory. Ash content of coals is an important parameter related to their eventual use as fuels, as feedstocks in liquefaction and gasification processes, and in metallurgical processes.

M-27   COMPOSITION AND PROPERTIES OF JET AND DIESEL FUELS DERIVED
FROM COAL AND SHALE

J. Solash and R. N. Hazlett
Naval Research Laboratory, Washington, D.C. 20375

ABSTRACT

Important properties controlling the availability and efficient use of fuels for Navy aircraft and ships are a) low temperature properties, b) stability, c) combustion behavior, and d) safety. In general, these critical properties are controlled by composition. Therefore, a variety of instrumental analyses - capillary gas chromatography, liquid chromatography, carbon-13 and proton nmr, and electron impact and field ionization mass spectrometry - have been applied to jet and diesel fuels made from coal and shale. The low temperature properties are controlled by n-alkanes and the combustion behavior is degraded by aromatics as well as partially saturated polynuclear aromatics. Fuel stability is degraded by sulfur and nitrogen compounds, both of which are prevalent in middle distillate fuels derived from alternate fossil fuel sources.

M-28 ELECTROCHEMICAL BEHAVIOR OF COALS, H-COAL LIQUIDS & Fe^{++} ION

R. P. Baldwin, K. F. Jones, J. T. Joseph, and J. L. Wong
Department of Chemistry, University of Louisville
Louisville, KY 40292

ABSTRACT

Recently, Coughlin et al. have shown that a redox couple involving the oxidation of coal and the reduction of H$^+$ at the cathode and primarily CO_2 at the anode. We have examined the elec-

trochemical behavior of various Kentucky coals and H-Coal liquids
at platinum electrodes under voltammetric and electrolytic conditions, and found that the reported electrochemical process is not
a general one, and that Fe^{++} may yield a similar phenomenon. Thus,
agueous o.1M LiClO4 slurries of most coals were found to yield
characteristic voltammograms containing anodic waves starting at
+0.4 volts vs. SCE and extending throughout the positive potential
range. The resulting oxidation currents were proportional in
magnitude to the coal concentration employed. Upon filtration, the
bulk of the electro-activity of the coal slurry was found to be
retained in the filtrate while the remaining coal residue showed
drastically diminished currents upon formation of a new slurry.
Atomic absorption analysis of the filtrate revealed iron concentrations in the parts-per-thousand range. It was determined that the
half-wave potential of Fe^{2+} occurred at +0.45 volts under the
conditions employed. Also, acetonitrile coal slurries and
acetonitrile solutions of H-Coal liquids exhibited no larger
electrolysis currents than were obtained with blank solutions.

M-29   MOSSBAUER EFFECT STUDY OF VICTORIAN BROWN COAL

J. D. Cashion and B. Maguire
Department of Physics, Monash University, Clayton
L. T. Kiss
S.E.C. Herman Research Labs., Richmond, Vic., Australia

ABSTRACT

[57]Fe Mössbauer spectra have been taken of bed-moist and dried
samples of Latrobe Valley brown coals. The samples contain iron
concentrations of 0.07-1.8% and cover the known variations in the
field. The bed-moist samples (67% water) typically gave several
poorly resolved doublets in the region -0.3 to + 0.6 mm s^{-1} w.r.t.
Fe metal, with maximum absorption dips of 0.1-0.3%. In contrast
to work on higher rank, higher sulphur U.S. coals, only one of
our coals had detectable amounts of pyrite. The remaining absorption lines could not be assigned to any known minerals or clays.
The isomer shift and quadrupole splittings could correspond to
either low-spin Fe^{II} or to trivalent Fe and we believe that they
are all due to hydrated organically bonded iron.
Measurements at 78 K on one sample which was allowed to dry in
air ($\sim$20% water) showed an increase of a factor of 10 in the
absorption (Debye-Waller factor) and a completely changed spectrum.
We take this as evidence that the iron atoms have principally
water ligands in the bed-moist state. Dried and briquetted
samples gave a large intensity (3%) doublet due to organically
bonded Fe and one or more washed-out hyperfine fields of 55 T,
48 T and 33 T probably corresponding to haematite, goethite and
organically bonded iron respectively. One sample also contained
szomolnokite as an oxidation product from pyrite.

T-2 DESULFURIZATION OF HOT COAL GASES BY REGENERATIVE SORPTION

V. M. Jalan and D. Wu
Giner, Inc., Waltham, MA 02154

ABSTRACT

With an objective to contribute to the integration of coal gasifier with advanced power generation systems, such as molten carbonate fuel cells, this study has investigated high temperature regenerable desulfurization processes in which the H_2S content of coal gases is reduced from 200 ppm to 1 ppm. Commercially available processes involved very low temperature scrubbing prior to use in the fuel cells and , consequently, introduce penalties in capital cost and system efficiency.

As a result of a systematic thermodynamic screening, four candidates (ZnO, V_2O_3, Cu, and WO_2) were identified for intermediate to high temperature (350-700°C) desulfurization of fuel gases derived from coal. Of these, ZnO was experimentally studied using a bench scale, isothermal packed bed reactor. It was demonstrated that ZnO can reduce the sulfur levels to less than 1 ppm from coal gases at 650°C, and it can be completely regenerated to ZnO. However, severe decrease in sulfur capacity at high temperatures and further degradation upon regeneration were observed.

Electron microscopy, microanalysis, and surface area measurements were obtained and examined in conjunction with a pore plugging model for this type of gas-solid reactions. Major causes for the observed decrease in sulfur capacity of the sorbent were shown to be the pore plugging during sulfurization and sintering during regeneration reaction. Further research to solve these two problems is continued.

T-3 AIR/WATER OXIDATIVE DESULFURIZATION OF COAL AND SULFUR-CONTAINING COMPOUNDS

R. P. Warzinski, S. Friedman and R. B. LaCount
U. S. Department of Energy, Pittsburgh Energy Technology Center
Pittsburgh, PA 15213

ABSTRACT

Air/water Oxydesulfurization has been demonstrated in autoclave experiments at the Pittsburgh Energy Technology Center for various coals representative of the major U.S. coal basins. The applicability at present of this treatment for producing an environmentally acceptable coal has been restricted by recently proposed SO_2 emission standards for utility boilers. The product would, however, be attractive to the many smaller industrial coal users

who cannot afford to operate and maintain flue gas desulfurization
systems. It is also possible that the utility industry could realize
a benefit by using chemically cleaned coal with partial flue gas
scrubbing. The higher cost of the cleaned coal would be offset by
the reduction in capital and operating costs resulting from decreased
FGD requirements. The susceptibility of sulfur in coal to oxidative
removal varies with the nature of the sulfur-containing species.
The inorganic sulfur compounds, primarily pyrite, marcasite, and iron
sulfate, are more amenable to treatment than the organically bound
sulfur which exhibits varying degrees of resistance depending on its
chemical environment. Air/water $O_x$ydesulfurization consistently
removes in excess of 90 percent of the pyritic sulfur; the extent
and efficiency of organic sulfur removal however, depends on the
type of coal and severity of treatment used. In general, the
organic sulfur of the higher rank coals exhibits more resistance to
treatment than that of the lower rank coals; however, the accompanying loss in heating value is greater for the latter. Similar treatment of sulfur-containing model compounds further illustrates the
relative susceptibilities of different chemical species to oxidation.
Application of these data to the understanding of the complex
chemistry involved in the treatment of coal is a preliminary step
toward improving the efficiency of Oxydesulfurization.

T-4   LIQUID SULFUR DIOXIDE TREATMENT OF COAL:
COMMINUTION, EXTRACTION, AND DESULFURIZATION

D. F. Burow and R. K. Sharma
Dept. of Chemistry, University of Toledo, Toledo, OH 43606

ABSTRACT

Preliminary observations suggest the utility of liquid sulfur
dioxide for desulfurization of coal. Initial exploratory work has
demonstrated that under mild condutions: (1) between 35 and 70% of
the organic sulfur can be removed from selected coals, (2) extraction
of a variety of organic materials from the coal occurs, (3) comminution of selected coals is extensive, and (4) pyrite can be oxidized.
Results obtained as functions of temperature and added reagents will
be presented. Hypotheses concerning the origins of these observations will be considered and present limitations of the methods will
be outlined.

The support of this research by the U.S. Department of Energy
and the samples supplied by the Coal Research Section, College of
Earth and Mineral Sciences, Pennsylvania State University are
gratefully acknowledged.

T-5   REACTION OF BITUMINOUS COAL WITH INORGANIC SOLVENTS

T. S. Miller and A. P. Hagen
Department of Chemistry
The University of Oklahoma, Norman, OK 73019

ABSTRACT

Anhydrous ammonia, methanol, and other protonic reagents have
been utilized for the chemical comminution of coal. We have found
that Oklahoma bituminous coal is readily comminuted by
triflouroacetic acid, difluorophosphoric acid, and the mixture
acetic acid – acid anhydride – sulfuric acid. Approximately 5% of
the original coal dissolves during the TFA treatment with a much
greater loss of usable fuel occurring with the acetic acid mixture.
TFA reduces the pyritic as well as the organic sulfur by 50% and
the ash by 50%. The acetic acid treatment leaves a solid product
which is high in ash and sulfur. Micrographs after treatment with
TFA dramatically show the layered structure of the coal.

T-6   BROMINATION OF ANTHRACITE

C. G. Woychik* and D. D. L. Chung†
Department of Metallurgy and Materials Science
Carnegie-Mellon University, Pittsburgh, PA 15213

ABSTRACT

Bromination of anthracite has been performed by exposure of
anthracite to bromine vapor. Bromine absorption was evident by
weight uptake. On further bromination, anthracite disintegrated
into a fine powder. Differential scanning calorimetry has been
used to characterize the thermal behavior of anthracite before and
after bromination. The measurement was made by using the Perkin-
Elmer Model DSC-2. Combustion of anthracite was achieved by
heating the sample in the presence of a flow of oxygen at 40 psi.
The exothermic peak which is associated with combustion was
observed to commence at $868^\circ K$ in pristine anthracite and at $890^\circ K$
in brominated anthracite. The height of the peak is independent
of bromine concentration, but the width of the peak decreases
significantly with increasing bromine concentration.

*Supported by Benedum Foundation

†Also of the Department of Electrical Engineering

T-8   CHARACTERIZATION OF THE OXIDATION OF
           BITUMINOUS COAL BY ^{13}C n.m.r.

J. A. MacPhee and B. N. Nandi
Energy Research Laboratories, Ottawa   K1A OG1

ABSTRACT

     The low-temperature air oxidation of coal has been studied
in relation to the changes produced in coking properties. For this
study a Cape Breton Development Corporation high-volatile bituminous
coal (Harbour 26 seam) was selected. During the course of the
oxidation in air (105°C) modification of coal properties and structure were monitored by means of the Ruhr dialatometric test, and
elemental and spectroscopic analyses. Micrographs of the fresh and
oxidized coals reveal the extent of the oxidation in a qualitative
manner. Infra-red spectra have been found to yield little information in this area. In contrast to this it was found that the
recently developed ^{13}C n.m.r. technique using cross polarization
and magic angle spinning is particularly applicable to the examination of oxidized coals.

T-9   REFRACTORY OXIDES FOR HIGH-TEMPERATURE
           COAL-FIRED MHD AIR HEATERS

Richard J. Pollina
Montana State University†, Bozeman, MT   59717
Ronald R. Smyth
FluiDyne Engineering Corporation*, Minn., MI   55422

ABSTRACT

     In a coal-burning MHD (Magnetohydrodynamic) power plant, the
use of a regenerative air heater is one option for producing the
required high plasma temperatures. However, thermal, mechanical
and chemical stresses which the MHD environment places on materials
are enormous and candidate materials must be carefully tested. We
report on the results of such testing for a variety of materials
studied in heater test facilities at both FluiDyne Engineering
Corporation and at Montana State University. We describe the
experimental program and test facilities for MHD heater development,
the rationale for the choice of materials, and their performance
(success or failure) under various environments (slagging and non-
slagging). Future materials needs are discussed in light of the
observed effects.

Supported by U. S. D.O.E. Contract

†ET-79-C-01-3087
*Et-79-C-01-3005

T-11   APPLICATION OF A MULTIPLE-BEAM LASER DOPPLER VELOCIMETER
TO MEASURE VELOCITY DISTRIBUTIONS OF A FLUIDIZED BED

E. J. Johnson and M. E. McDonnell
West Virginia University, Morgantown, WV 26506

ABSTRACT

Coal gasification is most efficiently performed in a fluidized
bed, but the inability to mathematically describe the process has
hampered scaling up the gasifiers. New techniques to measure the
velocity distribution of the fluidized particles may provide data
to critically evaluate models for scaling. This velocity profile
can be obtained with a laser Doppler velocimeter (LVD) that is
fast, simple, accurate and nonperturbing to the flow. The technique has been demonstrated with a multiple-beam LDV developed in
this lab on a two inch diameter by eight foot high fast fluidized
bed. In this instrument, a large number of beams intersect to form
sharp, intense spots well separated by dark area, rather than the
sinusoidally varying intensity pattern of the conventional two beam
LDV. This modification provides a less ambiguous description of
velocities. From first principles the correlation function of particles moving through the measuring volume is calculated and
experimentally verified with particles traveling at a uniform
speed. This technique is then used to map out the velocity profile
in the fluidized bed as a function of height in the bed, proximity
to the wall, and pressure drop over the bed.
(This work performed with support from the Energy Research
Center of West Virginia University.)

462

P. S. Virk and D. J. Ekpenyong
Department of Chemical Engineering
Massachusetts Institute of Technology, Cambridge, MA   02139

ABSTRACT

Modern interpretations of coal structure all suggest that it
possesses a molecular topology wherein sigma bonds invariably lie
in close proximity either to aromatic π systems or to electro-
negative hetero-atoms. We hypothesize that such a topology is
well suited to the occurrence of concerted molecular reactions
termed pericyclic which must therefore be intimately involved in
all coal processing. Selected pericyclic reactions of model coal
moieties are then examined by frontier orbital analysis. The
reactions studied include hydrogen transfers, cope re-arrangements,
retro-ene fragmentations, and cheleotropic extrusions, while the
coal model substrates include α,ω-diaryl n-alkanes and α-aryloxy,
ω-aryl n-alkanes both with and without hydrogen donors such as
molecular hydrogen and the two dihydronaphthalene isomers. The
present results accord with the few experimental reactions reported
with coal model compounds; further, the analyses indicate the
frontier orbital interactions of greatest strength and proffer
suggestions for possible catalysis of the reactions involved.

T-15   THE HYDROGASIFICATION OF LIGNITE AND SUB-BITUMINOUS COALS

B. Bhatt, P. T. Fallon and M. Steinberg
Brookhaven National Laboratory, Upton, New York   11973

ABSTRACT

A North Dakota lignite and a New Mexico sub-bituminous coal
have been hydrogenated at up to 900°C and 2500 psi hydrogen
pressure. Yields of gaseous hydrocarbons and aromatic liquids
have been studied as a function of temperature, pressure, residence
time, feed rates and H_2/coal ratio. Coal feed rates in excess of

10 lb/hr have been achieved in the 1 in. I.D. x 8 ft reactor and methane concentration as high as 55% have been observed. A four-step reaction model was developed for the production and decomposition of the hydrocarbon products. A single object function formulated from the weighted errors for the four dependent process variables, CH_4, C_2H_6, BTX, and oil yields, was minimized using a program containing three independent iterative techniques. The results of the nonlinear regression analysis for lignite show that a first-order chemical reaction model with respect to C conversion satisfactorily describes the dilute phase hydrogenation. The activation energy for the initial products formation was estimated to be 42,700 cal/gmole and the power of hydrogen partial pressure was found to be +0.14. The overall correlation coefficient was 0.83. The mechanism, the rate expressions, and the design curves developed can be used for scale-up and reactor design.

T-16   A DISCUSSION OF PHYSICAL AND CHEMICAL
PROBLEMS PERTINENT TO UCG

T. L. Eddy and C. I. Anekwe
Mechanical Engineering & Mechanics
West Virginia University, Morgantown, WV 26506

ABSTRACT

Recent field tests and modeling studies of the underground gasification of coal (UCG) have suggested a number of topical areas which are in need of further investigation. These include linking methods to control link zone geometry; appropriate methods to model reverse combustion link enhancement (2nd pass linking); the cause of plugging (tars or ash?) which occurs in both Eastern and Western coals; allowable flow rates as restricted by permeability, inter-well distance, cross sectional area, pipe restrictions, etc.; appropriate ignition and extinction temperatures for modeling; cavity wall reaction mechanisms; and cavity and link zone (or char) size and location methods. The characteristics and importance of these problems will be presented.

T-17   AN X-RAY DIFFRACTION STUDY OF THE REACTIONS OF ZINC
COMPOUNDS WITH COAL

H. Beall and R. J. Wadja
Department of Chemistry
Worcester Polytechnic Institute, Worcester, Massachusetts 01609

ABSTRACT

The importance of zinc chloride as a catalyst in conversion of

coal to liquid and gaseous fuels is well known and study of the
aspects of its reaction with coal are warranted. Recent studies
have reported the likelihood that certain transition metal
halides, both anhydrous and hydrated, will separate the graphite-
like planes of condensed aromatic carbon rings in the coal to
produce intercalation compounds. The work described in this
paper uses X-ray powder diffraction to investigate the question
of whether the efficacy of zinc chloride as a coal liquefaction
and gasification catalyst could possibly involve the formation of
intercalation compounds as an intermediate step thus providing
a method for reagents to penetrate the solid coal. Evidence is
presented that such compounds may, indeed, be formed. Furthermore, addition of zinc metal reduces the weight of zinc chloride
(or zinc oxide) needed to form the apparent intercalation compounds
with coal. Experiments with magnesium compounds indicate that the
results with zinc compounds and coal cannot be extended to the
analogous magnesium species.

T-18   MODEL COMPOUND STUDIES OF THE MINERAL MATTER CATALYSIS
                        IN COAL LIQUEFACTION

B. C. Bockrath and K. T. Schroeder
U. S. Department of Energy, Pittsburgh Energy Technology Center
                Pittsburgh, PA 15213

ABSTRACT

    The mineral matter naturally occurring in coal is recognized
as a coal liquefaction catalyst. Pyrite is considered to be the
primary catalytic mineral. It is known to convert to pyrrhotite
under reaction conditions. Pyrrhotite is believed to be the
active form of the mineral. Our investigations have shed some
light on the type of reactions which can be catalyzed by the
pyrite/phrrhotite system. The models studied include the
isomerization of cis- to trans-decalin, the reactions of tetralin
to give naphthalene, 1-methylindan and n-butylbenzene, the reduction of flourenone to flourene and the cleavage of 2-benzylphenol
to toluene and phenol. Significant differences in the mechanism
of formation of naphthalene and 1-methylindan from tetralin and
in the catalytic activity of pyrite versus pyrrhotite have been
found.

T-19   SHOCK ACTIVATION OF CATALYSTS*

    R. A. Graham, B. Morosin, P. M. Richards, F. V. Stohl,
                    and B. Granoff
        Sandia National Labs.,† Albuquerque, NM 87185

ABSTRACT

Scientists in the Soviet Union have demonstrated that high
pressure shock-wave loading can cause significant improvement in
the performance of catalysts. This increased catalytic activity
is apparently the result of the shock-induced defects, especially
vacancies, which act to facilitate atomic migration. We have
carried out shock activation experiments on a coal-derived pyrite
which has been previously used as a catalyst in coal liquefaction
studies. The pyrite powder was packed to a density of about 2.0
Mg/m^3 in a copper capsule and explosively loaded to a pressure of
about 15 GPa in the copper. The starting and shock-activated
samples were analyzed by x-ray diffraction and magnetization
measurements. The diffraction patterns of the shock-activated
samples were dominated by broadened pyrite lines indicative of a
significant increase in crystal defects. The diffraction patterns
also showed the presence of pyrrhotite ($Fe_{1-x}S$) in quantities of
a few percent. An iron carbide found in the shocked material was
apparently formed from carbon originating from either the calcite or
organic impurities in the starting material. Magnetic properties
of the sample were found to be substantially changed by the shock
loading. The study has demonstrated that shock loading can
significantly alter the crystalline order of pyrite and produce
measurable quantities of pyrrhotite. The effects of shock-
activated pyrite on the liquefaction of coal are being assessed by
means of tubing reactor experiments.

*This work sponsored by US DOE Contract DE-AC04-76-DPO0789.
†A U. S. Department of Energy facility.

T-20   IMPROVED METHANATION CATALYSTS

C. S. Brooks, G. S. Golden and F. D. Lemkey
United Technologies Research Center, East Hartford, CT

ABSTRACT

New procedures have been developed for preparation of improved
methanation catalysts. Supported metal catalysts have been
prepared, starting with controlled solidification (unidirectional or
gas   atomization) of eutectic or proeutectic alloys of transition
metals (rhodium and nickel) and aluminum, followed by chemical
dissolution in caustic of the aluminum component and acid precipi-
tation of the aluminum as alumina to provide a high surface area
oxide matrix for the dispersed catalytically active rhodium
(0.36 w/o) or nickel (6.1 w/o). Unsupported skeletal nickel
catalysts have been produced by gas   atomization and
rotary   atomization techniques from proeutectic nickel
(28.6 w/o Ni) aluminum alloy powders by caustic activation. Characterization of the supported and unsupported catalysts by x-ray

diffraction, electron microscopy, BET nitrogen adsorption and
hydrogen chemisorption demonstrated that the activated metals,
rhodium and nickel, were in a highly dispersed state and, in the
case of the alumina supported catalysts, present on a high surface
area stabilized substrate. The superior surface physical properties, pore structure, total surface areas and available metal
surface areas, and the superior initial catalytic activity of these
experimental catalysts compared with more conventional rhodium
and nickel supported and Raney^R catalysts establish the merits of
controlled solidification procedures as a new path for synthesis
of methanation catalysts.

T-21   MODELING OF FISCHER-TROPSCH SYNTHESIS

S. Novak
Exxon Research & Engineering Co., Corporate Research Science Labs
P. O. Box 45, Linden, New Jersey 07036

ABSTRACT

    The standard model for the growth and termination of hydro-
carbon chains in the Fischer-Tropsch (FT) reaction is based on a
classical polymer kinetic theory due to Flory. Unfortunately, the
Flory model predicts a low liquid product yield. Although the
weight distributions of hydrocarbon products observed with most
Fischer-Tropsch catalysts do, in fact, fit a Flory distribution,
to accept the Flory model as the definitive model for hydrocarbon
production is to admit that little can be done to improve the
liquid product yield. Consequently, we have postulated additional
chemical processes, which could conceivably occur in FT synthesis,
and we have theoretically determined the effect of these processes
on the weight distribution of products. Certain processes resulted
in a dramatic increase in the liquid product selectivity -- as
much as 2-3 times the selectivity predicted by the Flory model.
Thus, if catalysts can be tailored so as to include these chemical
processes, such catalysts could dramatically increase the liquid
product yield as compared to ordinary FT catalysts.

T-22   A FUNDAMENTAL CHEMICAL KINETICS APPROACH TO
                     COAL CONVERSION

R. E. Miller and S. E. Stein
Department of Chemistry, West Virginia University
Morgantown, WV 26506

ABSTRACT

    The ultimate goal of this research program is to discover

characteristic molecular reaction pathways, and their kinetic
behavior, in high temperature condensed-phase systems containing
aromatic substances. Such information can then be applied to the
thermal chemistry of coal system. We shall present recent progress
in which experimental determinations of product evolution rates
in well-defined "model" systems are analyzed using quantitative
models consistent with available thermochemical and kinetic data.
Two types of systems will be discussed, namely, bond breaking
in tetralin and liquid-phase pyrolysis of pure 1,2-diphenylethane.
Implications of these results for interpreting coal reactions will
also be discussed.

T-23 A GENERAL, QUANTITATIVE, KINETIC MODEL FOR COAL LIQUEFACTION*

Tom Gangwer
Brookhaven National Laboratory, Upton, NY 11973

ABSTRACT

A model for coal liquefaction has been developed which is
based on the concept of the thermal bond cleavage step being rate
limiting. The model provides rate laws which describe the
preasphaltene, asphaltene, oil and gas time/yield curves for the
coal liquefaction process. The quantitative agreement between
the derived rate laws and the kinetic data obtained from fifteen
publications will be presented. A general liquefaction reaction
scheme was developed and used to relate the proposed model to the
experimentally observed products. For the diverse coal liquefaction systems studied, the rate constants obtained from analysis
of the benzeme solubles plus gas/time data vary from ~1 to
46×10^3 min $^{-1}$ for data covering the later stages of conversion
at temperatures between 350° to 450°C. The rate constant ranges
for preasphaltene, asphaltene, oil and gas formations were found
to be 1.0 to 56.0 min^{-1}, 0.03 to 0.1 min^{-1}, 0.003 to 0.1 min^{-1} and
0.007 to 0.2 min^{-1} respectively over the 350° to 450° range. The
rate constant ranges for preasphaltene and asphaltene disappearance
were found to be 0.03 to 0.09 min^{-1} and 0.005 to 0.06 min^{-1} respectively. Based on the quality of the kinetic fit to the reported
coal liquefaction systems, which cover a diverse range of reaction
conditions, coal types and donor solvent compositions, it is
proposed that the thermal bond cleavage/hydrogen capping model
provides a good, quantitative description of the rate limiting
processes occurring during coal liquefaction.

*This work was supported by the Division of Chemical Sciences,
U.S. DOE, Washington, D.C., under Contract # DE-AC02-76Ch0016.

468

T-24

LIQUEFACTION OF KENTUCKY AND ILLINOIS COALS
IN A BATCH MICROREACTOR

D. C. Cronauer, R. G. Ruberto, and D. C. Young
Gulf Research and Development Company
P. O. Drawer 2038, Pittsburgh, PA 15230

ABSTRACT

Bituminous coals from the Illinois No. 6 and Kentucky No. 9
and 11 seams have been liquefied in a batch microreactor using
tetralin and selected catalysts. The levels of coal conversion
to toluene and pentane solubles were observed at a reaction
temperature of 450°C and times between 0 and 60 minutes. The
conversions of Kentucky coal to toluene solubles in the presence of
both Co/Mo and Ni/W catalysts were similar, while the Co/Mo appeared
to be more effective with the Illinois coal. However, in both
cases conversions were higher than those of non-catalytic runs.
Oil yields (pentane solubles) were increased in the presence of
catalyst. The amount of hydrogen transfer as determined by
naphthalene yield appeared to be somewhat independent of the presence
of catalyst. For comparison, product yields of experiments made
with octahydrophenanthrene were higher than those from the above
catalytic runs. This work was done as part of a DOE sponsored
project to study liquefaction with labeled and unlabeled donor
solvents.

T-25

BALL VALVE DESIGN FOR SOLID WITHDRAWAL SERVICE
IN COAL LIQUEFACTION

Alexander J. Patton
Gulf + Western Mfg. Co., Warwick, RI

ABSTRACT

Valves used for drawdown of the solids produced in the coal
liquefaction process must be designed to operate at high temperatures and must prevent the buildup of solids within the valve
internals to insure operability. Consideration must also be given
to materials used for seating surfaces in terms of being able to
withstand abrasion. Research into the design and materials for this
application resulted in a new ball valve design with a special inlet
port liner and O-ring seal. A prototype valve was built and put in
service in a small scale, experimental coal liquefaction plant. For
comparative purposes a standard design valve was also used, with
both valves being subjected to over 1000 open/close cycles, after
which both valves were removed, tested, and disassembled for inspection. The results of this experimental program will be presented,
including process parameters, service conditions, and material

evaluation. Particular emphasis will be placed on wear of seating
surfaces and the effects of process material on the physical
properties of seal and packing materials.

T-26

AGING OF SRC LIQUIDS

T. Hara, L. Jones, K. C. Tewari, and N. C. Li
Duquesne University, Pittsburgh, PA 15219

ABSTRACT

The viscosity of SRC-LL liquid increases when subjected to
accelerated aging by bubbling oxygen in the presence of copper
strip at 62°C. Precipitates are formed and can be separated from
the aged liquid by Soxhlet extraction with pentane. A 30-70 blend
of SRC-I with SRC-LL was subjected to oxygen aging in the absence
of copper, and the viscosity increased dramatically after 6 days
at 62°. The content of preasphaltene and its molecular size
increase with time of aging, accompanied by decrease of asphaltene
and pentane-soluble contents. For the preasphaltene fraction on
aging, gel permeation chromatography shows formation of larger
particles. ESR experiments show that with oxygen aging, spin concentration in the preasphaltene fraction decreases. Perhaps some
semiquinone, together with di- and tri-substituted phenoxy
radicals, generated by oxygen aging of the coal liquid, interact
with the free radicals already present in coal to yield larger
particles and reduce free radical concentration. We are currently
using the very high-field (600-MHz) NMR spectrometer at Mellon
Institute to determine changes in structural parameters before and
after aging of SRC-II and its chromatographically separated fractions.

T-27

LIQUEFACTION REACTIVITY OF EXTRUDED COAL

S. Mori, B. H. Davis, A. W. Fort and W. G. Lloyd
Institute for Mining and Minerals Research, Lexington, KY 40583

ABSTRACT

The reactivity towards liquefaction of a Kentucky #9 seam coal
and its extrudate (from the Jet Propulsion Laboratory's 1.5-inch
coal extruder) was measured in a micro-autoclave. A glass lined
10-cc reactor was designed so that gas pressure and reactor
temperature could be monitored continuously. The use of a
fluidized sand bath furnace, relatively coarse mesh coal, and
reactor geometry designed to promote ebullient mixing, provides
highly consistent results without mechanical agitation. Conversion
reproducibility for all runs has been better than $\pm$ 2%. The

470

thermomechanical history of the extrudate (about 1 min at 425°C) has only a slight effect upon reactivity: coal and extrudate afford similar conversion curves at 435°C, with maxima at less than 15 min. Conversions of coal and extrudate calculated on the basis of fixed carbon are virtually identical.

T-28   SUMMARY REPORT OF THE RAWLINS 1 FOR
         GASIFICATION OF STEEPLY DIPPING COAL BEDS

B. E. Davis and J. E. Miranda
Gulf Research and Development Company
P. O. Drawer 2038, Pittsburgh, Pennsylvania   15230

ABSTRACT

Gulf Research & Development Company has entered into a contract with the Department of Energy (DOE) for the gasification of steeply dipping coal beds. The first test of this program was completed in December of 1979. The test facility will be described and results of the first test will be discussed on the basis of defined test objectives. A brief description of Test 2 will also be presented.

T-29   HYDRATION PROCESS FOR ENHANCED CALCIUM UTILIZATION
           IN FLUIDIZED-BED COMBUSTION

D. S. Moulton, E. B. Smyk, and J. A. Shearer
Argonne National Laboratory, Argonne, IL   60439

ABSTRACT

Fluidized-bed combustion uses our coal reserves to produce steam or electric power in an environmentally acceptable manner, however, the calcium in the sorbent is only partially utilized. A method of increasing the calcium utilization has been discovered in which the spent sorbent is hydrated to reactivate it for additional use. Hydration produces calcium hydroxide which is subsequently dehydrated when the material reenters a combustor. The hydration-dehydration cycle alters the pore structure allowing further sulfation. The effectiveness of the method was verified by tests made in the ANL 15 cm atmospheric fluidized-bed combustor.
   A continuous-bed cooling and hydrating system is proposed to reactivate the spent sorbent and recover both the sensible heat and the heat of hydration. The kinetics of the hydration reaction using steam have been investigated, and test results from a fluidized-bed hydrator are presented.

AUTHOR INDEX

Alexander, C........444
Anekwe, C.I........463

Balanis, C.A.......175
Baldwin, R.P.......455
Beall, H...........463
Beck-Montgomery, S..417
Bhatt, B...........462
Bloomquist, C.A.A...443
Bluhm, D.D.........417
Bockrath, B.C......464
Brooks, C.S........465
Buckle, D.C. 446
Burow, D.F. 458

Cannon, S.E........444
Cashion, J.D. 456
Chiou, M. 450
Chung, D.D.L.......459
Cook, J.M..........445, 446
Cooper, B.R........438
Coughlin, R.W......388
Cronauer, D.C. 468
Curry, K. 154

Dalal, N.S.........454
Davies, T.D. 447
Davis, B.E. 470
Davis, B.H.........469
Davis, R.H.........451
Deno, N.C. 154
Duerbrouck, A.W. ...344
Drykacz, G.R. 443

Eddy, T.L..........463
Ekpenyong, D.J. 462
Essenhigh, R. 309

Fallon, P.T. 462
Fanslow, G.E. 417
Finney, W.C. 451
Fort, A.W. 469
Francis, H.E.......449
Friedman, S. 457
Fuchs, L.H. 443

Gangwer, T. 467
Gethner, J.S.......453
Gierl, D. 448

Given, P..........167
Golden, G.S.......465
Goodman, D.W......235
Graham, R.A. 464
Grady, W..........448
Grandy, D.W. 101
Granoff, B. 291, 464
Grant, G.C. 446

Hagen, A.P........459
Hamblen, D.G......121
Hara, T...........469
Hazlett, R.N. 455
Hickey, J. 388
Hicks, D.G. 454
Horwitz, E.P. 443

Jacobs, I.S.......452
Jagadeesh, M.S. ...448
Jalan, V.M. 457
Johnson, E.J......460
Jones, A.D........154
Jones, K.F........455
Jones, L. 469
Joseph, J.T.......455

Kalliat, M. 445
Kelland, D.R......452
Kelley, R.D.......235
Kiss, L.T. 456
Kopenaal, D.W. 454
Krishnamurthy, S...256
Kwak, C.Y. 445

LaCount, R.B. 457
Lalvani, S. 388
Larsen, J.W. 1
Larsen, R. 451
Lawson, W. 452
Lemkey, F.D. 465
Levine, H. 450
Levinson, L.M......209, 452
Li, N.C. 469
Lloyd, W.G. 449, 469
Lucht, L.M. 28

472

Maciel, G.W......66
MacPhee, J.A.....460
Madey, T.E......235
Maguire, B......456
Markuszewski, R..357
Maxwell, E.......452
McDonnell, M.E...461
McKee, D.W......236
Merten, U.......438
Miknis, F.P......66
Miller, K.J.....344
Miller, R.E.....466
Miller, T.S.....459
Minard, R.......154
Miranda, J.E....470
Miyagawa, I.....444
Montano, P.A....291
Mori, S.........469
Morosin, B......464
Moulton, D.S....470
Mraw, S. 332

Nandi, B.N.....460
Nelson, S.O....417
Novak, S.......466

O'Fallon, N.M...198
O'Reilly, J.E...454

Patton, A.J.....468
Pavlovic, A.S...445, 446
Peppas, N.A....28
Petrakis, L.....101
Pollina, R.J....451, 460
Potter, T......154

Quate, C.F.....141

Raj, S.........443
Rakitsky, W....154
Raymond, R. Jr..447
Renton, J.J....446
Retcofsky, H.L..167
Richards, P.M..464
Rose, K.D.......82
Ruberto, R.G....468

Schmidt, P.W...445
Schroeder, K.T..464
Schlosberg, R.H..167
Scouten, C.G....82

Seehra, M.S......448
Shah, Y.T........256
Sharma, R.K. 458
Shearer, J.A.....470
Shou, J.K........449
Silbernagel, B.G..332
Simonyi, T.......448
Smyk, E.B........470
Smyth, R.R.......460
Solash, J........455
Solomon, P.R.....121
Stein, S.E.......167, 466
Steinberg, M.....462
Stiegel, G.J.....256
Stiller, A.S.....445
Stohl, F.V.......464
Sullivan, M.J....66
Szeverenyi, N....66

Taylor, S.R......344
Tewari, K.C......469
Thanh, L.C.......451
Thompson, R.R....49
Treat, R.........452

Virk, P.S........462

Wadja, R.J.......463
Wagner, K........154
Wang, Z..........449
Warzinski, R.P...457
Wender, I........438
West, P.R........444
Wheelock, T.D....357
Wiser, W.H.......438
Wong, J.L........455
Woodmansee, D.E..438
Woychik, C.G.....459
Wu, D............457

Yates, J.T. Jr...235
Yates, L.P.......449
Yevak, R.J.......154
Young, D.C.......468

Zeller, F. III...452

AIP Conference Proceedings

		L.C. Number	ISBN
No.1	Feedback and Dynamic Control of Plasmas	70-141596	0-88318-100-2
No.2	Particles and Fields - 1971 (Rochester)	71-184662	0-88318-101-0
No.3	Thermal Expansion - 1971 (Corning)	72-76970	0-88318-102-9
No.4	Superconductivity in d-and f-Band Metals (Rochester, 1971)	74-18879	0-88318-103-7
No.5	Magnetism and Magnetic Materials - 1971 (2 parts) (Chicago)	59-2468	0-88318-104-5
No.6	Particle Physics (Irvine, 1971)	72-81239	0-88318-105-3
No.7	Exploring the History of Nuclear Physics	72-81883	0-88318-106-1
No.8	Experimental Meson Spectroscopy - 1972	72-88226	0-88318-107-X
No.9	Cyclotrons - 1972 (Vancouver)	72-92798	0-88318-108-8
No.10	Magnetism and Magnetic Materials - 1972	72-623469	0-88318-109-6
No.11	Transport Phenomena - 1973 (Brown University Conference)	73-80682	0-88318-110-X
No.12	Experiments on High Energy Particle Collisions - 1973 (Vanderbilt Conference)	73-81705	0-88318-111-8
No.13	π-π Scattering - 1973 (Tallahassee Conference)	73-81704	0-88318-112-6
No.14	Particles and Fields - 1973 (APS/DPF Berkeley)	73-91923	0-88318-113-4
No.15	High Energy Collisions - 1973 (Stony Brook)	73-92324	0-88318-114-2
No.16	Causality and Physical Theories (Wayne State University, 1973)	73-93420	0-88318-115-0
No.17	Thermal Expansion - 1973 (lake of the Ozarks)	73-94415	0-88318-116-9
No.18	Magnetism and Magnetic Materials - 1973 (2 parts) (Boston)	59-2468	0-88318-117-7
No.19	Physics and the Energy Problem - 1974 (APS Chicago)	73-94416	0-88318-118-5
No.20	Tetrahedrally Bonded Amorphous Semiconductors (Yorktown Heights, 1974)	74-80145	0-88318-119-3
No.21	Experimental Meson Spectroscopy - 1974 (Boston)	74-82628	0-88318-120-7
No.22	Neutrinos - 1974 (Philadelphia)	74-82413	0-88318-121-5
No.23	Particles and Fields - 1974 (APS/DPF Williamsburg)	74-27575	0-88318-122-3

No.24 Magnetism and Magnetic Materials - 1974
(20th Annual Conference, San Francisco)   75-2647   0-88318-123-1

No.25 Efficient Use of Energy (The APS Studies on
the Technical Aspects of the More Efficient
Use of Energy)   75-18227   0-88318-124-X

No.26 High-Energy Physics and Nuclear Structure
- 1975 (Santa Fe and Los Alamos)   75-26411   0-88318-125-8

No.27 Topics in Statistical Mechanics and Biophysics:
A Memorial to Julius L. Jackson
(Wayne State University, 1975)   75-36309   0-88318-126-6

No.28 Physics and Our World: A Symposium in Honor
of Victor F. Weisskopf (M.I.T., 1974)   76-7207   0-88318-127-4

No.29 Magnetism and Magnetic Materials - 1975
(21st Annual Conference, Philadelphia)   76-10931   0-88318-128-2

No.30 Particle Searches and Discoveries - 1976
(Vanderbilt Conference)   76-19949   0-88318-129-0

No.31 Structure and Excitations of Amorphous Solids
(Williamsburg, VA., 1976)   76-22279   0-88318-130-4

No.32 Materials Technology - 1975
(APS New York Meeting)   76-27967   0-88318-131-2

No.33 Meson-Nuclear Physics - 1976
(Carnegie-Mellon Conference)   76-26811   0-88318-132-0

No.34 Magnetism and Magnetic Materials - 1976
(Joint MMM-Intermag Conference, Pittsburgh)   76-47106   0-88318-133-9

No.35 High Energy Physics with Polarized Beams and
Targets (Argonne, 1976)   76-50181   0-88318-134-7

No.36 Momentum Wave Functions - 1976 (Indiana University)   77-82145   0-88318-135-5

No.37 Weak Interaction Physics - 1977 (Indiana University) 77-83344   0-88318-136-3

No.38 Workshop on New Directions in Mossbauer
Spectroscopy (Argonne, 1977)   77-90635   0-88318-137-1

No.39 Physics Careers, Employment and Education
(Penn State, 1977)   77-94053   0-88318-138-X

No.40 Electrical Transport and Optical Properties of
Inhomogeneous Media (Ohio State University, 1977)   78-54319   0-88318-139-8

No.41 Nucleon-Nucleon Interactions - 1977 (Vancouver)   78-54249   0-88318-140-1

No.42 Higher Energy Polarized Proton Beams
(Ann Arbor, 1977)   78-55682   0-88318-141-X

No.43 Particles and Fields - 1977 (APS/DPF, Argonne)   78-55683   0-88318-142-8

No.44 Future Trends in Superconductive Electronics
(Charlottesville, 1978)   77-9240   0-88318-143-6

No. 45	New Results in High Energy Physics - 1978 (Vanderbilt Conference)	78-67196	0-88318-144-4
No. 46	Topics in Nonlinear Dynamics (La Jolla Institute)	78-057870	0-88318-145-2
No. 47	Clustering Aspects of Nuclear Structure and Nuclear Reactions (Winnepeg, 1978)	78-64942	0-88318-146-0
No. 48	Current Trends in the Theory of Fields (Tallahassee, 1978)	78-72948	0-88318-147-9
No. 49	Cosmic Rays and Particle Physics - 1978 (Bartol Conference)	79-50489	0-88318-148-7
No. 50	Laser-Solid Interactions and Laser Processing - 1978 (Boston)	79-51564	0-88318-149-5
No. 51	High Energy Physics with Polarized Beams and Polarized Targets (Argonne, 1978)	79-64565	0-88318-150-9
No. 52	Long-Distance Neutrino Detection - 1978 (C.L. Cowan Memorial Symposium)	79-52078	0-88318-151-7
No. 53	Modulated Structures - 1979 (Kailua Kona, Hawaii)	79-53846	0-88318-152-5
No. 54	Meson-Nuclear Physics - 1979 (Houston)	79-53978	0-88318-153-3
No. 55	Quantum Chromodynamics (La Jolla, 1978)	79-54969	0-88318-154-1
No. 56	Particle Acceleration Mechanisms in Astrophysics (La Jolla, 1979)	79-55844	0-88318-155-X
No. 57	Nonlinear Dynamics and the Beam-Beam Interaction (Brookhaven, 1979)	79-57341	0-88318-156-8
No. 58	Inhomogeneous Superconductors - 1979 (Berkeley Springs, W.V.)	79-57620	0-88318-157-6
No. 59	Particles and Fields - 1979 (APS/DPF Montreal)	80-66631	0-88318-158-4
No. 60	History of the ZGS (Argonne, 1979)	80-67694	0-88318-159-2
No. 61	Aspects of the Kinetics and Dynamics of Surface Reactions (La Jolla Institute, 1979)	80-68004	0-88318-160-6
No. 62	High Energy e^+e^- Interactions (Vanderbilt , 1980)	80-53377	0-88318-161-4
No. 63	Supernovae Spectra (La Jolla, 1980)	80-70019	0-88318-162-2
No. 64	Laboratory EXAFS Facilities - 1980 (Univ. of Washington)	80-70579	0-88318-163-0
No. 65	Optics in Four Dimensions - 1980 (ICO, Ensenada)	80-70771	0-88318-164-9
No. 66	Physics in the Automotive Industry - 1980 (APS/AAPT Topical Conference)	80-70987	0-88318-165-7
No. 67	Experimental Meson Spectroscopy - 1980 (Sixth International Conference , Brookhaven)	80-71123	0-88318-166-5
No. 68	High Energy Physics - 1980 (XX International Conference, Madison)	81-65032	0-88318-167-3
No. 69	Polarization Phenomena in Nuclear Physics -- 1980 (Fifth International Symposium, Santa Fe)	81-65107	0-88318-168-1
No. 70	Chemistry and Physics of Coal Utilization - 1980 (APS, Morgantown)	81-65106	0-88318-169-X

Date Due

UML 735